MUSCLE RECEPTORS AND MOVEMENT

Symposium on muscle receptors and movement, Sherrington School of Physiology, St Thomas's Hospital Medical School, London, July 8th and 9th, 1980

BACK: S. Rossignol, J. Duysens, M. Hulliger, H. Forssberg, D. G. Stuart, E. D. Schomburg, H. Steffens, J. A. Hoffer, W. Z. Rymer, K. Fukushima, M. Kato, P. B. C. Matthews, S. Andreassen

MIDDLE: P. M. H. Rack, D. R. Westbury, B. Appelberg, V. Dietz, J. A. Stephens, C. Ghez, U. Proske, K. Appenteng, J. H. J. Allum, P. H. Ellaway, F. Emonet-Dénand, D. W. Harker, K.-E. Hagbarth, J. C. Houk, M. J. Stacey

FRONT: L. Jami, R. W. Banks, J. Noth, J. Paillard, P. Wand, A. Taylor, A. Prochazka, S. Grillner, Y. Laporte, I. A. Boyd, D. Barker, S. Homma, Å. B. Vallbo

MUSCLE RECEPTORS AND MOVEMENT

Proceedings of a Symposium held at the Sherrington School of Physiology, St Thomas's Hospital Medical School, London, on July 8th and 9th, 1980

Edited by

A. Taylor
and
A. Prochazka

Softcover reprint of the hardcover 1st edition 1981

First published 1981 by
Scientific and Medical Division
MACMILLAN PUBLISHERS LTD
London and Basingstoke
Companies and representatives throughout the world

ISBN 978-1-349-06024-5 ISBN 978-1-349-06022-1 (eBook)
DOI 10.1007/978-1-349-06022-1

The organisers gratefully acknowledge the following companies and organisations for their financial support, which made the symposium possible:

The Wellcome Trust
St Thomas's Hospital Research Endowments Committee
Medelec Ltd
Digitimer Ltd
Kodak Ltd
Cambridge Electronic Design Ltd
Niagara Therapy Ltd
The Parkinsons Society

Contents

The contributors

Dr J. H. J. ALLUM*, Institut für Hirnforschung, Universität Zürich, August-Forel-Strasse 1, CH-8029 Zürich, Switzerland
Dr O. ANDERSSON, Department of Physiology III, Karolinska Institutet, Lindingövägen 1, S-114 33 Stockholm, Sweden
Dr S. ANDREASSEN*, present address: Institute for Electronic Systems, Aalborg University Centre, DK-9100 Aalborg, Denmark
Dr B. APPELBERG*, Department of Physiology, University of Umeå, S-901 87 Umeå, Sweden
Dr K. APPENTENG*, Sherrington School of Physiology, St Thomas's Hospital Medical School, London, SE1 7EH, UK
Dr R. W. BANKS*, Department of Anatomy and Embryology, University College, Gower St, London, WC1E 6BT, UK
Prof D. BARKER*, Department of Zoology, University of Durham, Science Laboratories, South Road, Durham, DH1 3LE, UK
Dr H.-B. BEHRENDS*, Physiologisches Institut II, Der Universität, Humboldtallee 7, D-3400 Göttingen, W. Germany
Prof P. BESSOU, Laboratoire de Physiologie—CNRS ERA 846—Faculté de Médecine, 133 Route de Narbonne, 31062 Toulouse Cedex, France
Prof I. A. BOYD*, Institute of Physiology, University of Glasgow, Glasgow, G12 8QQ, UK
Dr F. J. CLARK, present address: Department of Physiology and Biophysics, Medical School, University of Nebraska, 42nd and Dewey St, Omaha, Nebraska, NE 68105, USA
Dr P. E. CRAGO, Sears Tower, RM EB-65, Case Western Reserve University, Cleveland, Ohio, OH 44106, USA
Dr V. DIETZ*, Neurologische Klinik, Der Universität, Hansastrasse 9, D-7800 Freiburg i. Br., W. Germany
Dr J. DUYSENS*, Laboratorium voor Neuro- en Psychofysiologie, Campus Gasthuisberg, Herestraat, B-3000 Leuven, Belgium
Dr F. R. EDWARDS, Department of Neurosurgery, State University of New York, Syracuse, New York, NY 13210, USA

* Symposium attendee

Dr P. H. ELLAWAY*, Department of Physiology, University College, Gower St, London, WC1E 6BT, UK

Dr F. EMONET-DÉNAND*, Laboratoire de Neurophysiologie, Collège de France, 11 Place Marcelin Berthelot, 75231 Paris Cedex 05, France

Dr H. FORSSBERG*, Department of Physiology III, Karolinska Institutet, Lidingövägen 1, S–114 33 Stockholm, Sweden

Dr L. GAUTHIER, Centre de Recherche en Sciences Neurologiques, Department of Physiology, Faculty of Medicine, University of Montreal, Case Postale 6128, Succursale A, Montreal, Quebec, Canada, H3C 3J7

Dr M. H. GLADDEN*, Institute of Physiology, University of Glasgow, Glasgow, G12 8QQ, UK

Prof S. GRILLNER*, Department of Physiology III, Karolinska Institutet, Lidingövägen 1, S–114 33 Stockholm, Sweden

Prof K.-E. HAGBARTH*, Kliniskt Neurofysiologiska, Central Laboratoriet, 750 14 Uppsala 14, Sweden

Dr K. HAYASHI, Department of Physiology, School of Medicine, Chiba University, Chiba 280, Japan

Dr J. A. HOFFER*, present address: Laboratory of Neural Control, NINCDS, Building 36, RM 5A 29, NIH, Bethesda, Maryland, MD 20205, USA

Dr S. HOMMA*, Department of Physiology, School of Medicine, Chiba University, Chiba 280, Japan

Prof J. C. HOUK*, Department of Physiology, The Medical School, Northwestern University Medical and Dental Schools, Ward Building 5–319, 303 East Chicago Avenue, Chicago, Illinois, IL 60611, USA

Dr M. HULLIGER*, Institut für Hirnforschung, Universität Zürich, August-Forel-Strasse 1, CH–8029 Zürich, Switzerland

Dr L. JAMI*, Laboratoire de Neurophysiologie, Collège de France, 11 Place Marcelin Berthelot, 75231 Paris Cedex 05, Paris, France

Dr M. JOFFROY*, Laboratoire de Physiologie—CNRS ERA 846—Faculté de Médecine, 133 Route de Narbonne, 31062 Toulouse Cedex, France

Dr H. JOHANSSON, Department of Physiology, University of Umeå, S–901 87 Umeå, Sweden

Dr S. W. JOHNSON, Department of Electrical and Electronic Engineering, University of Bristol, Bristol, BS8 1TD, UK

Dr C. JULIEN, Centre de Recherche en Sciences Neurologiques, Department of Physiology, Faculty of Medicine, University of Montreal, Case Postale 6128, Succursale A, Montreal, Quebec, Canada, H3C 3J7

Prof Y. LAPORTE*, Laboratoire de Neurophysiologie, Collège de France, 11 Place Marcelin Berthelot, 75231 Paris Cedex 05, Paris, France

Dr G. E. LOEB*, Laboratory of Neural Control, IRP, National Institute of Neurological and Communicative Disorders and Strokes, Building 36, RM 5A 29, NIH, Bethesda, Maryland, MD 20205, USA

Dr J. P. LUND, Centre de Recherche en Sciences Neurologiques, Department of Physiology, Faculty of Medicine, University of Montreal, Case Postale 6128, Succursale A, Montreal, Quebec, Canada, H3C 3J7

Dr P. A. LYNN, Department of Electrical and Electronic Engineering, University of Bristol, Bristol, BS8 1TD, UK

Dr P. B. C. MATTHEWS*, University Laboratory of Physiology, Parks Rd, Oxford, OX1 3PT, UK

Prof S. MILLER*, Department of Anatomy, The Medical School, The University, Newcastle upon Tyne, NE1 7RU, UK

Dr R. B. MUIR, present address: Department of Anatomy, Faculty of Medicine, Erasmus University, Rotterdam, The Netherlands

Dr P. R. MURPHY*, Department of Physiology, University College, Gower St, London, WC1E 6BT, UK

Dr K. S. K. MURTHY*, Division of Neurosurgery, Department of Surgery, University of Texas Medical School at Houston, 6400 West Cullen St, Houston, Texas 77030, USA

Dr Y. NAKAJIMA, Department of Physiology, School of Medicine, Chiba University, Chiba 280, Japan

Dr J. NOTH*, Neurologische Klinik, Der Universität, Hansastrasse 9, D–7800 Freiburg i. Br., W. Germany

Dr B. PAGÈS, Laboratoire de Physiologie—CNRS ERA 846—Faculté de Médecine, 133 Route de Narbonne, 31062 Toulouse Cedex, France

Dr J. PETIT, Laboratoire de Neurophysiologie, Collège de France, 11 Place Marcelin Berthelot, 75231 Paris Cedex 05, Paris, France

Dr E. M. POST, Department of Neurology, Northwestern University Medical School, 303 East Chicago Avenue, Chicago, Illinois, IL 60611, USA

Dr. A. PROCHAZKA*, Sherrington School of Physiology, St Thomas's Hospital Medical School, London SE1 7EH, UK

Dr U. PROSKE*, Department of Physiology, Monash University, Clayton, Victoria 3168, Australia

Dr P. M. H. RACK*, Department of Physiology, The Medical School, University of Birmingham, Birmingham, B15 2JT, UK

Dr G. A. L. REED, Department of Electrical and Electronic Engineering, University of Bristol, BS8 1TD, UK

Dr S. ROSSIGNOL*, Centre de Recherche en Sciences Neurologiques, Department of Physiology, Faculty of Medicine, University of Montreal, Case Postale 6128, Succursale A, Montreal, Quebec, Canada, H3C 3J7

Dr W. Z. RYMER*, Department of Physiology, Northwestern University Medical School, 303 East Chicago Avenue, Chicago, Illinois, IL 60611, USA

Dr K. SATO, Department of Physiology, School of Medicine, Chiba University, Chiba 280, Japan

Prof E. D. SCHOMBURG*, Physiologisches Institut II, Der Universität, Humboldtallee 7, D–3400 Göttingen, W. Germany

Prof T. A. SEARS*, Sobell Department of Neurophysiology, Institute of Neurology, The National Hospital, Queen Sq., London, WC1N 3BG, UK

Dr M. SHITO, Department of Physiology, School of Medicine, Chiba University, Chiba 280, Japan

Dr P. SOJKA, Department of Physiology, University of Umeå, S–901 87 Umeå, Sweden

Dr M. J. STACEY*, Department of Zoology, University of Durham, Science Laboratories, South Road, Durham, DH1 3LE, UK

Dr H. STEFFENS*, Physiologisches Institut II, Der Universität, Humboldtallee 7, D–3400 Göttingen, W. Germany

Dr J. A. STEPHENS*, Sherrington School of Physiology, St Thomas's Hospital Medical School, London SE1 7EH, UK

Prof A. STRUPPLER*, Neurologische Klinik der Technischen Universität, Möhlstrasse 28, 8 München 80, W. Germany

Prof D. G. STUART*, Department of Physiology, University of Arizona, Health Sciences Center, Tucson, Arizona, AZ 85724, USA

Prof A. TAYLOR*, Sherrington School of Physiology, St Thomas's Hospital Medical School, London, SE1 7EH, UK

Dr J. TROTT*, present address: Department of Physiology, The Medical School, University Walk, Bristol, BS8 1TD, UK

Prof Å. B. VALLBO*, Department of Physiology, University of Umeå, S–901 87 Umeå, Sweden

Dr B. WALMSLEY, Department of Physiology, Monash University, Clayton, Victoria 3168, Australia

Dr P. WAND*, Max Planck Institut für Experimentelle Medizin, 3400 Göttingen, Hermann–Rein Strasse 3, W. Germany

Dr D. R. WESTBURY* Department of Physiology, The Medical School, University of Birmingham, Birmingham, B15 2JT, UK

Foreword

For many decades of scientific endeavour the physiology of sensory receptors has been a major field of enquiry and one in which complexity of receptor structure is matched by a corresponding diversity in the experimental approaches used in its investigation. At one time it might have been thought that such complexity was the reserve of the special senses. However, that could hardly be accepted now, in the face of the continuous and increasing challenge to experimental skill, which questions concerning the structure and properties of muscle receptors and the role of these receptors in the control of muscular contraction pose for the investigator.

Accurately positioned within the three-dimensional space of the body image, the precise voluntary movements of the extremities occur on a postural stage, itself continuously and largely 'automatically' set through appropriately fine adjustment of the activities of the proximal and axial musculature. It is usually taken as axiomatic that the smooth precision of such movements is learnt and executed through mechanisms of muscular control which are highly dependent on the information supplied by the muscle receptors; only within the last decade however, has direct evidence become available as to the nature of the information transferred during voluntary movement, albeit, thus far, for a limited number of muscles and movements of the extremities. In 1965, on the occasion of the first symposium supported by the Nobel Foundation, which happened also to be on the topic of 'muscle afferents and motor control', the inaugural address was given by Adrian who thought that the subject was '. . . ripe for discussion . . . but still with some details to be filled in of the main outlines of the peripheral and spinal apparatus for controlling movement'. Although many symposia on the subject of motor control have been held since then, few have dealt so specifically with muscle receptors and their role in such control. Why then, it may be asked, is the time once again ripe for assembling a symposium of experts to discuss this subject; which key issues need to be debated and what extra details are still missing from the overall picture?

Each of us, according to age, training and experience, carries a very personal perspective of a given scientific topic, and with it, a highly individual view of what currently is important or at least interesting. With this in mind and by way of introduction, I outline my own perspective as I contemplated attending the symposium whose subject matter comprises this book. This converges to that First Nobel Symposium, a memorable and exciting occasion which I was fortunate enough to attend. A year beforehand, I had begun my research on the human intercostal muscle stretch reflexes. The experiments took into account the coactivation of alpha and gamma intercostal motoneurons which occurs in response to the natural, spontaneous and centrally initiated command for respiratory movements as had been recently discovered independently by von Euler and myself. With these human experiments, done in collaboration with J. Newsom Davis and one of the organisers of this symposium—A. Taylor—we were already confronted with the fundamental problem of distinguishing between reflex and voluntary behaviour in the conscious human subject. Not surprisingly, therefore, the topic which perhaps interested me most was Oscarsson's account of his work with Rosen (1963), which had convincingly demonstrated the projection of Group I muscle afferents to the cerebral cortex, a projection whose existence had previously been denied save for a preliminary, brief report of one by Amassian and Berlin (1959), that had been overlooked by many investigators. The belief that such a projection did not exist, reinforced the idea, based on behavioural and psychophysical experiments, that the information signalled by Group I muscle afferents (muscle spindles and Golgi tendon organs) does not project into consciousness. This idea occupied a central place in Merton's servo theory of muscular control for which he proposed that the spinal servo loop, subserved by the stretch reflex, has an insentient mode of operation, both when driven through fusimotor activity as conceived for the 'length follow-up' servo-mechanism, or, when simply responding to muscle stretch.

The absence of behavioural responses to stimulation of Group I afferents in animal experiments (by other authors) had led Oscarsson to conclude 'that the Group I projection represents a cortical mechanism as unrelated to conscious perception as the motor regulating mechanisms in the cerebellum.' Since the cerebral cortex, like the cerebellum, is concerned with the execution and co-ordination of movement, and was now also similar with regard to the afferent information it received from muscular and cutaneous afferents through the fastest paths available, Oscarsson suggested that 'these supraspinal pathways to the sensorimotor cortex constitute feedback channels used in the integration of motor activity'. Interestingly, Hammond had previously suggested that the long latency of the stretch reflex of the human biceps muscle might be due to a cerebellar loop. Thus by 1965 the anatomical and physiological basis for such a loop through the cerebral cortex was known to exist and knowledge of this pervaded our own thoughts as, like Hammond had done, we wrestled with the problem of the long latency of the intercostal muscle stretch reflex.

A few years later, Phillips (1969) in The Ferrier Lecture, linked the results from studying the strength and distribution of the monosynaptic connections of Group Ia muscle afferents to motoneurons of the primate hand to those revealing their projections to the cerebral cortex, to form his idea of a 'transcortical loop' in the conscious human subject, with a loop time for the stretch reflex shorter than the earliest voluntary response to a brief mechanical stimulation of a moving limb, i.e. shorter than a kinaesthetic reaction time. It will be seen that this proposal retains the idea of the insentient automatic compensation for unexpected variations in mechanical load inherent in the servo theory, but now the 'error' signal in the jargon of servo-mechanisms ultimately exerts its effects on the cells of origin of the cortico-motoneuronal tract. In effect, Sherrington's stretch reflex had been 'encephalised'. I emphasise this here because many authors use the term 'stretch reflex' with scant regard for its true meaning as originally embodied in Merton's theory, translated to Phillips's hypothesis, and having firm epistemological roots in Sherrington's stretch reflex, which is the sustained contraction of muscle (or motoneuron firing) in response to muscle stretch and dependent on the excitation of muscle receptors. During the last decade these key papers, particularly that of Phillips, have stimulated a great deal of human-based research on the 'long loop' reflexes. Furthermore, this effort has been paralleled by combined behavioural and electrophysiological studies aimed at deciphering the kind of information received by the primate cerebral cortex from muscle receptors during the course of normal and impeded 'voluntary' movements of the primate limb.

Equally importantly, the last decade has seen the full exploitation of microelectroneuronography, first introduced by Hagbarth and Vallbo in 1968. This allows direct recording from human muscle afferents during voluntary movement and so has allowed important inferences also to be drawn about alpha–gamma coactivation. Paralleling this remarkable achievement the convenors of our symposium—A. Taylor and A. Prochazka—have independently pioneered the recording of receptor nerve discharge for jaw and limb movements, respectively, in the awake animal.

Through the work of Smith, The First Nobel Symposium also saw the beginning of the exploration of living muscle spindle structure and function through direct visualisation, a method which Boyd and his collaborators have subsequently developed to a high degree of refinement.

These are only some of the topics critically discussed in this symposium. I have singled them out only for the purpose of illustrating the perspective we need to have on the continuously evolving nature of the concepts which we manipulate and the difficulties which thus arise in deciding, as Matthews asked in discussion, 'How much is new?', 'How much is true?' and 'How much is general?' While one might expect disagreement over the answers to such questions, which I encourage others to provide by reading this book, I am sure we can all agree how appropriate it is that this symposium on muscle receptors and movement should be held at the school where almost a century ago Sherrington not only received his medical training, but also as a lecturer embarked on a scientific career that was to lay the foundations of the subject debated in the following pages.

August, 1980

T. A. Sears
Sobell Department of Neurophysiology
Institute of Neurology
The National Hospital
Queen Square
London, WC1N 3BG

Preface and Acknowledgements

This book contains the Proceedings of the Symposium entitled *Changing Views of the Function of Muscle Receptors in Movement Control* held at the Sherrington School of Physiology, St Thomas's Hospital Medical School, London, UK on July 8th and 9th, 1980. The symposium was attended by 105 leading international scientists engaged in research into the control of movement in mammals.

An innovative feature of the symposium was the inclusion of short, formal critiques of presentations by recognised experts in the topics concerned. Each 'critic' had been sent copies of the relevant presentations two to three weeks prior to the meeting. Critiques varied from the very mild to the downright scurrilous, but were taken in good part by all concerned. Judging by the favourable (indeed almost gleeful) audience reaction, this formula might well reappear in symposia in the future.

We should like to make mention of the efforts of our publishers—Dr S. Sharrock and Mr R. M. Powell of Macmillan Publishers Ltd—in ensuring the availability of this book within nine months of the meeting. Finally, we wish to thank all of the contributors to the book for their enthusiasm and their willingness to act as subjects in this peer-review experiment.

The illustration on the front of the dust jacket is taken from the paper by R. W. Banks, D. Barker and M. J. Stacey, and we thank Professor Barker and his co-authors for permission to use it.

September, 1980

A. Taylor and A. Prochazka
Sherrington School of Physiology
St Thomas's Hospital Medical School
London, SE1 7EH

SECTION 1
MUSCLE SPINDLE STRUCTURE AND SENSITIVITY

Overview

Muscle spindles respond to stretch of muscle with a frequency-coded afferent discharge which contains information regarding length and velocity. However, despite some thirty years of research in what might be regarded as the present era, no completely satisfactory generalised quantitative description is yet available of the transducing properties of the primary and secondary endings. The desire to have a mathematical expression of 'transfer function' to describe spindle response was largely founded on the needs of models of feedback control inspired by the engineering approach originating in the 1950s. We may now reflect that these concepts were too restrictive and for long persuaded us to try to ignore the peculiar and non-linear properties of spindles or to restrict our observations to minute length changes in which an assumption of linearity was justified. We are now painfully aware of the deficiencies of this approach and are casting around for new ways of describing spindle transducing properties which incorporate non-linear length and velocity effects without the implication that they are regrettable defects. This session is notable for the clear description of one new approach to this problem, followed, however, by an exposure of its weakness.

The effects of fusimotor stimulation are complex and not yet fully worked out, even at the simple descriptive level, but the papers presented in this area certainly advance the subject significantly. At last the details and functional significance of the various intrafusal muscle fibres and their innervation are becoming clear, and the agreement reached here was most encouraging. Readers will however still be left wondering whether we have yet found the right way of looking at fusimotor effects on spindle responsiveness and whether the terms 'dynamic' and 'static' have been adequately defined in view of their constant use in describing data.

The papers in this section were all read at the symposium.

Structural aspects of fusimotor effects on spindle sensitivity

R. W. BANKS*, D. BARKER† AND M. J. STACEY†

SUMMARY

The distribution of motor and sensory axons to the three types of intrafusal muscle fibre have been determined using reconstructions of serially sectioned spindles and teased, silver-impregnated, whole spindles.

The results obtained from the analysis of fusimotor axons indicate that there is very little trail innervation, and a correspondingly low static input, distributed to bag_1 fibres.

Whereas the S_1 secondary ending is predominantly distributed to chain fibres, it almost invariably also innervates both types of bag fibre.

The variability of some histological features of primary and secondary innervation from four hindlimb muscles is described, and the significance of these findings discussed in relation to spindle sensitivity.

INTRODUCTION

The sensitivity of the sensory nerve endings in mammalian muscle spindles is controlled by the central nervous system via motor (γ and β) activation of intrafusal muscle fibres. It is now known that three types of intrafusal muscle fibre are normally present in each muscle spindle (Banks, Harker and Stacey, 1977), and it is generally agreed that their different mechanical properties are important in determining the characteristics of the sensory response. Thus, the bag_1 fibre mediates the dynamic responsiveness of the primary ending, whereas the bag_2 and chain fibres mediate its static responsiveness. Also, the secondary ending, which usually shows comparatively little dynamic sensitivity, predominantly innervates chain fibres. Finally, dynamic fusimotor

*Department of Anatomy and Embryology, University College London
†Department of Zoology, University of Durham

neurons innervate bag_1 fibres whereas static fusimotor neurons innervate bag_2 and chain fibres.

There remains one major area of disagreement and that is whether or not static fusimotor axons frequently innervate bag_1 fibres. Studies using the glycogen-depletion technique suggest that they do (Brown and Butler, 1973; Barker *et al.*, 1976; Emonet-Dénand *et al.*, 1980), but observation of living spindles during fusimotor stimulation indicates that bag fibres activated by dynamic axons (presumably bag_1) are not also activated by static axons (Bessou and Pagès, 1975; Boyd *et al.*, 1977).

This paper summarises recent results obtained from reconstructions of serially sectioned muscle spindles and from analysis of silver-impregnated, teased whole spindles. We discuss their relevance to spindle sensitivity and its fusimotor control. The results fall into two parts: first, the distribution of fusimotor axons to the three types of intrafusal muscle fibre, and second, some details of the histology of primary and secondary sensory endings. The preparative methods used in serial sectioning and reconstructions, and in the silver impregnation, are fully described elsewhere (Banks *et al.*, 1978; Banks, in preparation; Barker *et al.*, 1970) and only relevant details will be given here. All results were obtained from cat hindlimb muscles. The serial sectioning was performed on tenuissimus muscles only.

THE INTRAFUSAL DISTRIBUTION OF FUSIMOTOR AXONS

Serial, 1 μm thick transverse sections of three almost complete spindles and one half-spindle were used to make schematic reconstructions of their somatic motor innervation. The results of the half-spindle were typical, and will be described here.

Intrafusal muscle fibres were identified as nuclear-bag and nuclear-chain types by their lengths, diameters and equatorial nucleation. The bag fibres were further subdivided into bag_1 and bag_2 types by the dissociation of the bag_1 fibre from the others in the equatorial region and by the association of elastic fibres with the polar regions of the bag_2 fibre (figure 1). Neuromuscular junctions and their supplying axons, whether myelinated or unmyelinated, were easily recognisable (figure 1) and readily traceable through successive sections.

The reconstructed half-spindle was the proximal pole of spindle 5 in the study of Banks, *et al.* (1978). In this pole a bag fibre, presumably bag_1, was activated by two dynamic γ axons. Correspondingly, the bag_1 fibre possessed two motor endings, and each was supplied by an axon with no other intrafusal terminals in the pole (figure 1). Three other axons entered the pole, one ending in the periaxial space at some distance from the muscle fibres. Each of the remaining two axons supplied motor endings to the bag_2 fibre and to two of the four chain fibres (figure 1). They were presumably static axons and may be correlated with the trail innervation seen in silver-impregnated preparations (Barker *et al.*, 1973). In the example of this half-spindle, therefore, there was no common innervation of bag_1 with either bag_2 or chain fibres. In the three reconstructed whole spindles only one axon was found to supply a bag_1 fibre and another type of fibre (in this case a chain), whereas eight axons supplied endings to both bag_2 and chain fibres (Banks, in preparation). The distribution of all 26 axons traced is given in table 1, and is illustrated schematically in figure 2.

Table 1. Intrafusal distribution of fusimotor axons in three spindles and one half spindle (cat tenuissimus).

Muscle-fibre type	*Number of axons supplied to* *one pole*	*both poles*	*Totals*
bag_1	6	1	7
bag_2	5	—	5
chain(s)	4	1	5
bag_1 and chain	1	—	1
bag_2 and chain(s)	8	—	8
			26

If the glycogen-depletion results are an accurate reflection of the innervation of the bag_1 fibre, virtually each one should be innervated by at least one static axon in addition to any dynamic axons. This is because, on average, each static axon depletes three out of five bag_1 fibres (Barker *et al.*, 1976) and, since there are about three static axons to each spindle, the probability of at least one static axon innervating the bag_1 fibres is 0.94. Despite the small sample involved, our results clearly contradict this inference. It might be argued that some of the axons innervating bag_1 fibres in the reconstructed spindles also innervate bag_2 or chain fibres in other spindles; however, this complication is not suggested by the glycogen-depletion results, and it is known that most fusimotor axons have the same action on different primary endings (Emonet-Dénand *et al.*, 1977). It might also be argued that the axons from bag_1 and other fibres have not been traced sufficiently far back to reach their common origin; however, many such examples have been found for the bag_2/chain system, and this is in complete accord with both glycogen-depletion and direct observation of living spindles.

In the recent classification of sub-categories of static and dynamic fusimotor action (Emonet-Dénand *et al.*, 1977) the intermediate categories II to V were interpreted as arising from various degrees of common innervation of bag_1 with bag_2 or chain fibres. Since about 30% of the responses fell in these categories a fairly high proportion of such common innervation would be expected, and this would be more or less consistent with the glycogen-depletion results. However, in the light of our present findings we believe that

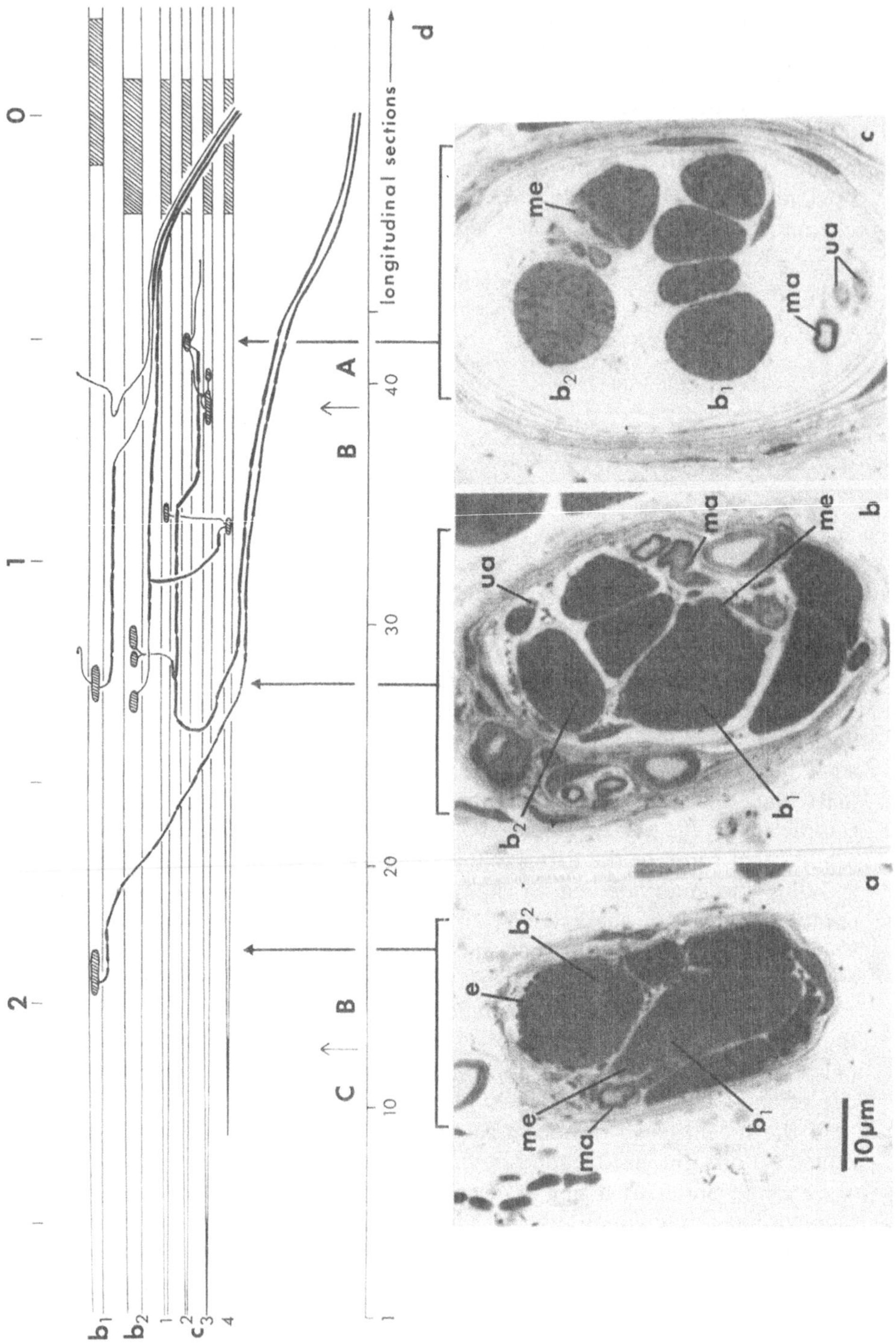
0
1
2
b1
b2
c 1 2 3 4
longitudinal sections
A
B
C
10
20
30
40
d
me
ma
ua
e
b1
b2
10µm
a
b
c

Figure 1 (opposite) (a)–(c). Representative 1μm thick transverse sections of the half-spindle described in the text. Bag_1 (b_1), bag_2 (b_2) and four chain fibres are present in all the sections. In (a) note the elastic fibres (e) associated with the bag_2 fibre, and the motor ending (me) on the bag_1 fibre. The myelinated axon (ma) supplying this ending is visible nearby. In (b) note the motor ending on bag_1. The section passes through the preterminal heminode of the supplying axon. An unmyelinated axon (ua) with its associated Schwann-cell nucleus is close to the bag_2 fibre, on which it terminated. This is a branch of the myelinated axon present in section (c). Other axons may be identified by reference to the reconstruction (d). In (c) note the motor ending on chain 2 and the three axons (one myelinated, two unmyelinated) within the periaxial space. The two remaining axons were present in the section but are not included in this field of view.

(d) Schematic reconstruction of the half-spindle described in the text. The locations of sections (a)–(c) are indicated by connecting lines. Only longitudinal dimensions are accurate, and are calibrated in slide numbers (lower scale), each slide containing 50 1 μm sections. The upper scale shows the approximate corresponding lengths (in mm) in the living spindle, allowing for 8% shrinkage. Vertical arrows indicate the limits of A, the periaxial space; B, the capsular pole; and C, the extracapsular pole. Muscle fibres are identified to the left of the drawing. The approximate location of the primary ending is indicated by the hatched rectangles to the right. Motor endings are indicated by hatched ovals. In the fusimotor axons unmyelinated regions and nodes of Ranvier are shown as thin lines, myelinated internodes as thick lines. One axon ends freely in the periaxial space, but perhaps had branches innervating muscle fibres in the other pole. Two other axons had freely ending ultraterminal branches.

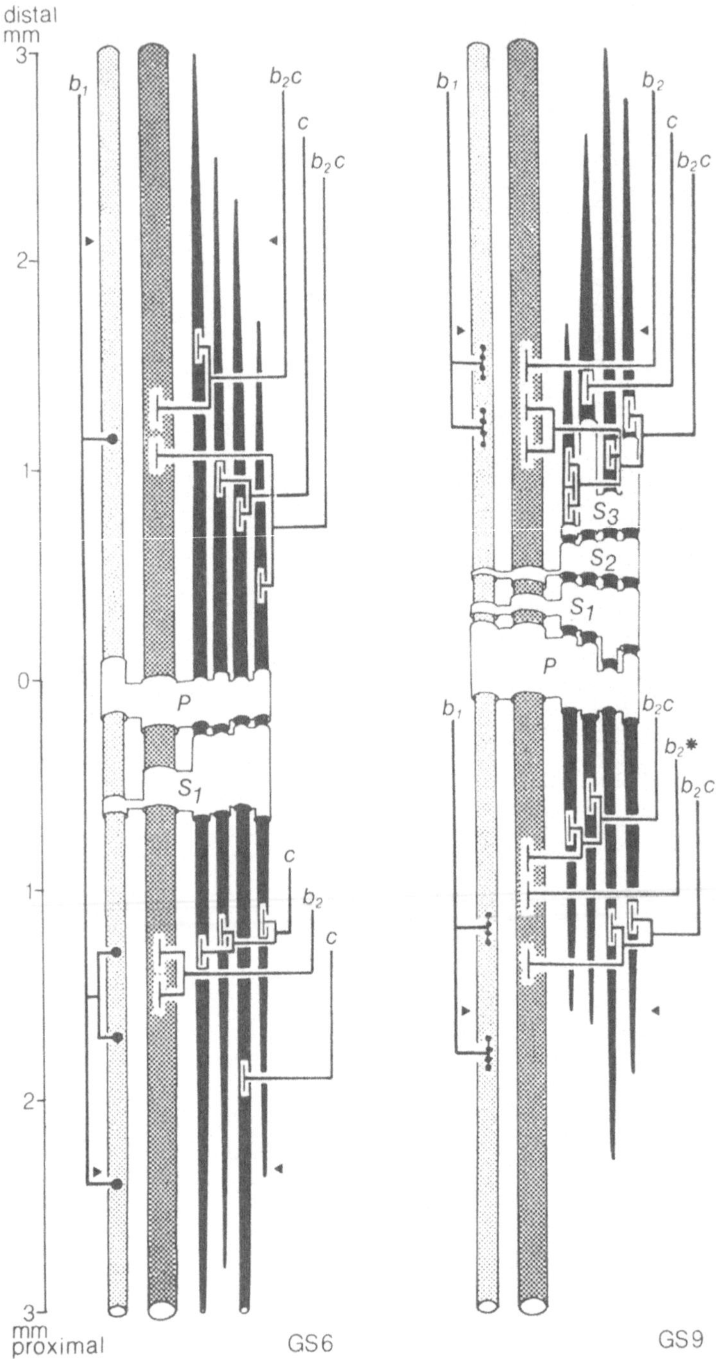

Figure 2 (continued opposite). Schematic representations of the innervation of the reconstructed cat tenuissimus spindles referred to in the text. Single asterisk in GS 9

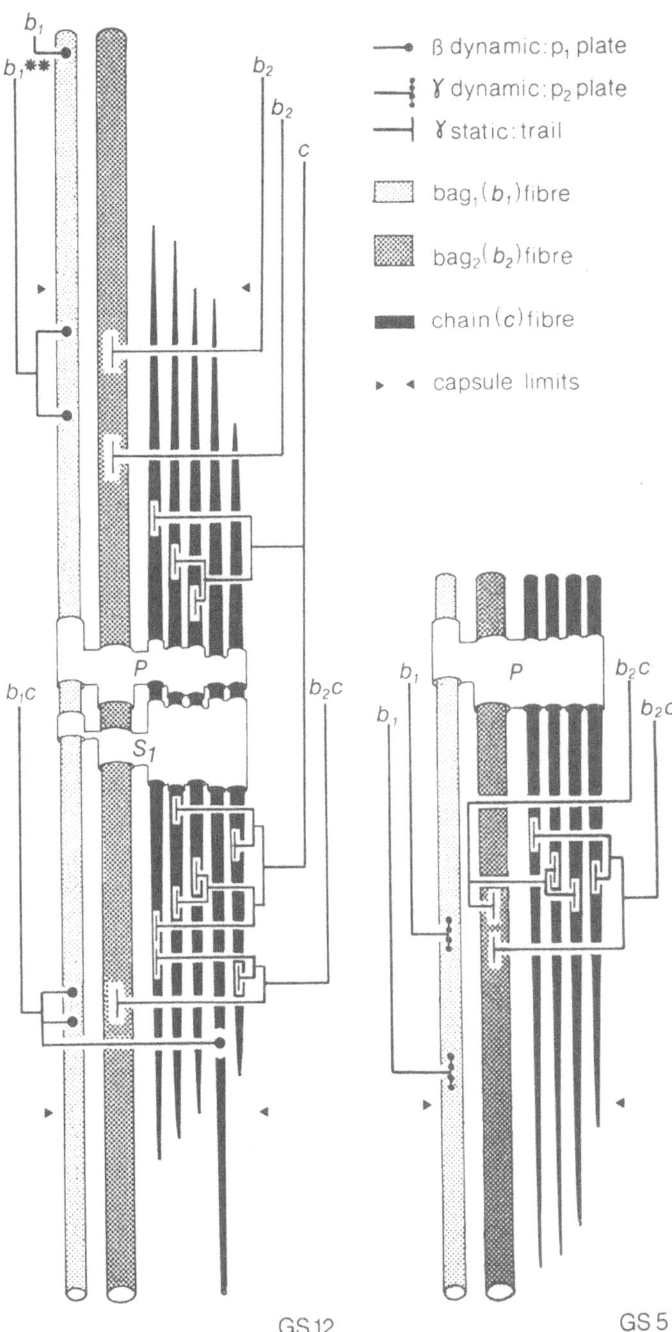

indicates uncertain identification of this motor axon and ending; double asterisk in GS 12 indicates possibility that this axon may be γ dynamic rather than β dynamic as shown.

only the category III responses of Emonet-Dénand *et al.* (1977) truly represent this occurrence, and if so this would indicate that about 5% of fusimotor axons have terminals on bag_1 as well as bag_2 or chain fibres. It seems likely that these axons are either predominantly dynamic or static when taking their total distribution into account (Emonet-Dénand *et al.*, 1977).

We conclude, therefore, that the incidence of static innervation (presumably via trail terminals) on bag_1 fibres is rather low, at least in tenuissimus muscle spindles. We are at present analysing the motor innervation of whole silver-impregnated spindles from a variety of hindlimb muscles in order to increase our sample size and to test the generality of the conclusion. It is our impression, at this preliminary stage, that trail endings rarely occur on bag_1 fibres.

THE SENSORY INNERVATION

Two primary endings and one S_1 secondary ending were reconstructed isometrically from serial, 1 μm thick transverse sections. Detailed descriptions of these together with the reconstructions themselves will be published elsewhere (Banks, Barker and Stacey, in preparation). The two primary endings were strikingly similar in several respects, namely the areas of contact between sensory terminals and each type of intrafusal muscle fibre; the form of the terminals, particularly the differences between those on bag_1 and bag_2 fibres; and the branching pattern of the afferent axons (Banks, Barker and Stacey, 1977). We then examined silver-impregnated, teased, whole spindles to see if these features were recognisable and, if so, how consistent they were. Spindles from four hindlimb muscles—peroneus brevis (PB), peroneus tertius (PT), superficial lumbrical (SL) and tenuissimus (T)—were used. The form of the terminals was the most consistent feature, and was readily interpretable in terms of that seen in the reconstructions. We were therefore able to identify bag_1 and bag_2 fibres in the silver preparations, relying on a constant relationship of sensory terminal form to muscle fibre type as one of several criteria.

We have previously reported that in four primary afferent axons from tenuissimus spindles the first-order branches supplied bag_1 fibres separately from bag_2 and chain fibres (Banks, Barker and Stacey, 1977). The branching pattern of many primary axons could be ascertained from silver-impregnated spindles and we were able to compare this feature in a number of muscles (table 2). Various degrees of segregation were present, and the proportion of spindles in which complete segregation occurred varied between the different muscles. Also shown in table 2 are the proportions of spindles that contained two bag_1 fibres in addition to a bag_2 fibre. Again this varied in the different muscles investigated.

All these features of the primary innervation are likely to have functional

Table 2. Primary endings in cat hindlimb muscles.

	Percentage with segregated distribution to bag_1 fibres		*Percentage innervating two bag_1 fibres (in spindles with three bag fibres)*	
Muscle	% segregated	total counted	% on two bag_1 fibres	total counted
Peroneus brevis	64	14	14	29
Peroneus tertius	54	11	5	22
Superficial lumbrical	30	23	29	28
Tenuissimus	80	30	7	62

correlates, but as yet the only one for which the relationship seems clear is that of the branching pattern of the axons and the occurrence of separate static and dynamic pacemakers (Hulliger and Noth, 1979).

The reconstructed S_1 secondary ending possessed terminals on all three types of intrafusal muscle fibre, though those on the four chain fibres dominated its appearance. Even so, 25% of the contact area between sensory terminal and muscle fibre occurred on the bag fibres (17% on bag_2 and 8% on bag_1). The branching pattern and distribution of four other S_1 secondary afferents were reconstructed and in each case the bag_1 fibre was included.

When we were able to identify bag_1 and bag_2 fibres confidently in silver-impregnated material, we began an analysis of the secondary endings in whole spindles from four hindlimb muscles (PB, PT, SL, T). In terms of the distribution to bag and chain fibres our results are similar to those of Boyd (1962). Most of the terminals of secondary endings were distributed to chain fibres, largely in the form of loose, widely spaced spirals and incomplete loops (figure 3(a)). On bag fibres the terminals were usually in the form of sprays (figure 3(a) and (b)) or claw-like configurations (figure 3(b) and (c)).

Of all the secondary endings, only those in the S_1 position consistently innervated one or other bag fibre in addition to chains. Thus among 144 S_1 endings, 130 had terminals on all three types of muscle fibre, 11 supplied bag_2 and chain fibres and three supplied bag_1 and chain fibres. It was apparent that the area of contact between sensory terminals and muscle fibres was very variable in different endings. As a simple estimate of this variability, we counted the numbers of terminal branches on the bag fibres. The counting was restricted to 72 of the best silver-impregnated endings from 58 spindles. The sample included 79 bag_1 fibres and 70 bag_2 fibres. In each case the majority of bag fibres had between eight and 12 S_1-terminal branches distributed to them. The variability between muscles is illustrated in table 3 where the proportions of bag_1 and bag_2 fibres possessing eight or more S_1-terminal branches are given for four hindlimb muscles (PB, PT, SL, T). The significance of these results is difficult to assess in terms of the proportions of S_1

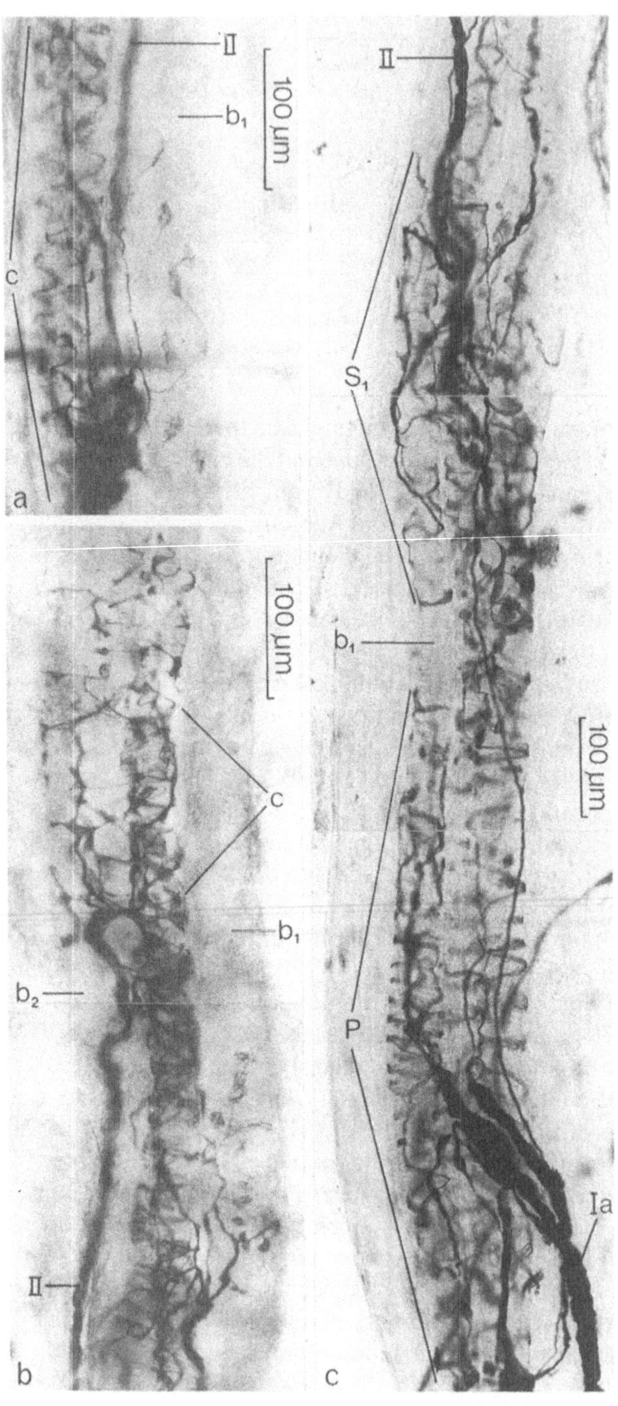
II
b1
100 µm
c
a
b2
b
S1
P
Ia

Table 3. Percentage of bag fibres receiving eight or more terminal branches from S_1 secondary endings.

Muscle	% bag_1 fibres	total counted	% bag_2 fibres	total counted
Peroneus brevis	53	19	100	15
Peroneus tertius	44	16	44	16
Superficial lumbrical	69	16	92	13
Tenuissimus	75	28	65	26

secondary endings supplied to bag fibres, since a similar count of terminal branches was not possible for chain fibres. However, as an approximate guide we may note that in the reconstructed ending described above there were about eight terminals on the bag_1 fibre and 18 on the bag_2 fibre. It seems, therefore, that in many S_1 secondary endings the bag_1 fibre alone may receive about 10% of the contact area of the ending.

It is tempting to see in these results the structural substrate for the variable dynamic component in the responses of secondary endings. If this is so it is at first surprising that dynamic fusimotor stimulation has so little effect on secondaries (Appelberg *et al.*, 1966). However, secondary terminals on bag fibres occur in regions of the muscle fibres that possess far more myofilaments than in the equatorial regions below the primary terminals. Dynamic fusimotor activation of bag fibres may not, therefore, have the same effect on secondary endings as on primary endings. This suggestion is supported by recent results of Jami and Petit (personal communication, 1980), who have found some secondary endings to be activated by dynamic axons, leading to an increase of static firing but without an increase in their dynamic indices.

Received on August 1st, 1980.

Figure 3 (opposite). Photographs of teased, silver preparations (method of Barker and Ip, 1963) of cat hindlimb muscle spindles illustrating features of sensory innervation.

(a) Part of an S_1 secondary ending supplied to a tenuissimus spindle showing spray of terminals supplied to the bag_1 (b_1) fibre on the right and loose spirals supplied to the chain fibres (c) on the left. II, secondary afferent.

(b) Secondary afferent (II) supplies an S_1 secondary ending to a superficial lumbrical spindle. The terminals are distributed to the bag_2 (b_2) fibre on the left, the bag_1 (b_1) fibre on the right, and the chain fibres (c) in the middle.

(c) Equatorial region of a superficial lumbrical spindle supplied with an S_1 secondary ending (upper half) and primary ending (lower half). Focus adjusted so as to pick out bag_1 fibre (b_1) and its primary (P) and secondary (S_1) terminals. Ia and II, primary and secondary afferents.

REFERENCES

Appelberg, B., Bessou, P. and Laporte, Y. (1966). Action of static and dynamic fusimotor fibres on secondary endings of cat spindles. *J. Physiol., Lond.*, **185**, 160–171

Banks, R. W., Barker, D., Bessou, P., Pagès, B. and Stacey, M. J. (1978). Histological analysis of cat muscle spindles following direct observation of the effects of stimulating dynamic and static motor axons, *J. Physiol., Lond.*, **283**, 605–619

Banks, R. W., Barker, D. and Stacey, M. J. (1977). Intrafusal branching and distribution of primary and secondary afferents, *J. Physiol., Lond.*, **272**, 66–67P

Banks, R. W., Harker, D. W. and Stacey, M. J. (1977). A study of mammalian intrafusal muscle fibres using a combined histochemical and ultrastructural technique, *J. Anat.*, **123**, 783–796

Barker, D., Emonet-Dénand, F., Harker, D. W., Jami, L. and Laporte, Y. (1976). Distribution of fusimotor axons to intrafusal muscle fibres in cat tenuissimus spindles, as determined by the glycogen depletion method, *J. Physiol., Lond.*, **261**, 49–70

Barker, D., Emonet-Dénand, F., Laporte, Y., Proske, U. and Stacey, M. J. (1973). Morphological identification and intrafusal distribution of the endings of static fusimotor axons in the cat, *J. Physiol., Lond.*, **230**, 405–427

Barker, D. and Ip, M. C. (1963). A silver method for demonstrating the innervation of mammalian muscle in teased preparations, *J. Physiol., Lond.*, **169**, 73–74P

Barker, D., Stacey, M. J. and Adal, M. N. (1970). Fusimotor innervation in the cat, *Phil. Trans. R. Soc. B*, **258**, 315–346

Bessou, P. and Pagès, B. (1975). Cinematographic analysis of contractile events produced in intrafusal muscle fibres by stimulation of static and dynamic fusimotor axons, *J. Physiol., Lond.*, **252**, 397–427

Boyd, I. A. (1962). The structure and innervation of the nuclear bag muscle fibre system and the nuclear chain muscle fibre system in mammalian muscle spindles, *Phil. Trans. R. Soc. B*, **245**, 81–136

Boyd, I. A., Gladden, M. H., McWilliam, P. N. and Ward, J. (1977). Control of dynamic and static nuclear bag fibres and nuclear chain fibres by γ- and β-axons in isolated cat muscle spindles, *J. Physiol., Lond.*, **265**, 133–162

Brown, M. C. and Butler, R. G. (1973). Studies on the site of termination of static and dynamic fusimotor fibres within muscle spindles of the tenuissimus muscle of the cat, *J. Physiol., Lond.*, **233**, 553–573

Emonet-Dénand, F., Jami, L., Laporte, Y. and Tankov, N. (1980). Glycogen depletion of bag_1 fibres elicited by stimulation of static γ axons in cat peroneus brevis muscle spindles, *J. Physiol., Lond.*, **302**, 311–321

Emonet-Dénand, F., Laporte, Y., Matthews, P. B. C. and Petit, J. (1977). On the subdivision of static and dynamic fusimotor actions on the primary ending of the cat muscle spindle, *J. Physiol., Lond.*, **268**, 827–861

Hulliger, M. and Noth, J. (1979). Static and dynamic fusimotor interaction and the possibility of multiple pace-makers operating in the cat muscle spindle, *Brain Res.*, **173**, 21–28

The action of the three types of intrafusal fibre in isolated cat muscle spindles on the dynamic and length sensitivities of primary and secondary sensory endings

I. A. BOYD*

SUMMARY

The effect of separate activation of the dynamic bag_1 fibre, the static bag_2 fibre and the nuclear chain fibres on the discharge of primary and secondary sensory endings in isolated cat muscle spindles is described. The static bag_2 fibre decreases, or does not change, the length sensitivity of primary endings and does not change, or slightly increases, the length sensitivity of secondary endings. The static bag_2 fibre decreases, or does not change, the dynamic sensitivity of both primary and secondary endings. The input from the static bag_2 fibre to secondary endings is about one fifth of that from the nuclear chain fibres.

Chain fibres drive the primary ending up to about 60 Hz and give a highly irregular discharge at higher frequencies; the length sensitivity of the primary ending is little altered. Chain fibres rarely drive the secondary ending but powerfully excite a secondary ending in the active pole, at the same time decreasing its dynamic sensitivity and increasing its length sensitivity, sometimes by a large amount.

The dynamic bag_1 fibre rarely affects the dynamic sensitivity of secondary endings, but may increase their length sensitivity slightly. It increases both the dynamic sensitivity and the length sensitivity of primary endings. The characteristic slow phase of adaptation during the stretch plateau is related to the mechanical creep which occurs in an active dynamic bag_1 fibre when it is stretched and held stretched. The marked increase in length sensitivity *during* a ramp stretch of primary endings, and of the occasional secondary ending whose dynamic sensitivity is increased, is probably velocity dependent and contributes appreciably to the increased dynamic response. Thus the input from the dynamic bag_1 fibre is a function of length and velocity.

* Institute of Physiology, University of Glasgow

INTRODUCTION

Cat muscle spindles usually contain one dynamic bag_1 intrafusal fibre, one static bag_2 fibre, and four or five nuclear chain fibres. About one spindle in five contains three nuclear bag fibres, of which two may be static bag_2 fibres or dynamic bag_1 fibres. The dynamic bag_1 fibre is supplied selectively by a dynamic γ axon(s), and, in about one spindle in three in the tenuissimus muscle, by a dynamic β axon as well. The static bag_2 fibre and the nuclear chain fibres are supplied by static γ axons, either selective to each type, or non-selective to both types. In many spindle poles, and some complete spindles, in our experience the innervation is entirely selective. Fast β axons which innervate chain fibres and extrafusal fibres are known to exist but we have not studied any of these. Glycogen-depletion studies suggest that some static axons may supply the dynamic bag_1 fibre as well as the static bag_2 fibre or chain fibres, but we have not encountered any of these to date. We have, however, encountered one axon which supplied both the dynamic bag_1 fibre and static bag_2 fibre and had a powerful mixed dynamic and static action, and one axon which innervated the dynamic bag_1 fibre and one chain fibre in a spindle studied by electron microscopy.

Since the great majority of fusimotor axons selectively innervate one type of intrafusal fibre we have been able to determine the action of each type of fibre separately on 29 primary sensory endings and six secondary sensory endings. The conclusions of these experiments are summarised elsewhere (Boyd *et al.*, 1979; Boyd, 1980); particular attention is focused here on the effect of activation of each type of intrafusal fibre, at different frequencies of stimulation, on the dynamic and length sensitivities of the primary and secondary sensory endings. Since the action of most axons was tested when the muscle was still in the cat (*see* Methods), and it was not known at that stage which spindle would be isolated successfully, it was not practical to test changes in dynamic or length sensitivity systematically at different amplitudes and velocities of stretch for all the available group Ia and group II discharges in different spinal root filaments.

The 'dynamic response' in this paper is the difference between the peak frequency at the end of the ramp stretch and the almost fully adapted frequency 2–2.5 sec later at the commencement of the release of stretch, corrected for receptor adaptation (*see* below). Changes in 'dynamic sensitivity' are expressed as the percentage increase or decrease in this corrected dynamic response.

The 'static response' is the difference between the discharge frequency at the commencement of the stretch and the almost fully adapted value at the start of the release, corrected for receptor adaptation. The 'length sensitivity' is this corrected static response divided by the amplitude of stretch and the effect of fusimotor activity is expressed as the percentage increase or decrease in this length sensitivity measured under static conditions. In some cases the length sensitivity of the primary sensory ending was also measured under dynamic conditions, in which case it was given by the slope of the linear increase in frequency which occurred during the dynamic phase of a slow stretch.

METHODS

Group Ia (and sometimes group II) afferents from several spindles in the tenuissimus muscle were isolated in dorsal root filaments, and as many as possible of the fusimotor axons which supplied them were isolated in ventral root filaments. The motor axons were classified as dynamic or static according to whether or not they increased the dynamic index of primary sensory endings during a ramp and hold stretch. Then a 4 cm length of muscle, with its nerve and blood supply intact, was reflected into a bath. One spindle supplied by at least three, and sometimes as many as seven, of the classified fusimotor axons was isolated and the behaviour of its individual intrafusal fibres when each of the axons was stimulated was recorded on videotape. On a number of occasions the action of each axon on the Ia or II discharge was tested again after the muscle was in the bath, and in a few cases after the spindle was isolated if the blood supply was very good and there was a well-maintained static discharge from the primary ending. The method is described in detail by Boyd (1980). The cats were fully anaesthetised with pentobarbitone sodium (Sagatal—May and Baker Ltd; 45 mg/kg I.P. supplemented when required).

RESULTS

The static bag$_2$ fibre

Primary sensory ending

Activation of one or both poles of the static bag$_2$ fibre by stimulation at 100 Hz of a static γ axon supplying it selectively results in an initial peak increase in Ia frequency of from 25/sec to 150/sec. This initial sharp peak is followed by pronounced receptor adaptation (figure 1a). The time course of this adaptation in the absence of the ramp stretch is shown by the upper broken line in figure 1a. In this figure the response to stretch during fusimotor stimulation, less the fusimotor response, is shown by the dotted line. Clearly, the dynamic response to stretch, corrected for receptor adaptation, is appreciably reduced when the static bag$_2$ fibre is active. The dynamic sensitivity of the primary ending was always decreased by activity in the static bag$_2$ fibre, and the effect for a sample of 10 such fibres in different spindles is shown in figure 4a (----).

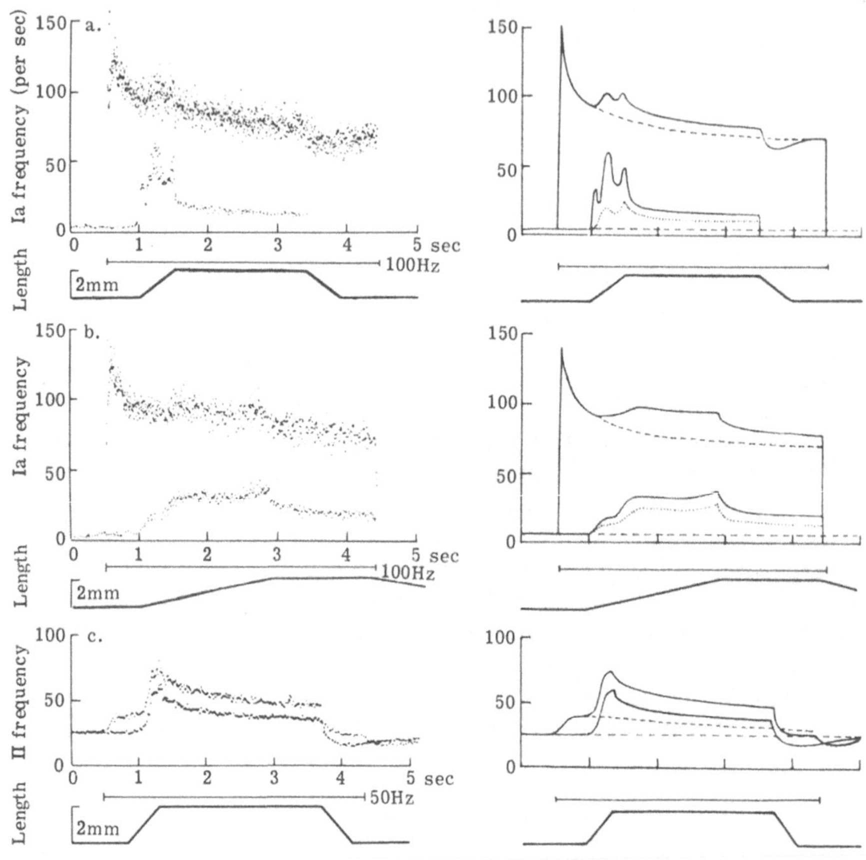

Figure 1. The static bag_2 fibre. Records of instantaneous frequency, and tracings of mean frequency, of the response of a primary sensory ending (a and b) and a secondary sensory ending (c) to a ramp stretch in the presence of (upper trace), and in the absence of (lower trace), activation of the static bag_2 fibre for the period shown by the continuous line beneath each trace.

(a) Primary ending. Static γ axon to static bag_2 fibre only at 100 Hz. Stretch velocity 4 mm/sec. Three recordings superimposed.
(b) Same primary ending and static γ axon as in (a). Stretch velocity about 1 mm/sec. Three recordings superimposed.
(c) Secondary ending. Static γ axon to static bag_2 fibre only at 50 Hz. Stretch velocity 4.5 mm/sec.

Dashed curves indicate receptor adaptation in the absence of the ramp stretch. Dotted curves indicate the response to stretch during γ stimulation less the fusimotor response.

The decrease at 100 Hz varied from 25% to 80% and was, in general, maximal at this frequency.

The static response of the primary ending in figure 1a, corrected for receptor adaptation, was also reduced by activity in the static bag_2 fibre. All such fibres either decreased, or did not change, the length sensitivity of the primary ending, and the values for a sample of 10 fibres are shown in figure 5a (----). Since the correction for receptor adaptation is large, and the discharge shows quite a large degree of irregularity, the measurements at different frequencies of stimulation should be regarded as approximate, only. However, it is clear that activity in the static bag_2 fibre never *increases* the length sensitivity of the primary ending. Confirmation of this is provided by the change in slope of the response during a slow stretch (figure 1b). The slope is definitely not increased, and is, in fact, slightly decreased during activation of the static bag_2 fibre at 100 Hz in this case (figure 1b, · · · ·).

Activity in the static bag_2 fibre does not cause driving of the primary sensory discharge, except very occasionally, and then only at very low frequencies of stimulation.

Secondary sensory ending

Activation of the static bag_2 fibre has a relatively small action on the secondary sensory ending, the group II frequency increasing by an amount varying from 0 to 50/sec, usually about 20/sec (figure 1c), reaching this maximum at between 50 Hz and 100 Hz. The dynamic sensitivity is unchanged, or decreases by up to 50% (figure 4b, ----), and in general any effect increases with the frequency of stimulation. In view of the small excitatory effect of most static bag_2 fibres it is doubtful whether any change of less than 20% is significant. The length sensitivity of the secondary ending may be unchanged or slightly increased (figure 5b, ----). It is doubtful whether these increases are significant, except in one case.

The nuclear chain fibres

Primary sensory ending

Nuclear chain fibres always drive the Ia discharge, usually 1:1 with the fusimotor frequency, up to about 60 Hz (figure 2b), unless the spindle is very slack. They rarely drive above 75 Hz, however, and at 100 Hz or above the discharge is usually very irregular (figure 2c) and has a lower mean frequency than the highest value at which driving occurs. It is impossible to assess the effect of chain fibre activity on the dynamic or length sensitivity of the primary ending, but neither is increased appreciably. Due to the mechanical oscillatory contraction of chain fibres at quite high frequencies of stimulation, the length signal is disrupted, as is the dynamic signal in most cases, though sometimes the dynamic peak at the end of the stretch has a constant value despite irregularity or driving elsewhere in the response.

Secondary sensory ending

In contrast, the chain fibres rarely drive the secondary sensory ending which gives a good length signal at all frequencies of stimulation. Activity in the chain fibres at the pole in which the secondary ending lies increases the group II afferent discharge by 70/sec to 100/sec (figures 2d and 2e), so that the input

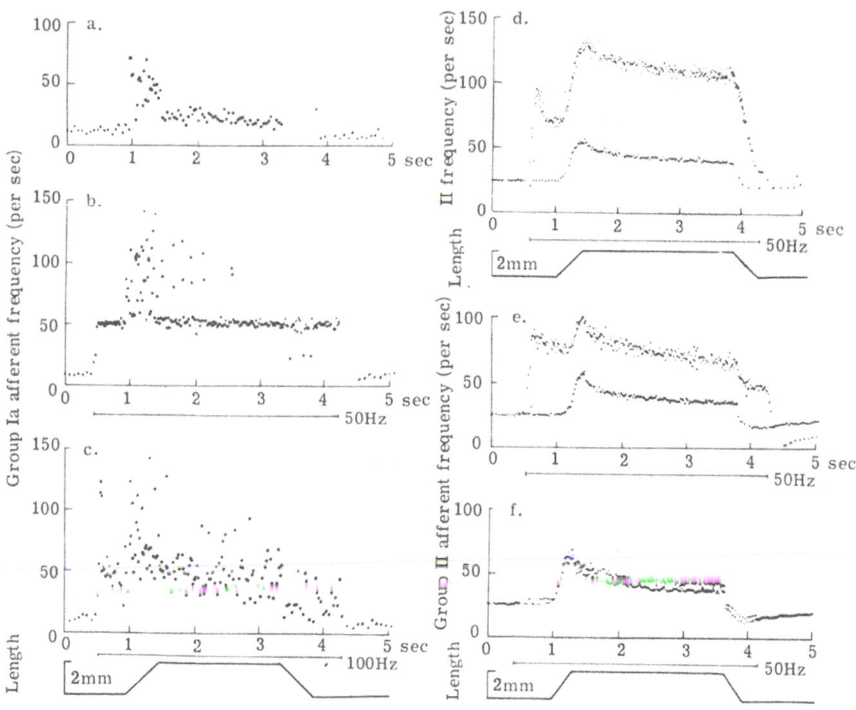

Figure 2. The nuclear chain fibres.

(a)–(c). Primary sensory ending. Static γ axon to nuclear chain fibres alone in one spindle stimulated at (a) 0 Hz, (b) 50 Hz and (c) 100 Hz. Stretch velocity 4 mm/sec.

(d) and (e). Secondary sensory endings. Static γ axon to nuclear chain fibres alone at the pole containing a secondary ending in two spindles. Lower trace, no stimulation. Upper trace, 50 Hz. Stretch velocity 4.5 mm/sec.

(f). Secondary sensory ending. Static γ axon to nuclear chain fibres alone at the pole opposite to that containing the secondary ending. Same spindle as (e). Stretch velocity 4.5 mm/sec.

from this source is, on average, about five times that from the static bag_2 fibre. The secondary sensory response to chain fibre activation adapts steadily after an initial peak, necessitating correction of measurements of the dynamic response and static response as before.

Activation of chain fibres at the same pole always decreases the dynamic sensitivity of the secondary ending by up to 50%, occasionally more (figures 2d and 2e; 4b, ———). The length sensitivity, on the other hand, is usually increased (figure 5b, ———), sometimes by a small amount (figure 2e), sometimes by a very large amount (figure 2d). Further, the largest increase may occur at a frequency less than that which produces maximum excitation of the ending.

Activation of the chain fibres at the pole opposite to that containing the secondary ending usually has very little effect on it (figure 2f), though there may be a modest effect if the spindle is held in a well-extended state.

The dynamic bag_1 fibre

Primary sensory ending

Activation of one or both poles of the dynamic bag_1 fibre by stimulation of a dynamic axon at 100 Hz results in a transient increase in the Ia discharge of between 5 and 100/sec, but not more. After this transient peak, the response settles to a value, varying from 2 or 3/sec (figure 3b) to 50/sec above the initial discharge, this 'on-response' usually being about 20/sec. There is little receptor adaptation thereafter (figure 3d).

The response to stretch follows the well-documented pattern which results from dynamic fusimotor activity (e.g. Crowe and Matthews, 1964; Lennerstrand and Thoden, 1968a). There is an initial transient stiction (or acceleration) effect (figure 3c) which tends to disappear in the increased response during the dynamic phase of the stretch (figure 3d). This increased response consists of a rapid rise, mirrored by a rapid fall immediately after the end of the stretch, and which is dependent on the velocity of stretching. During the dynamic phase of the stretch there is an almost linear increase in frequency which is related to the actual extent of the applied stretch at any point in time. The slope of this component increases, as shown by the continuous lines in figure 3, by an amount dependent on the dynamic fusimotor frequency. Finally, there is a period of slow adaptation with a time constant of about 0.5 sec, which is of small amplitude in the absence of fusimotor activity, but which is pronounced when the dynamic bag_1 fibre is active. These three features—the fast rise and fall, the increased length sensitivity during the ramp, and the slow decay—are still in evidence in the Ia discharge from a fully isolated spindle (figure 3b), even though the frequency of discharge may be lower than it was when the muscle was in the cat.

The slow decay follows the same time course as the creep which occurs in an active dynamic bag_1 fibre when it is held stretched (Boyd, 1976; Boyd *et al.*, 1977), and is apparently mechanical in origin. The increase in length sensitivity during the stretch is explained by the increase in stiffness of the active pole(s) of the dynamic bag_1 fibre so that a greater proportion of the stretch is transmitted to its primary sensory spiral. The origin of the fast rise and fall is unknown though it may be associated with a change in the potas-

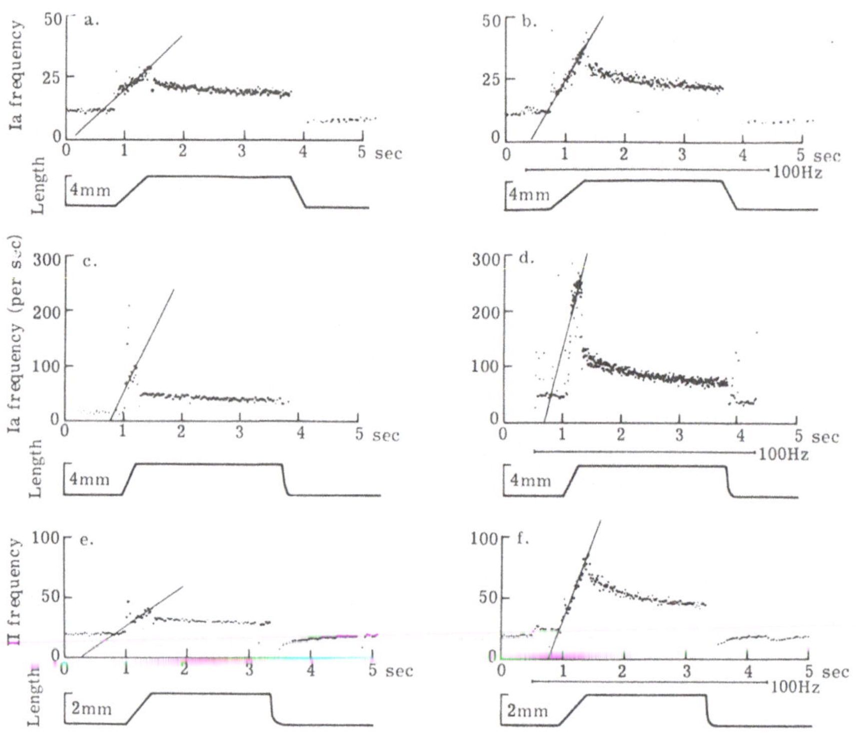

Figure 3. The dynamic bag_1 fibre.

(a) and (b). Primary sensory ending in a fully isolated spindle. Dynamic γ axon to dynamic bag_1 fibre alone, stimulated at (a) 0 Hz and (b) 100 Hz. Three recordings superimposed. Stretch velocity 8 mm/sec.

(c) and (d). Primary sensory ending in cat before isolation of spindle. Dynamic γ axon to dynamic bag_1 fibre alone, stimulated at (c) 0 Hz and (d) 100 Hz. Stretch velocity 16 mm/sec.

(e) and (f). Secondary sensory ending, muscle in bath but spindle not isolated. Dynamic γ axon to dynamic bag_1 fibre alone stimulated at (e) 0 Hz and (f) 100 Hz. Stretch velocity 4.5 mm/sec.

sium conductance of the receptor membrane or first node of Ranvier, as is the post-dynamic undershoot which occurs at high stretch velocities (Hunt *et al.*, 1978). If mechanical in origin it is too rapid to be detected by visual observation of the spiral with the techniques presently available.

The increase in dynamic sensitivity of the primary ending when the dynamic bag_1 fibre was active varied greatly from one spindle to another in the present experiments. For the samples of 10 dynamic bag_1 fibres in figure 4a, the increase in dynamic sensitivity varied from 30% to 250%, that is, the corrected dynamic response (or dynamic index) increased up to three or four times when the dynamic bag_1 fibre was maximally activated at 75–100 Hz. At the velocities used in the present study, however, the increase was often only about 100% and was due, mainly, to an increase in length sensitivity during the ramp (figures 3a and 3b) rather than to an increase in the initial fast component which is marked at high velocities of stretch.

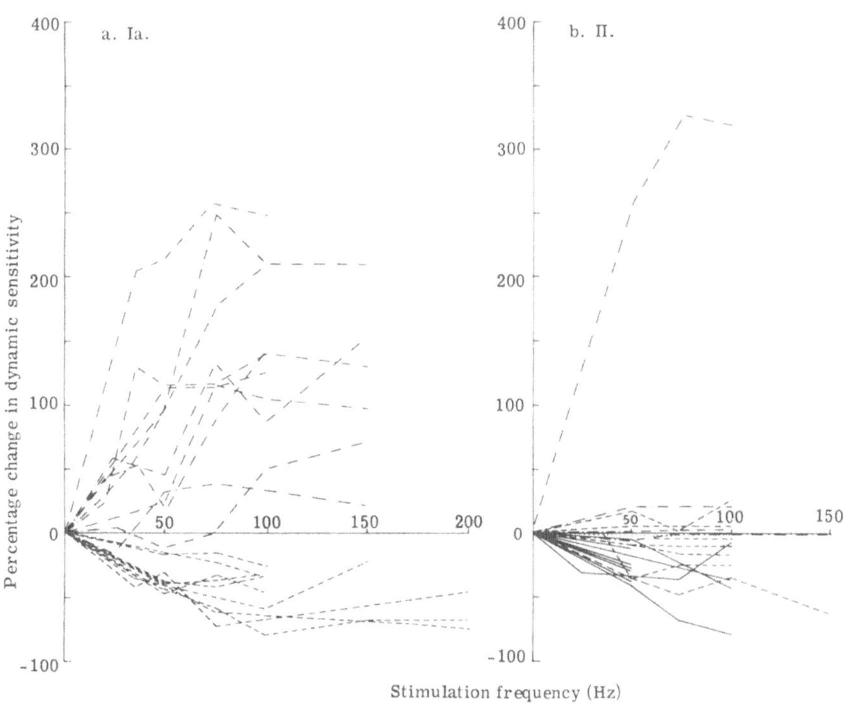

Figure 4. The effect of frequency of fusimotor stimulation on the dynamic sensitivity of (a) primary sensory endings and (b) secondary sensory endings. Dynamic bag_1 fibres, — · — · — ·. Static bag_2 fibres, - - - -. Nuclear chain fibres, ———. Dynamic bag_1 + static bag_2 fibres, ·····.

The length sensitivity of the primary ending, derived from the difference between the discharge frequency at the start of stretch and that at the commencement of release, was increased during activation of the dynamic bag_1 fibre by an amount which varied widely from 0 to 200% but was often close to 100% (figure 5a, — · — · — ·).

Secondary sensory ending

Of the six secondary sensory endings studied three were unaffected by activity in the dynamic bag_1 fibre, whereas the other three were excited to a small degree. In one case only there was a marked increase in both the dynamic and length sensitivities (figures 3e and 3f; figure 4b; figure 5b, — · — · — ·); the response showed all the characteristics of a primary ending, except that it was more regular, there was a higher initial discharge frequency and a less obvious

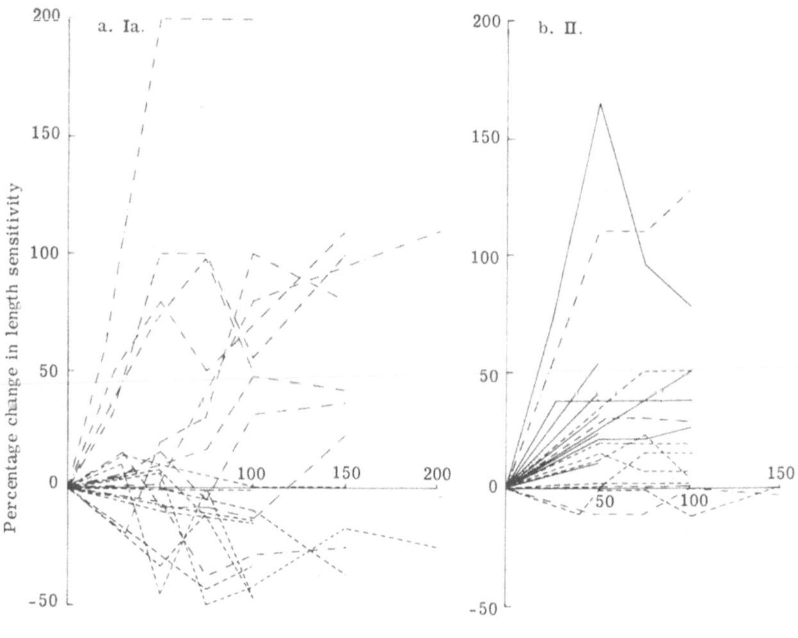

Figure 5. The effect of frequency of fusimotor stimulation on the length sensitivity of (a) primary sensory endings and (b) secondary sensory endings. Dynamic bag_1 fibres, — · — · — ·. Static bag_2 fibres, - - - -. Nuclear chain fibres, ———. Dynamic bag_1 fibre + static bag_2 fibre, ····. All primary and secondary endings were discharging steadily prior to the ramp stretch.

fast rising phase, than for the primary ending in the same spindle. It should be emphasised that it is very rare for the dynamic sensitivity of a secondary ending to be increased by dynamic fusimotor axons. However, one other secondary ending whose dynamic sensitivity was not significantly increased when the dynamic bag_1 fibre was active, did show a significant increase in length sensitivity. Thus, the possibility that the dynamic bag_1 fibre may modify the length sensitivity, but not the dynamic sensitivity, of some secondary endings must be borne in mind.

The high frequency and regularity of the discharge of the secondary ending whose dynamic sensitivity was increased when the dynamic bag_1 fibre became active, made it very suitable for a study of the effect of changes in the velocity of stretch in the presence and absence of dynamic fusimotor activity. The response to the same amplitude of stretch at six different velocities, both at 0 Hz and 100 Hz, is shown in figure 6a, and the effect of a very slow stretch in figure 6b. The increase in the slope of the response during the ramp when the dynamic bag_1 fibre was active is very obvious at all velocities (figures 3e and 3f; figure 6b), as is the increase in slope as the velocity increases (figure 6a). It is not immediately apparent, however, whether the increase in slope is velocity dependent. The frequency at any point in time during the ramp could merely reflect the degree of the applied stretch at that time, and the increase in the peak frequency could be due entirely to an increase in the amplitude of the initial fast component. It is shown in figure 6d, however, that the increase in slope *is* velocity dependent. The closed circles (●) show the length sensitivity during the dynamic phase of stretching when the dynamic bag_1 fibre was activated at 100 Hz; the sensitivity was not only much greater than when the fibre was inactive (▲) but also definitely increased with velocity. The length sensitivity measured after adaptation was complete was independent of velocity as expected, though it was considerably greater when the dynamic bag_1 fibre was active (○) than when it was not (△). The amplitude of the fast component of the dynamic response during stimulation at 100 Hz is shown in figure 6c (■, fast rise; □, fast fall). There is no significant fast component until the velocity exceeds 2 mm/sec, and it increases with velocity thereafter. The amplitude of the initial stiction effect at 100 Hz is also shown (×); this is constant until the velocity exceeds 2 mm/sec and then increases thereafter until it is overtaken by the fast component of the response.

DISCUSSION

The present work confirms the long-accepted view that the secondary sensory ending is a length-measuring device which derives most of its input from the nuclear chain fibres as expected since the secondary sensory terminals lie predominantly on these fibres (Boyd, 1962; Barker and Cope, 1962). Contraction of the chain fibres at the pole containing the ending greatly excites it

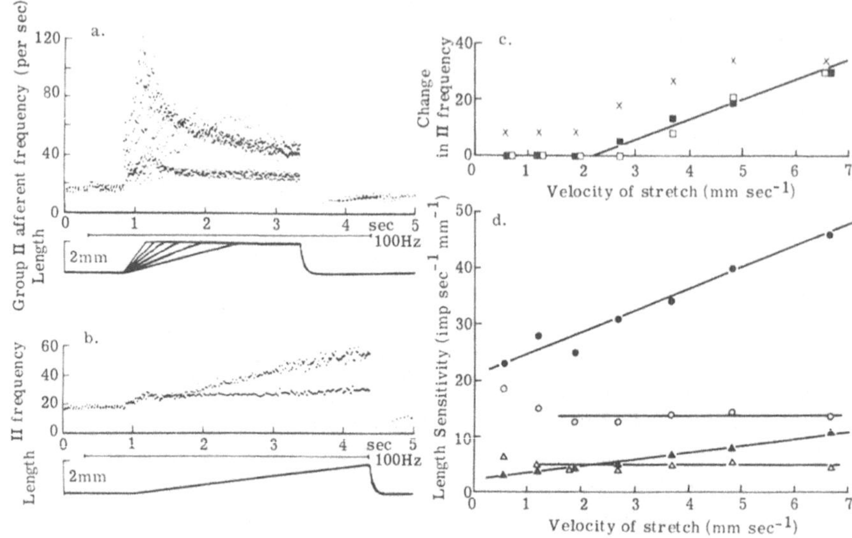

Figure 6. Unusual secondary ending (same as in figures 3(e) and 3(f) whose dynamic sensitivity was increased by dynamic bag$_{\text{I}}$ fibre contraction).

(a) and (b). Response to ramp stretches at seven different velocities during stimulation of a dynamic γ axon at 100 Hz (upper traces) and in the absence of stimulation (lower traces).

(c). Amplitude of fast rise (■) and fast fall (□) of the response of the secondary ending in (a) and (b) during ramp stretches at different velocities. ×, amplitude of initial stiction effect.

(d). Length sensitivity of the same secondary ending, derived from the slope of the linear response during the ramp (filled symbols) or from the difference between the discharge frequency immediately preceding the stretch and that preceding the release (open symbols). Circles, during activation of the dynamic bag$_{\text{I}}$ fibre at 100 Hz. Triangles, in the absence of stimulation. First two open symbol values high because adaptation was incomplete at end of ramp plateau after low velocity stretches.

and at the same time increases its length sensitivity, sometimes substantially. However, a majority of secondary endings have a significant proportion of terminations on nuclear bag fibres (Banks *et al.*, 1977) and it has been shown here that they receive a small but significant physiological input from the static bag_2 fibre. Activity in this fibre does not, however, modify the length sensitivity of the ending appreciably. These effects of the chain fibres and static bag_2 fibre on length sensitivity are consistent with the glycogen-depletion study of Jami *et al.* (1980). It is probable that simultaneous activity in several static axons to a spindle always results in an increase in the length sensitivity of its secondary endings. Some secondary endings receive a small input from the dynamic bag_1 fibre which may also increase the length sensitivity of the ending to some degree, but very rarely increases its dynamic sensitivity. The exceptional secondary ending in figures 3e and 3f, which did have its dynamic sensitivity increased by the bag_1 fibre, possibly had an appreciable number of its terminals on the nuclear bag region of the fibre. In general, activity in any of the three types of intrafusal fibre reduces, or does not alter, the dynamic sensitivity of the secondary ending.

The purpose of the input from the nuclear chain fibres to the primary ending is obscure. The Ia length signal is disrupted by driving or irregularity. It seems likely that the input from the chain fibres is minimal during extrafusal contraction because of the ease with which they are unloaded and kinked. The input may be substantial, however, when the muscle is extended and it is perhaps this dependence of the system on muscle length which is important.

Activity in the static bag_2 fibre provides a powerful positive bias to the Ia discharge without increasing the length sensitivity of the primary ending and, in fact, usually decreases it. Lennerstrand and Thoden (1968b) found that many static axons increased the length sensitivity of the primary ending, though many others decreased it. Further, Dutia (1980) found that in phase III of excitation of the primary ending by intra-arterial succinylcholine, which he reasonably attributed to recruitment of the static bag_2 fibre, the length sensitivity was greatly increased. Crowe and Matthews (1964), however, found that most static axons did not change the length sensitivity of the primary ending, and only a few increased it. Since static axons which selectively innervate the static bag_2 fibre decrease (or do not change) the length sensitivity, and the nuclear chain fibres do not change it in any meaningful way, some other explanation of the observed increases is needed.

The static levels of discharge frequency recorded by Dutia (1980) during scoline infusion were very much higher than those normally attained by stimulation of single static axons. Probably scoline depolarises and causes contraction of the whole length of both poles of the static bag_2 fibre except for the equatorial region, whereas static γ stimulation causes localised contractions at fusimotor endings. The occasional propagated action potential in a static bag_2 fibre would likewise cause the whole of both poles to contract and hence increase the length sensitivity of the primary ending. Barker *et al.*

(1978) recorded a propagated potential intracellularly in one static bag_2 fibre out of a total of seven which were impaled. Gladden (1981), on the other hand, has consistently recorded only local potentials in 11 static bag_2 fibres. Finally, it should be noted that scoline infusion activates both the static bag_2 fibre and the dynamic bag_1 fibre. Thus, a non-selective axon which activates both types of nuclear bag fibre could conceivably increase the length sensitivity of the primary ending. One, and only one, such axon was encountered in the present study. We believe such axons to be rare in the tenuissimus muscle though they may be more common in other muscles, and are dubious as to whether they can rightly be classified as static in any case. The possibility that an increase in length sensitivity of the primary ending results from activity in the dynamic bag_1 fibre is, however, very real.

The length sensitivity of the primary ending, assessed in the present work from the difference between the discharge at the start of the stretch and the value 2–2.5 sec after the stretch was completed, was usually increased by 50–100% by maximal activation of the dynamic bag_1 fibre. At first sight this finding is in conflict with the observations of Crowe and Matthews (1964) and Lennerstrand and Thoden (1968a) that stimulation of dynamic axons did not alter, or in the latter study slightly reduced, the length sensitivity of the primary ending. It could be argued that the apparent increase in length sensitivity in the present work was due simply to the fact that the discharge had not adapted fully at 2.5 sec, or that the response of the ending is so non-linear that a single 'two-point' assessment of it gives a result not applicable over the full range of stretch. It can be seen from figure 3, however, that the discharge frequency is almost constant 2.5 sec after the peak response, and in some cases it certainly was so. Further, the fact that the length sensitivity was nearly always increased for all frequencies of stimulation of almost all dynamic bag_1 fibres (figure 5) makes it unlikely that non-linearity of the response was responsible. The effect is probably due to the plasticity of the dynamic bag_1 fibre. If the muscle is stretched rapidly and held stretched when the bag_1 fibre is active, it may give way and creep, so that the final steady extension of the sensory spiral is greater than if the muscle is slowly extended to the same final length as was done by Lennerstrand and Thoden (1968a). Also, Crowe and Matthews (1964) found that dynamic fusimotor stimulation produced approximately the same increase in the discharge of the ending before and after a ramp stretch was applied. Dynamic fusimotor stimulation with the muscle held at any length causes contraction of the dynamic bag_1 fibre without any subsequent creep. Again, the frequency attained during stimula-tion at a particular length may be less than that when the fusimotor axon is first stimulated and then the muscle is stretched rapidly to the same final length.

If this hypothesis is correct, and we have not yet studied the origin of creep in the bag_1 fibre in sufficient detail to know that it is so, then variations in the degree of this plasticity could explain why the length sensitivity as measured in the present work was increased greatly for some dynamic bag_1 fibres and to a much smaller extent in others. Further, the interesting possibility arises that some contracting dynamic bag_1 fibre poles may fail to creep at all. If this were the case then the final discharge frequency at the end of the ramp plateau would be very much greater than that at the commencement of the stretch. The dynamic index would then be relatively small even though the increase in frequency during the ramp was large. The axon responsible might then be classified as static even though it selectively innervated the dynamic bag_1 fibre.

Crowe and Matthews (1964) did obtain a marked increase in the length sensitivity of primary endings *during* the ramp stretch when dynamic axons were stimulated. The slope of the response during the ramp increased with stimulation frequency, just as it did for the two primary endings, and one secondary ending, in figure 3 of this paper. Crowe and Matthews (1964) do not appear to have tested whether this effect was velocity dependent. If the dynamically measured length sensitivity of the primary ending increases with the velocity of stretch as well as with stimulation frequency, as has been shown to be the case for a secondary ending in the present work (figure 6), then this could be a contributory factor to the increase in dynamic response which is related to the velocity of stretch. At low stretch velocities it may, in fact, be more important than any increase in the fast initial rise. Relatively low velocities of extension are possibly more typical of some normal movements than high velocities, in which case the principal action of the dynamic bag_1 fibre is to increase the length sensitivity of the primary ending during the dynamic phase of stretch. Since this length sensitivity change is velocity dependent, the contribution to the Ia discharge which arises in the dynamic bag_1 fibre is a function of both length and velocity.

ACKNOWLEDGEMENT

This work was supported by a grant from The Wellcome Trust.

Received on July 3rd, 1980.

REFERENCES

Banks, R. W., Barker, D. and Stacey, M. J. (1977). Intrafusal branching and distribution of primary and secondary afferents, *J. Physiol., Lond.*, **272**, 66–67P

Barker, D., Bessou, P., Jankowska, E., Pagès, B. and Stacey, M. J. (1978). Identification of intrafusal muscle fibres activated by single fusimotor axons injected with fluorescent dye in cat tenuissimus spindles, *J. Physiol., Lond.*, **275**, 149–166

Barker, D. and Cope, M. (1962). The innervation of individual intrafusal muscle fibres. In *Symposium on Muscle Receptors* (ed. D. Barker), Hong Kong, Hong Kong University Press, 263–269

Boyd, I. A. (1962). The structure and innervation of the nuclear bag muscle fibre system and the nuclear chain muscle fibre system in mammalian muscle spindles, *Proc. R. Soc. B*, **245**, 81–136

Boyd, I. A. (1976). The mechanical properties of dynamic nuclear bag fibres, static nuclear bag fibres and nuclear chain fibres in isolated cat muscle spindles, *Prog. Brain Res.*, **44**, 32–50

Boyd, I. A. (1980). The isolated mammalian muscle spindle, *Trends Neurosci.*, **3**, 258–265

Boyd, I. A., Gladden, M. H. and Ward, J. (1977). The contribution of intrafusal creep to the dynamic component of the Ia afferent discharge of isolated muscle spindles, *J. Physiol., Lond.*, **273**, 27–28P

Boyd, I. A., Gladden, M. H. and Ward, J. (1979). The effect of contraction in the three types of intrafusal fibre in isolated cat muscle spindles on the Ia discharge during stretch, *J. Physiol., Lond.*, **296**, 41P

Crowe, A. and Matthews, P. B. C. (1964). The effects of stimulation of static and dynamic fusimotor fibres on the response to stretching of the primary endings of muscle spindles, *J. Physiol., Lond.*, **174**, 109–131

Dutia, M. B. (1980). Activation of cat muscle spindle primary, secondary and intermediate sensory endings by suxamethonium, *J. Physiol., Lond.*, **304**, 315–330

Gladden, M. H. (1981). The activity of intrafusal muscle fibres during central stimulation in the cat, this publication

Hunt, C. C., Wilkinson, R. S. and Fukami, Y. (1978). Ionic basis of the receptor potential in primary endings of mammalian muscle spindles, *J. gen. Physiol.*, **71**, 683–698

Jami, L., Lan-Couton, D. and Petit, J. (1980). A study with the glycogen-depletion method of intrafusal distribution of γ axons that increase sensitivity of spindle secondary endings, *J. Neurophysiol.*, **43**, 16–26

Lennerstrand, G. and Thoden, U. (1968a). Position and velocity sensitivity of muscle spindles in the cat. II: Dynamic fusimotor single-fibre activation of primary endings, *Acta physiol. scand.*, **74**, 16–29

Lennerstrand, G. and Thoden, U.(1968b). Position and velocity sensitivity of muscle spindles of the cat. III: Static fusimotor single-fibre activation of primary and secondary endings, *Acta physiol. scand.*, **74**, 30–49

Nature of the dynamic response and its relation to the high sensitivity of muscle spindles to small changes in length

J. C. HOUK*, W. Z. RYMER* AND P. E. CRAGO†

SUMMARY

The dynamic responses of spindle receptors to large ramp changes in length cannot be broken down into length, velocity and acceleration components. Instead, it is shown that, except for a brief transient phase associated with the initial burst, response is proportional to the product of a velocity-dependent and a length-dependent term. The velocity-dependent term is approximately $v^{0.3}$ for both primary and secondary endings, both in the presence and in the absence of spontaneous fusimotor activity. For the majority of receptors, length dependence is well characterised by an upwardly offset straight-line relation between discharge rate and muscle length. It seems likely that this peculiar form of relation derives from the mechanical properties of intrafusal muscle fibres. Regardless of the mechanism, the relation places many interesting constraints on theories of motor control.

INTRODUCTION

Background on dynamic response

The responses to ramp and hold stretches look very much as if they were made up of the sum of length, velocity and acceleration components as portrayed schematically in figure 1. There is a transient at stretch onset (the 'initial burst') that resembles an acceleration response, followed by a step increase in discharge rate that resembles a velocity response, both of which are superimposed upon a progressive increase in discharge rate that resembles a

* Northwestern University, Chicago
† Case Western Reserve, Cleveland, Ohio

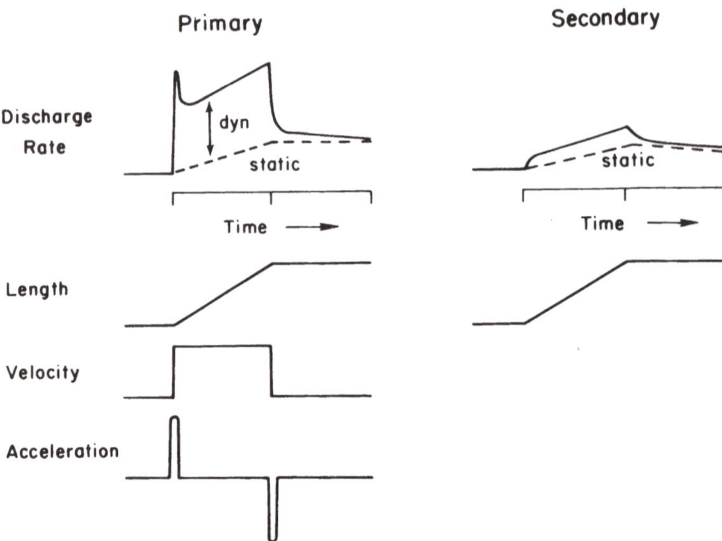

Figure 1. Dynamic responses of primary and secondary endings. The response of the primary ending looks as if it were made up of the sum of length, velocity and acceleration components, but this appearance is fortuitous.

position (or length) response. In line with this interpretation, the abrupt fall in discharge rate which occurs at ramp plateau (at which point velocity returns to zero) might be expected to provide an ideal measure of the velocity-dependent component of response, which is the main justification for the use of the 'dynamic index' as a measure of large-signal velocity sensitivity (Crowe and Matthews, 1964).

More detailed analyses of the dynamic responses to ramp and triangular stretches indicate that the model described in the previous paragraph is not generally valid and, in many instances, is quite misleading. Specifically, the initial burst appears to depend largely upon the occurrence of a certain amplitude of stretch rather than relating to acceleration (Lennerstrand and Thoden, 1968a; Brown *et al.*, 1969; Houk *et al.*, 1973). A second problem is that the magnitude of the step increase in rate discussed in the previous paragraph may not be proportional to stretch velocity (Lennerstrand and Thoden, 1968b; Hasan and Houk, 1975a), in contrast to expectations based on figure 1. Finally, the slope of the relation between muscle length and discharge rate (i.e. the positional sensitivity) turns out not to be a constant; it varies with stretch velocity and it typically exceeds the static sensitivity of the receptor (Crowe and Matthews, 1964; Lennerstrand and Thoden, 1968a,b). Andersson (in an appendix to Lennerstrand, 1968) attributed the variable positional sensitivity to the existence of a linear term with a long time constant in the transfer function (*see also* Crowe and Matthews, 1964; Lennerstrand

and Thoden, 1968a). However, this interpretation is inconsistent with the magnitude and speed of the initial decay in rate which accompanies cessation of constant velocity stretch (Hasan and Houk, 1975b).

Background on small-signal responsiveness

An equally puzzling problem in spindle receptor physiology concerns the radical differences in responsiveness that are observed when length changes are restricted to amplitudes less than about 0.5% of the physiological range. A variety of recent work has shown that primary endings are an order of magnitude more sensitive to these minute changes in length than to stretches of larger amplitude (Matthews and Stein, 1969; Poppele and Bowman, 1970; Matthews, 1972; Hasan and Houk, 1975a,b). The transition from high to low sensitivity has been studied in detail using very slow ramp stretches of small to intermediate amplitude (Hasan and Houk, 1972; Houk, Harris and Hasan, 1973; Hasan and Houk, 1975a,b), as illustrated in figure 2. This transition was

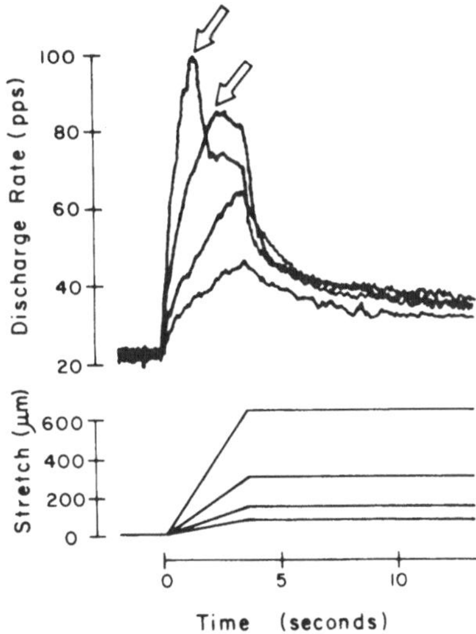

Figure 2. Evolution of the dynamic response. The family of responses of a primary ending to four different amplitudes of stretch illustrates the transition from high to low sensitivity and the manner in which the dynamic response evolves from this transition (modified from Hasan and Houk, 1975b). Note that the response to the smallest ramp is disproportionately large. Responses to larger ramps show severe saturation in the static phase and a lesser saturation in the dynamic phase. In fact, the initial portions of the responses show no saturation; they scale in proportion to the size of the ramp. Failure to scale (i.e. the departure from linear behaviour) begins at the time of the abrupt discontinuities in response marked by the arrows. These discontinuities become initial bursts in responses to larger and faster ramps.

shown to be a relatively abrupt event (marked by the arrow in figure 2) that occurs after a certain amplitude of stretch (0.1–0.2 mm in cat soleus spindles) rather than being related to a particular value of, or increment in, discharge rate. It was also shown that the decay of the initial burst marks the time during which this transition is in rapid progress. Prior to the peak of the initial burst, spindle receptors are well characterised by a linear dynamic model, whereas after the burst the response properties are highly non-linear.

A second important difference between large- and small-signal responsiveness concerns the overall appearance of the dynamic response, as is quite noticeable from a comparison of responses to different-sized ramps in figure 2. Responses to small ramps lack several of the characteristic features of the responses to larger ramps; namely, the initial burst (the acceleration-like component in figure 1), the step increase in rate (the velocity-like component in figure 1), and the large, fast component of adaptation at ramp plateau. It is the combination of these differences that is responsible for the rather dynamic appearance of responses to large ramps. While all of the named differences are attributable to non-linear features of the response properties of primary spindle receptors (Hasan and Houk, 1975b), a specific non-linear relation that characterises these features is wanting.

METHODS

These various deficiencies in our understanding of responses to stretches extending beyond the small-signal linear region prompted us to reinvestigate the velocity dependence and other features of the dynamic response. Our experiments were conducted on decerebrate cats with intact ventral roots in order to preserve a level of spontaneous fusimotor activity approximating that present in the intact animal. Except for the few filaments that were dissected for recording, dorsal roots were also left intact in order to preserve reflex function. Control observations in de-efferented spindles showed reduced overall sensitivity and lower initial discharge, but the responses were still adequately characterised by the equations developed later (cf. Rymer *et al.*, 1981). We studied a population of 95 primary and secondary endings in soleus muscles of 18 animals. More details concerning methods can be found in a recent report (Rymer *et al.*, 1979).

RESULTS

Our data on each receptor consisted of 5–10 responses to 10 mm ramps at each of eight velocities in the range 0.4–100 mm/sec. Figure 3A shows response averages constructed from the trials at 1 mm/sec and 50 mm/sec for a typical

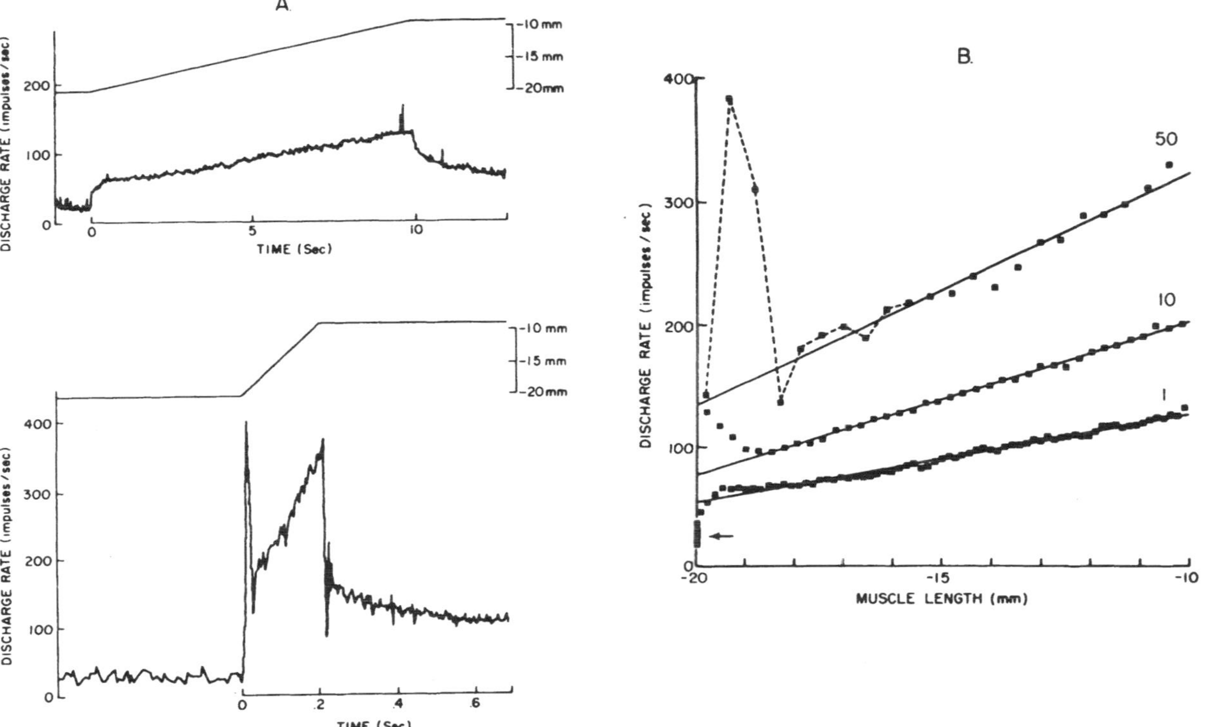

Figure 3. Dependence of dynamic responses on stretch velocity. Part A shows ensemble averages of the responses of a primary ending to a 10 mm stretch applied at two velocities (1 mm/sec and 50 mm/sec). Dynamic trajectories of these and one other response (10 mm/sec) are shown superimposed in part B, on a plot of discharge rate versus muscle length. The arrow indicates the initial discharge rate. Note that both the magnitude and the slope of the dynamic response increase with velocity, but that the increases are only about two-fold for a ten-fold change in velocity.

primary ending, and part B shows dynamic trajectories (plots of discharge rate versus muscle length during the constant-velocity phase of the ramp) constructed from response averages at three different velocities (1 mm/sec, 10 mm/sec and 100 mm/sec). The results obtained with primary endings will be described first, and the differences obtained with secondary endings will be noted in a separate section.

Velocity sensitivity

Superposition of dynamic trajectories as in figure 3B highlights the specific dependence of response on velocity independent of any length-related contribution. At any given length, corresponding to a particular amplitude of stretch, discharge rate is clearly greater at higher stretch velocities, but the increases are much less than proportional to the changes in velocity. Using the increment in discharge rate above the initial value as the measure of response, we found typically that a ten-fold increase in velocity results in less than a two-fold increase in response, which is a rather weak dependence. The exception is the response at ramp onset which is approximately proportional to velocity, up until the peak of the initial burst.

Plots of log (response) versus log (velocity) fell along straight lines with slopes less than one which indicated a fractional power relation (*see also* Schaefer, 1973). Response was taken as the increment in discharge rate measured after 7 mm of stretch at each of the eight velocities. Comparing different receptors, the slopes of regression lines had a narrow distribution about a mean value of 0.31. Thus, our velocity data are well summarised by the equation

$$r - r_o = Kv^{0.3} \quad (1)$$

where r represents discharge rate, r_o the initial value and v the velocity. The values of the parameter K (a scaling factor) correlated reasonably well with dynamic indices; however, the latter varied with velocity whereas the former have the advantage of being independent of velocity.

Relation between dynamic trajectories and length dependence

Figure 4 is a composite summary of our findings concerning dynamic trajectories. The solid trajectories are analogous to those shown in figure 3, and the dashed extrapolations illustrate the tendency for regression lines fitted to the data, excluding the initial burst, to intersect in the vicinity of a common point, designated r*, x*. This concept of trajectories following the course of lines that radiate from a common point is supported by other data we compiled.

The fact that our summary equation required no additive term dependent

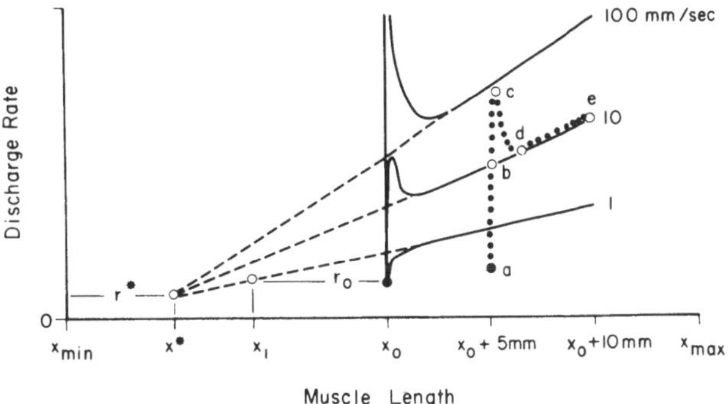

Figure 4. Composite summary showing the dependence of the dynamic response on muscle length and velocity for a typical primary ending. The initial burst divides responses to constant velocity stretch into two phases: (1) an initial transient characterised by high sensitivity (steep slope) and linear dynamic responsiveness, and (2) a steady state in velocity characterised by a family of straight lines (dashed) having slopes proportional to $v^{0.3}$ and tending to intersect at a common point (r*,x*). The dotted trajectory illustrates how a typical response to a constant velocity stretch (10 mm/sec) displays high sensitivity prior to the initial burst (a–c) but then rapidly converges upon the dashed line characterising the steady state at a constant velocity of 10 mm/sec. Note that the offset of the response (the segment a–b) is proportional to the product of muscle length and velocity raised to the 0.3 power, which contrasts with the previous view that it represents a velocity component.

on muscle length means that the obvious length dependence of the response must be accounted for either in terms of a slow transient (to be excluded later) or by allowing K in equation 1 to be a function of length (x), i.e.

$$r - r_o = K(x)v^{0.3} \qquad (2)$$

For cases in which the trajectory follows a linear course (true for the majority of endings), K(x) takes the form of an offset straight line

$$r - r_o = K_1(x - x_1)v^{0.3} \qquad (3)$$

where K_1 and x_1 are constants. If one treats velocity as a parameter in equation 3, one then generates a family of straight lines that radiate from the point r_o,x_1 (figure 4 shows the point, but not this particular family of lines).

The values of r* estimated by backward extrapolation of the trajectories were generally similar to the initial discharge rates r_o at the initial lengths x_o that we happened to choose. Correspondingly, values of x_1 and x* were similar. In any case, the velocity data do clearly support the concept that a length-dependent and a velocity-dependent term multiply to determine overall response. Added evidence for this was our finding that the slopes of dynamic trajectories depended on the same low fractional power of velocity as did the incremental responses.

Trajectories initiated from different starting lengths

A second type of evidence for the radiating line model was obtained by comparing responses to a given velocity of stretch, starting from two different initial lengths. The consistent result was that the trajectory initiated from the longer length (dotted trajectory in figure 4) had a larger offset (the distance from a to b) but the same slope as the control trajectory initiated from the shorter length. As a consequence the two trajectories promptly converged upon a common straight line. This result supports the empirical validity of the radiating line model and it also serves to define the duration of the transient phase of the response. The transient phase extends only from points a to d in the illustrated example, since beyond point d the response no longer depends on the past history of length change. This demonstrates that the transient phase is completed as the initial burst dies away, which is a very useful finding. It proves that the dependence of slope on velocity is not caused by a slow transient and it provides strong justification for our use of an algebraic expression rather than a differential equation to characterise the post-burst phase of the dynamic response (equations 2 and 3).

Effect of small-signal sensitivity in promoting the transition to a characteristic function

The radiating lines in figure 4, or more generally the family of curves $K(x)v^{0.3}$ in equation 2, appear to represent characteristic functions that define steady states in velocity. Thus, actual dynamic trajectories converge towards a given characteristic function whenever velocity is held constant. The course of the transition from a resting to a constant-velocity state is determined by the small-signal sensitivity of the receptor together with the stretch parameters. It is clear from figure 4 that high stretch velocities and large values of small-signal sensitivity combine to yield steep transitions on rate versus length co-ordinates, and this translates into rapid transitions on plots of discharge rate versus time. Thus, a major consequence of the high small-signal sensitivity of primary endings appears to be one of promoting transitions to characteristic functions corresponding to any given velocity of stretch.

Secondary endings

Although the dynamic responses of secondary endings are smaller than those of primary endings, our results indicate a basic qualitative similarity. Thus, secondary-ending dynamic responses are proportional to the same low fractional power of velocity and their dynamic trajectories show an initial steep rise (frequently culminating in an initial burst) followed by a convergence upon characteristic functions just as in the case of primary endings. One might expect that the lower small-signal sensitivities of secondary endings might delay the transitions to the characteristic functions, but this effect tends to be counterbalanced by the fact that the characteristic functions are also attenuated. In terms of figure 4, both the slope and the vertical extent of the line segment a to b are smaller in the case of secondary endings. In essence, the dynamic response of secondary endings is simply a scaled-down version of the dynamic response of primary endings. The lesser dynamic appearance results from the fact that the plateau phase of the response is not scaled down, and hence the dynamic to static ratio is much smaller.

An important consequence of these findings is that they counter the notion that one could derive a signal better related to velocity by subtracting responses of secondary endings from those of primary endings. Since both are closely related to $K\nu^{0.3}$, the result of subtraction would only be to modify the scaling parameter K.

DISCUSSION

Similarity between spindle dynamic responsiveness and muscle mechanical properties

Previous authors have pointed out many of the striking similarities between muscle mechanical and spindle receptor response properties (cf. Matthews, 1972 and Hasan and Houk, 1975b for reviews). In addition we wish to point out here the obvious resemblance between the characteristic functions described above and muscle length–tension curves. While further studies on muscles will be required to demonstrate the specifics, there is already present in the literature some suggestion that muscle force may under appropriate circumstances obey a multiplicative relationship between length and velocity similar to equation 3. The study of the cat soleus muscle by Joyce *et al.* (1969) is particularly helpful since the results are presented in terms of force–length trajectories (figure 3 in the quoted study) that are analogous to the rate–length trajectories described here. In particular, the trajectories they obtained during stimulation at 35 Hz appear to trace out a characteristic function that resembles a scaled version of the isometric length–tension curve. Furthermore, their figure 5 suggests that the increment in force above the isometric value during constant-velocity stretch superimposed on 35 Hz stimulation is a negatively accelerating function of velocity similar to the low fractional power of velocity reported here. The extra tension developed by isolated frog muscle fibres that are stimulated tetanically appears also to show a low fractional power dependence on stretch velocity (David Morgan, personal communication).

Functional significance of the dynamic response

It is clear that the information supplied to the CNS by spindle receptors is a highly complex representation of muscle length and velocity. What remains unclear is the use to which this information is put. Length and velocity information might be extracted by some complex algorithm, but we believe that this is unlikely. Instead, we favour a much simpler hypothesis.

The fact that the dynamic responses of primary endings typically have substantial offsets (due to short values of x^* in figure 4) when stretch begins from intermediate muscle lengths means that the dynamic response is large at relatively low velocities. The fact that the dynamic response depends on $v^{0.3}$ means that it increases relatively little at higher velocities. These features taken together indicate that the dynamic response is large when the muscle undergoes stretch, with only a weak sensitivity to the actual velocity. This suggests that the large dynamic response of primary endings is better suited for the detection of movement than for signalling the precise velocity at which movement occurs. In a later article in this book (Houk *et al.*, 1981), we relate this suggestion and other non-linear features of the dynamic response to the hypothesis for stiffness regulation. In addition, however, we would like to point out that these response features of primary endings would be expected to generate a reflex force in opposition to movement that is analogous to friction, as in fact has been observed by Roberts (1963) and confirmed by us (Houk *et al.*, 1977). The potential advantages of this type of damping remain to be fully evaluated, but may include a degree of damping that is relatively independent of the inertial load. Finally, if sensations of movement result from primary ending responses, as suggested by Goodwin *et al.* (1972), the limitations imposed by equation 2 should be demonstrable psychophysically.

ACKNOWLEDGEMENT

This research was supported in part by NIH grants NS06828, NS07226 and NS14703.

Received on June 9th, 1980.

REFERENCES

Brown, M. C., Goodwin, G. M. and Matthews, P. B. C. (1969). After-effects of fusimotor stimulation on the response of muscle spindle primary afferent endings, *J. Physiol., Lond.*, **205**, 667–694

Crowe, A. and Matthews, P. B. C. (1964). The effects of stimulation of static and dynamic fusimotor fibres on the response to stretching of the primary endings of muscle spindles, *J. Physiol., Lond.*, **174**, 109–131

Goodwin, G. M., McCloskey, D. I. and Matthews, P. B. C. (1972). The contribution of muscle afferents to kinaesthesia shown by vibration induced illusions of movement and by the effects of paralyzing joint afferents, *Brain*, **95**, 705–748

Hasan, Z. and Houk, J. C. (1972). Nonlinear behavior of primary spindle receptors in response to small, slow ramp stretches, *Brain Res.*, **44**, 680–683

Hasan, Z. and Houk J. C. (1975a). Analysis of response properties of deefferented mammalian spindle receptors based on frequency response, *J. Neurophysiol.*, **38**, 663–672

Hasan, Z. and Houk, J. C. (1975b). Transition in sensitivity of spindle receptors that occurs when muscle is stretched more than a fraction of a millimeter, *J. Neurophysiol.*, **35**, 673–689

Houk, J. C., Crago, P. E. and Rymer, W. Z. (1981). Function of the spindle dynamic response in stiffness regulation—a predictive mechanism provided by non-linear feedback, this publication

Houk, J. C., Harris, D. A. and Hasan, Z. (1973). Non-linear behaviour of spindle receptors. In *Control of Posture and Locomotion* (edited by R. B. Stein, K. B. Pearson, R. S. Smith and J. B. Redford), New York, Plenum, 147–163

Houk, J. C., Rymer, W. Z. and Crago, P. E. (1977). Complex velocity dependence of the electromyographic component of the stretch reflex, *Proc. XXVII Int. Cong. Physiol. Sci.*, Paris

Joyce, G. C., Rack, P. M. H. and Westbury, D. R. (1969). The mechanical properties of the cat soleus muscles during controlled lengthening and shortening movements, *J. Physiol., Lond.*, **204**, 461–474

Lennerstrand, G. (1968). Position and velocity sensitivity of muscle spindles in the cat. I: Primary and secondary endings deprived of fusimotor activation, *Acta physiol. scand.*, **73**, 281–299

Lennerstrand, G. and Thoden, U. (1968a). Dynamic analysis of muscle spindle endings in the cat using length changes of different length–time relations, *Acta physiol. scand.*, **73**, 234–250

Lennerstrand, G. and Thoden, U. (1968b). Position and velocity sensitivity of muscle spindles in the cat. II: Dynamic fusimotor single-fibre activation of primary endings, *Acta physiol. scand.*, **74**, 16–29

Matthews, P. B. C. (1972). *Mammalian Muscle Receptors and Their Central Actions*, Baltimore, Williams and Wilkins, 183–185

Matthews, P. B. C. and Stein, R. B. (1969). The sensitivity of muscle spindle afferents to small sinusoidal changes in length, *J. Physiol., Lond.*, **200**, 723–743

Poppele, R. E. and Bowman R. J. (1970). Quantitative description of linear behavior of mammalian muscle spindles, *J. Neurophysiol*, **33**, 59–72

Roberts, T. D. M. (1963). Rhythmic excitation of a stretch reflex, revealing (a) hysteresis and (b) a difference between responses to pulling and to stretching, *Q. Jl exp. Physiol.*, **48**, 328–345

Rymer, W. Z., Houk, J. C. and Crago, P. E. (1979). Mechanisms of the clasp-knife reflex studied in an animal model, *Brain Res.*, **37**, 93–113

Rymer, W. Z., Post, E. M. and Edwards, F. R. (1981). Functional roles of fusimotor and skeletofusimotor neurons studied in the decerebrate cat, this publication

Schaefer, S.-S. (1973). The characteristic curves of the dynamic response of primary muscle spindle endings in the absence and presence of stimulation of fusimotor fibers, *Brain Res.*, **59**, 395–399

A critique of the paper by Houk, Rymer and Crago

P. B. C. MATTHEWS*

We can all agree with Houk, Rymer and Crago that, 'the information supplied to the CNS by spindle receptors is a highly complex representation of muscle length and velocity'. Moreover there is no point in attempting to defend the old simplifications, which have always been recognised as such. And, of course, it can be accepted that a good fit to their particular experimental data is normally provided by their characteristic equation, namely

$$r - r_0 = K_1(x - x_1)\, v^{0.3}$$

But we should ask of their iconoclastic views, 'How much is new?', 'How much is true?', and 'How far can their findings be generalised?'

It has been clear from the very outset of such studies that the frequency of spindle firing during a long ramp stretch is the greater, the greater the velocity of stretching. But, equally, it has been obvious to all concerned that, however it is measured, the response to a dynamic stimulus increases far more slowly than in direct linear proportion to the value of the velocity, measured in millimetres per second. This is, for example, immediately apparent on simple inspection of the almost universally employed displays of instantaneous frequency. A specific statement about the lack of linearity of the velocity response was originally made with respect to the dynamic index of de-efferented endings (Matthews, 1963) and applies equally to the responses observed during electrical stimulation of single fusimotor axons, as has recently been reiterated for the primate (Cheney and Preston, 1976). Thus there is nothing new about the recognition of this lack of linearity in the velocity responsiveness of the primary ending, and its acceptance as an essential feature of spindle signalling.

The use of a power function to describe such a non-linearity, in which the response increases progressively more slowly with increasing stimulus, has plenty of precedents in physiology. It then almost invariably provides a better

*University Laboratory of Physiology, Oxford

fit than a linear relation, and in so doing performs a helpful descriptive simplification. It has already been specifically employed by Schäfer (1973) to cover the relation between the dynamic index and the velocity of stretching. But it would not invariably appear to provide a good fit of spindle firing during ramp stretches applied during stimulation of isolated fusimotor axons, since it does not appear to do so for all of the data published by Crowe and Matthews (1964) and by Cheney and Preston (1976). One difficulty with a power relation is to know precisely which measure of response it should be applied to, whether to the absolute frequency of firing, the increment above the resting discharge, the dynamic index, or what have you; this matters because the addition of a constant to a power relation degrades it, so that the data no longer fall on a straight line on a log–log plot. Thus it must be noted, as a matter of algebra, in relation to Houk, Rymer and Crago's figure 4 that the power relation cannot apply strictly on subtracting r_0 rather than r^* from the firing frequency at the end of the ramp; moreover, the discrepancy would get worse and worse with increasing initial muscle stretch, since any positional static response of the ending will increase r_0. Further, it remains problematical as to what would happen if r_0 at a given muscle length were to be increased by fusimotor action. Thus the coincidental agreement in their experiments between r_0 and r^* is a matter of some importance for the success of their curve fitting, and one is left wondering what would have happened if the two values had been more different. Likewise, it deserves note that the length at which they initially set the muscle again coincidentally turned out to have been nearly the same as the point from which their curves radiated.

However, what Houk, Rymer and Crago are stating which is quite new is that once it has settled down, outside the small signal range, the spindle obeys an equation in which the length and velocity terms are *multiplied* together rather than being *added*. And although this seems to fit their data quite reasonably with long ramps, we must ask how well their equation would hold up when we move outside this situation, if it is to be used as the basis for any wider generalisation. Now, life doesn't end with the end of a ramp and the transition to the hold phase of the stretch, but the equation simply predicts that we immediately return to the same 'resting' frequency whatever the size of the stretch, because the velocity becomes zero. To introduce a new length term during the dynamic phase is to get back to playing addition rather than multiplication, and thus attacks the essence of the scheme. But forgetting this aspect of the matter we have not been told anything about what happens when the muscle is released, and whether the same dynamic relation with the same velocity exponent applies. Yet from the point of view of motor control, release, corresponding to active muscle shortening, is of quite as much interest as stretch. Equally, it would be useful to know what happens in their preparation when stretch and release are combined in sinusoidal stretching. The equation predicts that the phase advance of the response on the stretching should be constant with variation of frequency. The current experimental position is, however, equivocal. Hunt and Wilkinson (1980) found such constancy of phase on studying the receptor potential, but a progressive phase advance with frequency was found by Hulliger *et al.* (1977) on studying the spike discharge.

Returning to the conditions for which the equation was developed, it may be noted that one of its crucial features is that, once it has equilibrated, the firing during a given velocity is related to the same absolute reference length (x_l/x^*, cf. equation 3, and figure 4) whatever the starting length of the stretch. This, however, does not immediately tally with the observations made by Jansen and Matthews (1962) on stretching at the single velocity of 3 mm/sec with a succession of 2–3 mm stretches. The incremental response above the pre-existing discharge then remained approximately constant with variation of muscle length, and this was so for endings with both large and small static positional responses (cf. Jansen and Matthews, 1962, figures 4, 6 and 8). This would be a remarkable coincidence if the spindle was thinking solely in terms of multiplication rather than addition, but the evidence is not final either way since a wide range of velocities was not then studied. Two differences should, however, be pointed out between Jansen and Matthews' preparations and those of Houk, Rymer and Crago, although both sets of experiments were on decerebrate cats the ventral roots of which had been left intact to provide continuous fusimotor activation of the spindles. First, Jansen and Matthews stretched the muscles up to approximately the physiological maximum, whereas in the recent experiments the muscles still seem to have been a good deal shorter when the stretch reached its maximum. From the point of view of the applicability of a universal equation, this should be quite immaterial, for to have any generality an equation should hold at all lengths. But it could well make an important difference if the behaviour of the spindles depends in large part upon the dynamic length–tension properties of the intrafusal muscle fibres, as we would all appear to believe; such a relation could well rise steeply at short muscle lengths, to flatten near the physiological maximum. Second, in Houk, Rymer and Crago's preparations the dorsal roots had been left virtually intact, as well as the ventral roots. Thus any stretch-evoked reflexes on to the fusimotor neurons would have been operative, and so there is no knowing as to whether or not the level of fusimotor activity remained constant throughout the course of a single stretch; moreover, the level of fusimotor activity at a given length could have been different for the different velocities of stretching. From what we are told elsewhere in this book such effects seem unlikely to have been large, but their possible occurrence creates a further element of uncertainty. From the logical point of view it is strange indeed to use such data for propounding a general mathematical description of spindle behaviour. It must be presumed that the present work was originally aimed at quite another target, such as the relation between spindle activity and the resulting stretch reflex, when all that would have mattered would have been the actual level of afferent firing, irrespective of how it were produced.

Now, a final thought. Does the CNS know the characteristic equation of the spindle and act upon it? If it does, then it would appear rather difficult for it to disentangle the length and velocity signals from each other since they are so compounded by multiplication, and in a way which is held to be the same for both primary and secondary endings. This probably wouldn't matter from the point of view of reflex control. But in the realm of sensation it is now well known that muscle vibration, which puts in an excess of Ia firing relative to spindle group II firing, leads to the conscious experience of continuous movement. It would be interesting to be told how this is thought to come about, rather than just that the limitations imposed by the equation should be demonstrable psychophysically.

Thus, while I welcome these attempts at a synthetic description of spindle behaviour I remain to be convinced that the present characteristic equation can in any way be regarded as a definitive basis for generalisation. As a beginning, just using ramp stretches, it requires to be tested with a variety of levels of constant fusimotor activity and over a wider range of muscle lengths. Additive relations still offer the advantage that they can include a term for the static response of an ending, albeit that any superadded velocity response is inevitably non-linear; but multiplicative terms between length (whether incremental or absolute) and velocity may well also be required as suggested by Lennerstrand and Thoden's (1968) analysis of the situation. There is the continuing problem that during dynamic fusimotor activation what appears to be a simple length response during the rising phase of a ramp is quite different from the length response seen under truly static conditions (Crowe and Matthews, 1964), and does appear to be in some degree velocity dependent. Given the known complexity of spindle behaviour the most appropriate analytical model to employ could well depend upon the use to which it was to be put. The present characteristic equation with its domination by the multiplicative term seems quite reasonable as an empirical description of certain experimental data obtained under semi-steady-state conditions. But as a basis for wide generalisation I suspect it of being stillborn. If so, I offer the epitaph that, 'It was too simple to be true and too complex to be helpful'.

Received on August 1st, 1980.

REFERENCES

Cheney, P. D. and Preston, J. B. (1976). Effects of fusimotor stimulation on dynamic and position sensitivities of spindle afferents in the primate, *J. Neurophysiol.*, **39**, 20–39

Crowe, A. and Matthews, P. B. C. (1964). The effects of stimulation of static and dynamic fusimotor fibres on the response to stretching of the primary endings of muscle spindles, *J. Physiol., Lond.*, **174**, 109–131

Hulliger, M., Matthews, P. B. C. and Noth, J. (1977). Effects of static and of dynamic fusimotor stimulation on the response of Ia fibres to low frequency sinusoidal stretching covering a wide range of amplitudes, *J. Physiol., Lond.*, **267**, 811–838

Hunt, C. C. and Wilkinson, R. S. (1980). An analysis of receptor potential and tension of isolated cat muscle spindles in response to sinusoidal stretch, *J. Physiol., Lond.*, **302**, 241–262

Jansen, J. K. S. and Matthews, P. B. C. (1962). The central control of the dynamic response of muscle spindle receptors, *J. Physiol., Lond.*, **161**, 357–378

Lennerstrand, G. and Thoden, U. (1968). Position and velocity sensitivity of muscle spindles in the cat. II: Dynamic fusimotor single-fibre activation of primary endings, *Acta physiol. scand.*, **74**, 16–29

Matthews, P. B. C. (1963). The response of de-efferented muscle spindle receptors to stretching at different velocities, *J. Physiol., Lond.*, **168**, 660–678

Schäfer, S. S. (1973). The characteristic curves of the dynamic response of primary muscle spindle endings in the absence and presence of stimulation of fusimotor fibres, *Brain Res.* **59**, 395–399

REPLY TO MATTHEWS'S CRITIQUE BY HOUK, RYMER AND CRAGO

We agree, as acknowledged briefly in our Introduction, that a lack of proportionality between the dynamic response and velocity is apparent from the older literature (although this fact is frequently ignored). What our data add is that the specific relation is approximately $v^{0.3}$ throughout a broad and functionally relevant range of velocities. Although Schafer fitted dynamic index versus velocity plots with power functions, it is important to note that a large range of exponents (some greater than 1) was required to fit data from different endings under different conditions of gamma stimulation. Our results suggest that increment in discharge rate may be a more useful response measure than dynamic index, since the associated exponents all cluster about the single value of 0.3. This holds for both primary and secondary endings both in the presence and absence of fusimotor activity.

We agree that $(r-\mathrm{r}^*)$ measures are likely to provide better fits to power functions than $(r-r_0)$ measures. We used the latter not only because of past tradition, but also since initial rate r_0 can be measured directly, whereas r* is a derived quantity subject to errors in extrapolation. Fortunately, this did not matter since r_0 and r* had similar values under the conditions of our experiments. These conditions, namely the use of intermediate muscle lengths within the physiological range and a presence of intact reflexes and spontaneous fusimotor activity, were selected because of their presumed relevance to motor control in intact animals. Thus, under physiologic conditions the distinction between r* and r_0 may not be particularly crucial, although this point warrants further investigation.

While Matthews acknowledges the multiplicative relationship between muscle length and velocity as a novel finding, he expresses considerable doubt as to the generality of this relation. The equation given represents what mathematicians call an 'asymptotic' solution. While asymptotic solutions are precise only for responses under restricted conditions, responses under more general conditions often will show many of the features predicted by the asymptotic equation. In fact, the increase in the magnitude of the initial burst that we regularly observed at longer initial lengths is well accounted for as a multiplicative effect, even though the overshoot associated with the initial burst is not predicted by the equation we provide. In other words, the multiplicative effect appears to be more general than the conditions under which the equation holds. Further studies will be required to define the limits of this generality.

Two other limitations of our characterisation commented upon by Matthews are (1) the absence of a static response to maintained stretch ($v=0$) and (2) the inadequate testing of predictions regarding response to shortening. Regarding the former, we have found that the emergence of a static response as velocity is decreased towards zero is not well explained by an additive term. We currently believe that a pacemaker switch, modelled by the substitution of a new equation rather than the introduction of an additive term, may be the proper way to characterise the static state. Our rather preliminary observations on responses to shortening indicate a definite decrease in dynamic sensitivity (cf. asymmetry discussed by Houk, Crago and Rymer, this publication) with perhaps some preservation of the multiplicative feature, but this requires further study.

In an untimely burial ceremony, Matthews suggests that the multiplicative relation is too complex to be useful. Actually, it is not nearly as complex as many of the linear transfer functions that have been proposed earlier. Perhaps the real issue is that we are so accustomed to linear models that the mind is resistant to consideration of non-linear alternatives. Our own philosophy is that we should approach biological non-linearities in a forthright manner.

Received on November 13th, 1980.

Fusimotor actions on the sensitivity of spindle secondary endings

L. JAMI* and J. PETIT*

SUMMARY

The effects of stimulating single γ axons on the responses of spindle secondary endings to low-frequency sinusoidal stretching have been examined in cat peroneus tertius muscle. The action of each axon during sinusoidal stretches (0.1–2 mm extent at 0.5–1 Hz) was systematically compared with its action during large ramp stretches (3 mm at 0.4–0.6 mm/sec).

Fusimotor axons that enhance the sensitivity of a secondary ending to slow ramp stretch can also increase by 50–100% the sensitivity of this ending to low-frequency sinusoidal stretching, provided two conditions are satisfied: (1) that the extent of sinusoidal stretching exceeds 0.1 mm and (2) that the sinusoidal stretch is applied at muscle lengths shorter than 1–2 mm below the physiological maximum.

The same γ axons can enhance the sensitivity of secondary endings during stretches at either constant or variable velocity. Earlier work showed that these axons have an extensive distribution to intrafusal chain muscle fibres. Their action might therefore be equivalent to that expected when several γ axons with more restricted distributions are activated simultaneously.

INTRODUCTION

The sensitivity of secondary endings to changes in muscle length, expressed in impulses sec^{-1} mm^{-1}, is defined as the increase in firing rate of an ending elicited by a 1 mm increase in muscle length. It can be determined under static conditions, by measuring the firing frequency of an ending for different muscle lengths, after complete adaptation at each length. The sensitivity of

* Laboratoire de Neurophysiologie, Collège de France, Paris

the ending is then represented by the slope of the linear frequency–length relationship obtained in these conditions; it is usually termed *position sensitivity* (Lennerstrand, 1968; Brown *et al.*, 1969). Under dynamic conditions, the sensitivity of a secondary ending can be determined during movements either at constant velocity (e.g. ramp stretching of the muscle) or at variable velocity (e.g. sinusoidal stretching of the muscle). In the former instance, the instantaneous frequency of discharge of a secondary ending, recorded during a slow ramp stretch, provides a display of the frequency–length relationship, and the slope of this relationship gives the sensitivity of the ending. In the latter instance, as sinusoidal changes of muscle length elicit approximately sinusoidal modulation of the firing frequency of secondary endings, the ratio of this modulation to the extent of the applied stretch represents a measure of the ending sensitivity. In de-efferented spindles, different values of sensitivity are found in these three conditions and the fact that higher values are found under dynamic conditions demonstrates the existence of dynamic components in the responses of secondary endings to changes in muscle length (Renkin and Vallbo, 1964; Brown *et al.*, 1969; Jami and Petit, 1978).

In the first two conditions (i.e. static conditions and movement at constant velocity), individual static fusimotor axons exert the same action: among the 3–6 axons acting on a secondary ending, only one or two will significantly increase the ending sensitivity (Jami and Petit, 1978). The action has been shown to depend on the pattern of distribution of the axon to intrafusal muscle fibres, the relevant factor for enhancing sensitivity being the number of activated chain fibres. Stimulation of a γ axon which innervates all the chain fibres in one pole of a spindle can elicit a two- or three-fold increase in the sensitivity of a secondary ending of that spindle. This axon will have the same action on all the secondary endings of a spindle (Jami *et al.*, 1980). On the other hand, since a great variety of patterns is known to exist in the distribution of individual static axons to intrafusal muscle fibres in the different spindles they supply (*see* Barker *et al.*, 1978), it is not surprising that an individual γ axon does not increase the sensitivities of all the secondary endings it excites. But when large samples of secondary ending responses to stimulation of single γ axons are examined, significant increases of sensitivity are found in 25–30% of instances (Jami and Petit, 1978).

In the third condition, namely during sinusoidal stretching of the muscle, fusimotor stimulation has been reported to decrease regularly the sensitivity of secondary endings to stretches of 0.2–30 Hz with amplitudes ranging from 20 μm to 1 mm (Chen and Poppele, 1973, 1978; Cussons *et al.*, 1977). In these studies however, the action of the studied γ axons on the sensitivity of secondary endings measured under static conditions or during ramp stretches was not investigated. The conclusion that γ axons decrease the sensitivity of secondary endings to sinusoidal stretch may thus have been reached simply because the examined samples did not include instances of axons enhancing the sensitivity of secondary endings. But there remained the possibility that a

γ axon exerts opposite actions on the sensitivity of a secondary ending during sinusoidal stretch and during ramp stretch.

In the present study, the effects of stimulating single γ axons on the responses of secondary endings to low-frequency sinusoidal stretching have been examined in cat peroneus tertius muscle. The action of each γ axon during sinusoidal stretches of various extents was systematically compared with its action during large ramp stretches applied at a slow constant velocity.

METHODS

The experiments were performed on 10 adult cats anaesthetised with pentobarbital sodium (Nenbutal, Abbott Laboratories, 40 mg/kg i.p.). The techniques used have been fully described elsewhere (Jami and Petit, 1978). The tendon of the peroneus tertius muscle was dissected free and attached to an electromagnetic puller giving either ramp or sinusoidal stretches. Single afferent fibres from secondary endings of peroneus tertius spindles were functionally isolated in dorsal root filaments and single γ motor axons were isolated in ventral root filaments. The effect of stimulating each axon at 30–100/sec on the discharge of each activated secondary ending was observed under the following conditions: (1) during stretches of the muscle at 0.4–0.6 mm/sec over 3 mm; (2) during faster stretches (5 mm/sec) of the same amplitude; and (3) during sinusoidal stretches of 0.1–2 mm extent (peak-to-peak movement) applied at 0.5 Hz or 1 Hz. The position of the puller was adjusted so that the final length of the muscle after a ramp was 1–2 mm shorter than its maximal physiological length. Sinusoidal stretches started from different muscle lengths ranging between 4 mm and 1 mm from the maximal physiological length, so that the shorter position was well beyond the length for which the muscle became slack. *In situ*, the amplitude of possible movements for the peroneus tertius was 4–7 mm, depending chiefly on the size of the muscle.

The actions of 44 single γ axons (conduction velocities 23–45 m/sec) on 50 secondary endings (conduction velocities of the afferent fibres 22–58 m/sec) were studied. In total, 88 examples of the action of a γ axon on the response of a secondary ending were observed. In 24 (27%) instances, stimulation of the axon at 50/sec elicited 100–300% increases in the sensitivity of the ending measured during slow ramp stretches. The studied γ axons were presumed to be static since dynamic axons only very exceptionally activate secondary endings (Appelberg *et al.*, 1966; Brown *et al.*, 1967; Durkovic and Preston, 1974).

The method used to calculate the sensitivity of secondary endings during ramp stretches has been described earlier (Jami and Petit, 1978). The modulation of secondary ending discharge during sinusoidal stretching of the muscle was assessed by measuring the difference between the mean maximal and mean minimal instantaneous frequencies occurring during a cycle. The measures were averaged over 5–10 successive cycles.

RESULTS

A typical example of the action of a γ axon that increased the sensitivity of a secondary ending during a slow ramp stretch is illustrated by figure 1. Records 1a and 1b show that stimulation of this axon at 50/sec increased the slope of the response from 24 to 83 impulses sec^{-1} mm^{-1} (245% increase). Records 2 show that during sinusoidal stretching the effect exerted by this axon depended on the extent of the applied stretch. The initial muscle length was set 1 mm further for sinusoidal stretch than for the ramp stretch, so that the movement occurred in the middle of the range covered during the ramp.

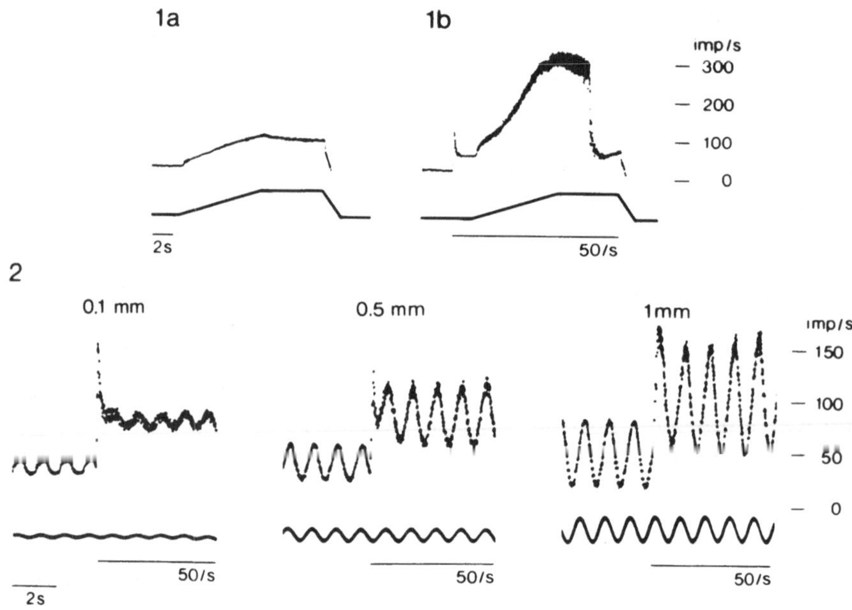

Figure 1. Actions exerted by a γ axon which increased the sensitivity of a secondary ending.

1a: Discharge of the passive secondary ending (conduction velocity of the afferent fibre 29 m/sec) recorded with an instantaneous frequencymeter (upper trace) during a ramp stretch of 3 mm at 0.4 mm/sec. The lower trace shows the muscle length.

1b: Response of the ending to the same stretch during stimulation of the γ axon (conduction velocity 28 m/sec). Note that stimulation is 'driving' the afferent discharge at the initial muscle length and at the onset of stretch.

2: Responses of the ending to sinusoidal stretches of increasing extent at 1 Hz (*see* text for further comments).

During 0.1 mm sinusoidal stretching, fusimotor stimulation at 50/sec elicited irregularity of the ending discharge together with an increase of its mean level, but the average modulation of the discharge appeared reduced by about 25% and the sensitivity of the ending decreased from 140 impulses sec^{-1} mm^{-1} in the passive state to 105 impulses sec^{-1} mm^{-1} during fusimotor stimulation. For larger stretches (0.5 mm and 1 mm) the effect of fusimotor stimulation was, on the contrary, to enhance the modulation of the response so that the sensitivity of the ending increased by about 63% (from 66 to 110 impulses sec^{-1} mm^{-1} for 0.5 mm stretches and from 65 to 105 impulses sec^{-1} mm^{-1} for 1 mm stretches). The same axon thus apparently exerted three different actions on the sensitivity of the secondary ending, namely (1) a very large increase during the slow ramp stretch, (2) a slight decrease during small sinusoidal stretches and (3) an appreciable increase during larger sinusoidal stretches, although this increase amounted to one quarter only of that observed during the slow ramp stretch.

Figure 2 illustrates the action of a fusimotor axon that did not increase the sensitivity of a secondary ending during slow ramp stretch (figure 2, records 1a and 1b). Stimulation of this axon at 50/sec elicited reductions of 33% and 20% in the modulation of the ending discharge during sinusoidal stretches of 0.1 mm and 0.5 mm, respectively, but for 1 mm stretch there was no change in modulation, and for 1.5 mm stretch a slight increase (15%) was apparent (figure 2, records 2). Similar effects were observed by Cussons *et al.* (1977) in soleus muscle spindles.

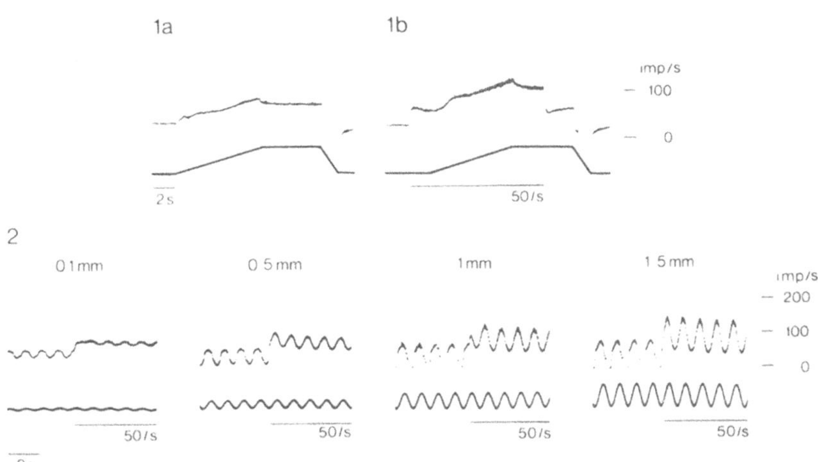

Figure 2. Actions exerted by a γ axon which did not increase the sensitivity of a secondary ending. Same conventions as in figure 1. Conduction velocity of the afferent fibre 39 m/sec. Conduction velocity of the γ axon 35 m/sec. From the same experiment as figure 1.

In peroneus tertius spindles, fusimotor actions on the sensitivity of secondary endings to low-frequency sinusoidal stretching were consistently found to change with the extent of stretch: there was a reduction of sensitivity for small stretches and an increase for large stretches. But the extent of sinusoidal stretching for which the change occurred was much smaller for γ axons that enhanced the sensitivity of secondary endings to ramp stretches (between 0.1 mm and 0.5 mm in figure 1) than for γ axons that did not exert this effect (between 1 mm and 1.5 mm in figure 2). Also, the increase was much higher in the former cases (63% for 0.5 mm in figure 1) than in the latter (15% for 1.5 mm in figure 2). Mean values calculated from the pooled observations made in a single experiment (figure 3) confirm the qualitative consistency of the effects, although they smooth out the quantitative differences. In this experiment, 18 actions of a single γ axon on a secondary ending were studied (they were elicited by five axons activating 10 endings) and significant

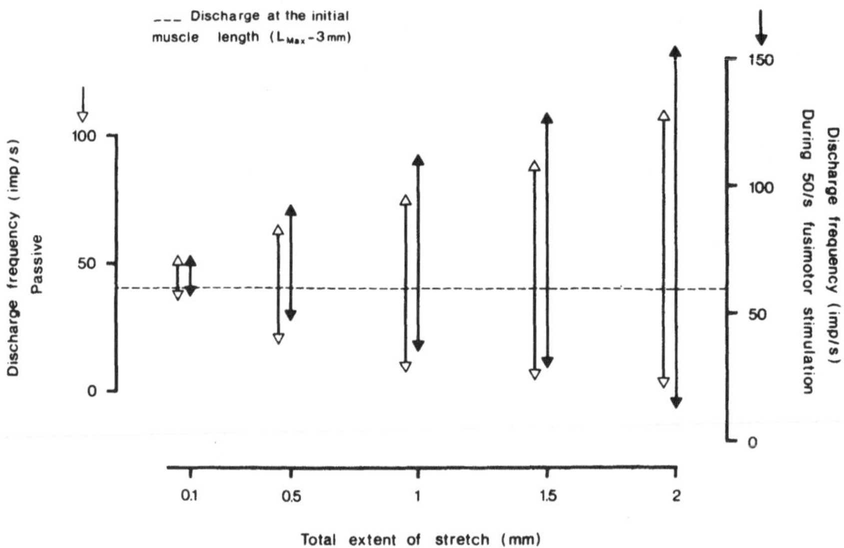

Figure 3. Modulation of secondary ending discharges in response to sinusoidal stretches of various extent at 1 Hz.

The modulation is represented by the height of the vertical bars. Bars limited by open triangles (left-hand scale) give the average modulation of 10 passive secondary endings studied in a single experiment. Bars limited by filled triangles (right-hand scale) give the average modulation of the same endings when activated by single γ axons (18 instances of the action of a γ axon on a secondary ending). The measures were taken from records such as those shown in figures 1 and 2. The left-hand scale has been shifted upwards so that the average level of discharge of the passive endings at the initial muscle length coincides with the average level of discharge at the same length during fusimotor stimulation. The average increase in static discharge frequency was 20 impulses/sec.

increases of sensitivity were observed in four instances only. Stimulation of individual γ axons at 50/sec produced on average a reduction of 20% in modulation of secondary endings discharges during sinusoidal stretches of 0.1 mm; for stretches of 0.5 mm, the reduction was only 5%, and from 1 mm upwards the modulation was increased during fusimotor stimulation by an amount that augmented with the extent of stretch: 11%, 16% and 27% for 1 mm, 1.5 mm and 2 mm, respectively. Stretches of 1 mm represented 14–25% of the maximal possible extension of peroneus tertius *in situ*. For stretches of this extent and above, fusimotor stimulation induced either no change or, more frequently, an increase in the sensitivity of secondary endings.

There was no relation between the degree of excitation of an ending by a γ

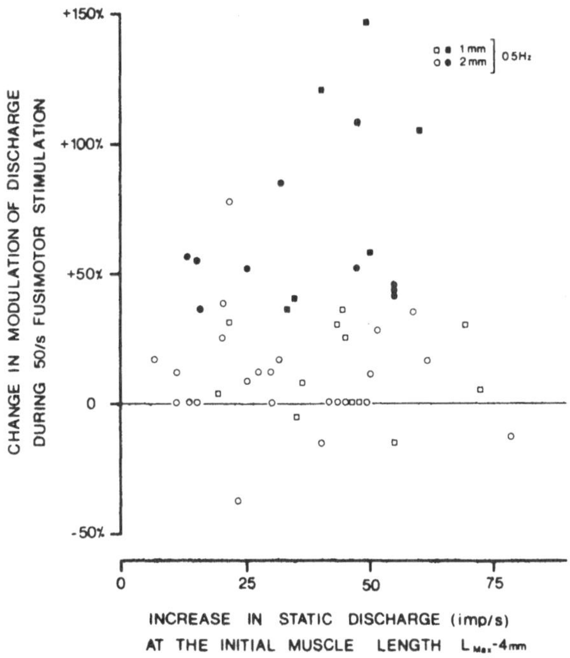

Figure 4. Lack of relation between the degree of excitation of secondary endings by individual γ axons and the change elicited by these axons in the modulation of the ending discharges in response to sinusoidal stretches of 1 mm (squares) and 2 mm (circles) at 0.5 Hz.

Each symbol represents the action of a single γ axon on a secondary ending (N=52). Filled symbols are instances in which the axon elicited an increase in the sensitivity of the ending during a slow ramp stretch. There were 10 such instances in the sample of 2 mm stretches (N=34) and six in the sample of 1 mm stretches (N=18). The average increase in static discharge frequency was 31 impulses/sec in the total sample. The average passive modulations were 69 and 46 impulses/sec for 2 mm and 1 mm stretches, respectively.

axon and the action exerted by this axon on the modulation of the ending discharge during sinusoidal stretches. This is illustrated in figure 4 for a sample of 52 instances of the action of a γ axon on a secondary ending, studied in five experiments. The changes elicited by stimulation of γ axons at 50/sec in the modulation of the discharge of secondary endings during 1 mm or 2 mm sinusoidal stretches are plotted against the increases elicited in the discharge of these endings when the muscle was maintained at the initial length. Reduction of modulation was observed in five cases only and in 10 cases there was no change in modulation, whereas in 11 other cases the increase was less than 20%. In the 26 other instances (i.e. half of the sample), increases ranging from 25% to 145% were observed, so that the mean effect for the whole sample was an increase of 32%. The largest increases were of course elicited by axons which enhanced the sensitivity of the endings during ramp stretches (filled symbols in figure 4). There was no relation between the sizes of increase in sensitivity observed during movements at constant velocity and those observed during sinusoidal stretching, but the latter were usually the weaker (figure 1).

In de-efferented spindles, the sensitivity of primary and secondary endings to sinusoidal stretching of the muscle is known to depend on the initial length of the muscle (Matthews and Stein, 1969; Poppele, 1973). For primary endings, static fusimotor stimulation tends to remove the dependence (Goodwin *et al.*, 1975) and this was found, in the present study, to occur also for secondary endings when activated by γ axons that *did not* enhance their sensitivity to ramp stretches. A typical example is illustrated in figure 5 (row 1): in the passive state, the modulation of this secondary ending discharge almost doubled when the initial muscle length was extended from 4 mm to 1 mm short of the maximal physiological length, but during fusimotor stimulation the modulation of discharge remained constant at all lengths. As a consequence, the slight increase in modulation (less than 15%) observed during fusimotor stimulation when the muscle was at the shorter length was replaced by a clear reduction (33%) of modulation at the longer length. The same occurred, although through a different evolution, for secondary endings activated by γ axons that did enhance their sensitivity (figure 5, row 2): during fusimotor stimulation the modulation of the secondary ending discharge tended to decrease as the muscle length came close to its physiological maximum (compare the first and last records in figure 5, row 2). In the instance illustrated the increase in modulation of discharge elicited by the γ axons was about 75% when the initial muscle length was 4 mm or 3 mm from the maximal physiological length; it fell to 25% when the muscle length came within 2 mm of the maximum and it was finally replaced by a reduction of about 25% after the muscle had been stretched 1 mm further.

Fusimotor axons that enhanced the sensitivity of secondary endings to low-frequency sinusoidal stretching often produced their maximal effects for relatively low rates of stimulation (30–50/sec), as was observed for the action

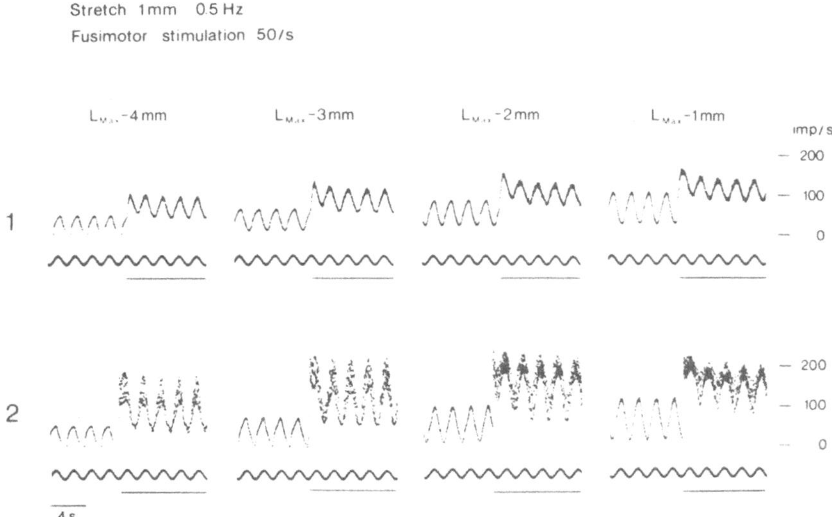

Figure 5. Effects of varying the initial muscle length on the modulation of secondary ending discharges in response to sinusoidal stretch, in the passive state and during the stimulation of single fusimotor axons. The periods of stimulation are indicated by the horizontal bars below each record. 1 and 2 are from the same experiment.

1: Action of an axon (conduction velocity 30 m/sec) which did not enhance the sensitivity of the secondary ending (conduction velocity of the afferent fibre 29 m/sec) during a slow ramp stretch.

2: Action of another axon (conduction velocity 28 m/sec) which enhanced the sensitivity of the secondary ending (conduction velocity of the afferent fibre 25 m/sec) during a slow ramp stretch.

of γ axons on the sensitivity of secondary endings measured during ramp stretches. Stimulation at higher rates rarely elicited a further increase in sensitivity and could even produce a relative decrease (*see* Jami and Petit, 1978).

DISCUSSION

These experiments show that in peroneus tertius spindles, stimulation of a γ axon that enhances the sensitivity of a secondary ending to slow ramp stretch can also increase by 50–100% the sensitivity of this ending to low-frequency sinusoidal stretching. This effect is exerted under two conditions: (1) extent of sinusoidal stretch exceeding 0.1 mm, that is only 2.5% of the possible extension of the muscle *in situ*; and (2) sinusoidal stretch applied at muscle lengths shorter than 1–2 mm below the physiological maximum, that is at muscle lengths up to 50–85% of the maximal extension of peroneus tertius *in situ*. The two conditions might explain the difference between the present findings and those of Cussons *et al.* (1977) in soleus muscle. It is for small stretches of 0.5–30 Hz, falling within the linear range (i.e. mostly under 0.5 mm extent) that they observed 50% reductions of secondary endings sensitivity during fusimotor stimulation. Moreover, they systematically worked at muscle lengths that were 1–2 mm short of the physiological maximum. Occasionally however, they met γ axons that slightly increased the sensitivity of a secondary ending to stretches of 1 mm amplitude at 1 Hz, although they reduced this sensitivity for smaller stretches. Chen and Poppele (1978) also mentioned that the effect of fusimotor stimulation on secondary endings sensitivity can be reversed for small- and large-amplitude stretches (cf. their figure 4C). This was regularly observed in the present study, but the effects exerted on the responses of secondary endings to large sinusoidal stretches by axons which did not increase the sensitivity of these endings to ramp stretches were small enough (figure 2) to be considered as insignificant.

Comparison between fusimotor actions on secondary endings and dynamic actions on primary endings

The γ axons that were able to increase the sensitivity of secondary endings to sinusoidal stretch exerted this effect in a manner which resembles the action of dynamic axons on primary endings (Hulliger *et al.*, 1977). These authors termed *dynamic paradox* the fact that, with increasing amplitudes of stretch, dynamic fusimotor action first reduces and then increases the sensitivity of primary endings to low-frequency sinusoidal stretching. Moreover, the sizes of both reductions (−25% for 0.1 mm stretches) and increases (+60% for 1 mm stretches), as well as the amplitude of stretch for which the change occurred (below 0.25 mm), were similar to those observed in this study (cf. their figure 12).

However, this similarity should not be taken as an evidence of dynamic fusimotor action exerted on secondary endings. Refutation of any such assumption deserves emphasis since it was recently reported that secondary endings lay on all the types of intrafusal muscle fibres, dynamic effectors as well as static effectors (Banks *et al.*, 1979). Moreover, in de-efferented spindles of peroneus tertius, the secondary endings have higher dynamic indices than secondary endings of other muscles (Jami and Petit, 1979), suggesting that their sensitivity includes an appreciable dynamic component.

A close inspection of the effect exerted by an axon increasing the sensitivity of a secondary ending, already indicates that it is essentially an effect on length sensitivity and not on dynamic sensitivity. Data from figure 1 have been rearranged in figure 6 to show that, in the range of muscle length covered by the sinusoidal stretch of 1 mm extent, the sensitivity of the passive ending was 24 impulses sec^{-1} mm^{-1} for the ramp (figure 6, record 1a) and 65 impulses

$sec^{-1} mm^{-1}$ for the sinusoidal stretch (figure 6, left part of record 2). Such an increase might be ascribed to the dynamic component in the sensitivity of the passive ending since the mean velocity of the sinusoidal stretch was higher than that of the ramp. But upon fusimotor activation (figure 6, record 1b and right part of record 2) the values of the sensitivity became almost identical during both types of stretch (95 impulses $sec^{-1} mm^{-1}$ for the ramp and 105 impulses $sec^{-1} mm^{-1}$ for the sinusoidal stretch), as if the dynamic component

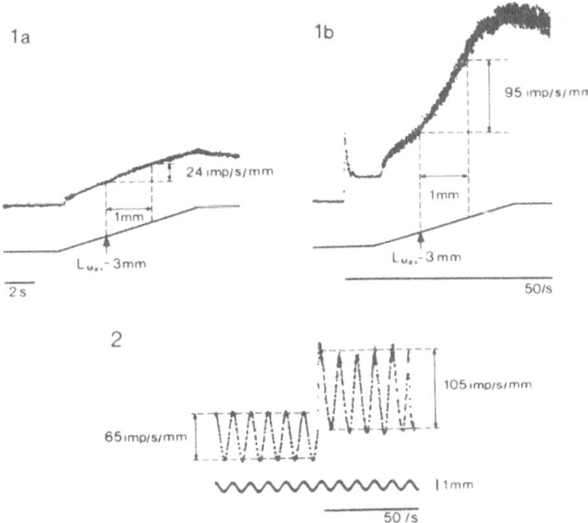

Figure 6. Same data as in figure 1. *See* text for comments.

in the sensitivity of the ending failed to express itself when the ending was excited by this axon.

At any rate, it can be ruled out that axons exerting effects such as are illustrated in figure 6 might be dynamic axons, since their intrafusal distribution is known to involve mainly chain fibres (Jami *et al.*, 1980) that are static effectors.

Another item of evidence against the possibility that these static axons could exert some kind of dynamic action on secondary endings is given by the test of their effect on dynamic indices (figure 7). Axons that enhance the sensitivity of secondary endings (filled circles in figure 7) can either reduce the dynamic indices of these endings, or leave them unchanged, or increase them by less than 20 impulses/sec. The figure shows that the effect of these axons is not different from the effect of axons that do not increase the sensitivity of secondary endings. As in soleus muscle (Brown *et al.*, 1967), fusimotor actions in peroneus tertius do not increase the dynamic sensitivity of secondary endings.

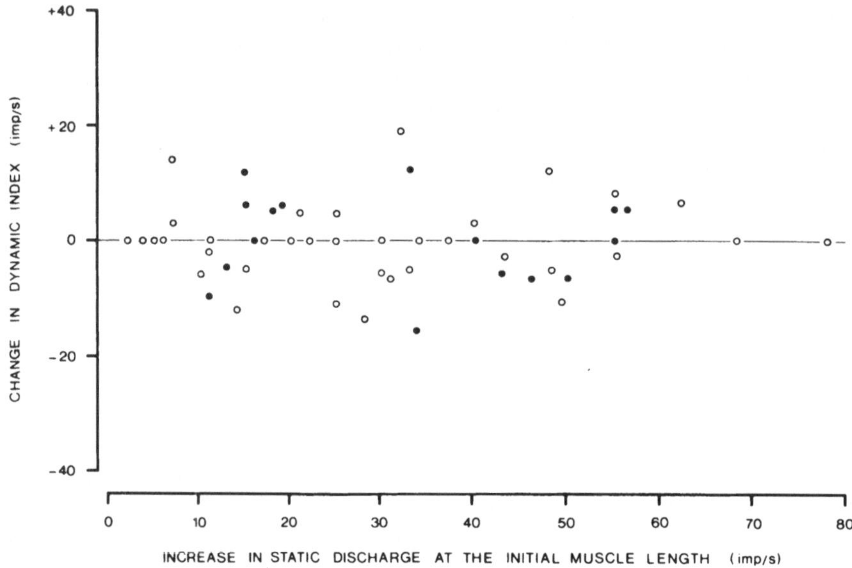

Figure 7. Lack of relation between the changes in dynamic indices of secondary endings and their increase in static discharge at the initial muscle length, both produced by 50/sec fusimotor stimulation.

The dynamic indices were measured at the end of 3 mm ramp stretches applied at 5 mm/sec. Each symbol represents the action of a single axon on a secondary ending (N=52; the sample is not exactly the same as in figure 4). Filled circles are instances in which the axon enhanced the ending sensitivity during both slow ramp and sinusoidal stretches. The average dynamic index of the passive endings was 36 impulses/sec. The average increase in static discharge frequency at the initial muscle length was 30 impulses/sec in the total sample.

Possible intrafusal mechanisms

The stiffness of de-efferented spindles is not uniformly distributed along intrafusal muscle fibres. Measurements made under static conditions by Poppele *et al.* (1979) have shown that the sensory area is stiffer than the intracapsular portion of muscle fibres, which in turn is stiffer than their extracapsular region. Contraction of chain fibres elicited by γ axons has been described as extensive but nevertheless predominant in the polar half of these fibres (Boyd, 1976). When contracted, polar regions of chain fibres are likely to become stiffer, which should improve the transmission of stretch to the sensory region, thereby contributing to increase the sensitivity of secondary endings. An axon distributing branches to all the chain fibres should be more efficient in this action than any axon with a more restricted distribution.

The fact that axons with such an extensive distribution can nevertheless reduce the sensitivity of secondary endings to small sinusoidal stretches might also result from the diffusion of the contraction they induce. Secondary endings lay on striated portions of intrafusal muscle fibres (Cooper, 1961; Boyd, 1962) and the extension of contraction to the region bearing a secondary ending would probably increase the stiffness of the sensory area, which might increase its resistance to deformation, and consequently reduce its sensitivity. If it were the case, the contracted sensory area would probably resist small stretches more effectively than large ones. Actual contraction of the secondary sensory area was rarely seen (Boyd, 1976) but this does not preclude that some changes in the mechanical properties of the region occur during diffuse activation of chain fibres.

Alternative explanations would imply that, although the chain fibres are contracted, their stiffness is reduced so that they can partly take up the applied stretch and interfere with its transmission to secondary endings. It would be interesting to know whether some decrease in the stiffness of chain fibres may occur when they are activated by fusimotor axons. In fact, a marked decrease in the stiffness of frog muscle fibres has been observed during caffeine-induced activation of the contractile system (Lännergren, 1971).

A further possibility is related to the innervation of the bag_2 fibre by static γ axons. In tenuissimus spindles there is a high incidence of static axons distributed exclusively to this fibre (Boyd *et al.*, 1977). Such axons excite scondary endings but reduce their sensitivity to slow ramp stretch (cf. figure 5 in Jami *et al.*, 1980) and might be expected also to reduce the modulation of secondary endings during low-frequency sinusoidal stretching. This effect could be due to the unloading of chain fibres by the contraction of the bag_2 fibre (Boyd *et al.*, 1977). Kinked chain fibres would be able to 'absorb' part of the applied stretch as long as its amplitude would allow persistence of the slack.

Functional implications

Evidence suggesting that, in physiological conditions, fusimotor activation increases the sensitivity of secondary endings may be summarised as follows. The action of γ axons that enhance the sensitivity of secondary endings has been found consistent in all the conditions in which this sensitivity was tested, that is under static conditions and during slow ramp stretch (cf. figure 5 in Jami and Petit, 1978; *see also* Lennerstrand and Thoden, 1968) as well as during low-frequency sinusoidal stretches of sufficient extent (figure 1). In each of these conditions relatively low rates of fusimotor stimulation (50/sec) elicited significant (i.e. at least 50–100%) increases of secondary ending sensitivity, although the absolute values of the increases were not necessarily the same in all conditions (figure 6). This pattern of fusimotor action is correlated with a pattern of fusimotor innervation involving all the chain fibres in a spindle pole (Jami *et al.*, 1980). The actions exerted by single static γ axons having this extensive distribution may be considered as representative of the actions exerted by the static fusimotor system as a whole, since when all the static axons supplying a spindle are acting together, they are likely to elicit contraction of all chain fibres in this spindle. Indeed, Brown *et al.* (1969) observed that stimulation of several static fusimotor fibres contained in coarse ventral root filaments markedly increased the sensitivity of secondary endings under both static and dynamic conditions, and notwithstanding the simultaneous activation of extrafusal motor units (*see also* Harvey and Matthews, 1961).

Fusimotor axons can strongly excite secondary endings even before any change occurs in muscle length. Under the assumption that secondary endings function as muscle length sensors, the physiological significance of their increased sensitivity during fusimotor activation might be either to improve their accuracy in measuring muscle length or simply to preserve their capacity for providing information about muscle length in spite of an excitation that is not related to length changes. It is difficult to decide between these two possibilities as long as very little is known about the use made by the central nervous system of the informations provided by secondary endings. The reduction of sensitivity that occurs for small sinusoidal stretches, as well as the fact that increases in sensitivity are observed only in a limited range of muscle lengths, are not in favour of an improved accuracy of secondary endings. The observations of Cody *et al.* (1975) in normal awake cats rather suggest that fusimotor action can adjust the sensitivity of secondary endings, according to the type of movement that is being performed.

ACKNOWLEDGEMENTS

The technical assistance of Mrs D. Lan-Couton is gratefully acknowledged. This work was supported by a grant from the Fondation pour la Recherche Médicale Française.

Received on July 9th, 1980.

REFERENCES

Appelberg, B., Bessou, P. and Laporte, Y. (1966). Action of static and dynamic fusimotor fibres on secondary endings of cat spindles, *J. Physiol., Lond.*, **185**, 160–171

Banks, R. W., Barker, D. and Stacey, M. J. (1979). Sensory innervation of cat hind-limb muscle spindles, *J. Physiol., Lond.*, **293**, 40–41P

Barker, D., Banks, R. W., Harker, D. W., Milburn, A. and Stacey, M. J. (1976). Studies of the histochemistry, ultrastructure, motor innervation and regeneration of mammalian intrafusal muscle fibres. In *Understanding the Stretch Reflex* (ed. S. Homma), Elsevier, Amsterdam (*Prog. Brain Res.*, **44**, 67–87)

Boyd, I. A. (1962). The structure and innervation of the nuclear bag muscle fibre system and the nuclear chain muscle fibre system in mammalian muscle spindles, *Phil. Trans. R. Soc. B*, **245**, 81–136

Boyd, I. A. (1976). The response of fast and slow nuclear bag fibres and nuclear chain fibres in isolated cat muscle spindles to fusimotor stimulation, and the effect of intrafusal contraction on the sensory endings, *Q. Jl exp. Physiol.*, **61**, 203–254

Boyd, I. A., Gladden, M. H., McWilliam, P. N. and Ward, J. (1977). Control of dynamic and static nuclear bag fibres and nuclear chain fibres by γ and β axons in isolated cat muscle spindles, *J. Physiol., Lond.*, **265**, 133–162

Brown, M. C., Engberg, I. and Matthews, P. B. C. (1967). Fusimotor stimulation and the dynamic sensitivity of the secondary ending of the muscle spindle, *J. Physiol., Lond.*, **189**, 545–550

Brown, M. C., Lawrence, D. G. and Matthews, P. B. C. (1969). Static fusimotor fibres and the position sensitivity of muscle spindle receptors, *Brain Res.*, **14**, 173–187

Chen, W. J. and Poppele, R. E. (1973). Static fusimotor effect on the sensitivity of mammalian muscle spindle, *Brain Res.*, **57**, 244–247

Chen, W. J. and Poppele, R. E. (1978). Small-signal analysis of response of mammalian muscle spindles with fusimotor stimulation and a comparison with large-signal responses, *J. Neurophysiol.*, **41**, 15–27

Cody, F. W. J., Harrison, L. M. and Taylor, A. (1975). Analysis of activity of muscle spindles of the jaw-closing muscles during normal movements in the cat, *J. Physiol., Lond.*, **253**, 565–582

Cooper, S. (1961). The responses of the primary and secondary endings of muscle spindles with intact motor innervation during applied stretch, *Q. Jl exp. Physiol.*, **46**, 389–398

Cussons, P. D., Hulliger, M. and Matthews, P. B. C. (1977). Effects of fusimotor stimulation on the response of the secondary ending of the muscle spindle to sinusoidal stretch, *J. Physiol., Lond.*, **270**, 835–850

Durkovic, R. G. and Preston, J. B. (1974). Evidence of dynamic fusimotor excitation of secondary muscle spindle afferents in soleus muscle of the cat, *Brain Res.*, **75**, 320–323

Goodwin, G. M., Hulliger, M. and Matthews, P. B. C. (1975). The effects of fusimotor stimulation during small amplitude stretching on the frequency-response of the primary ending of the mammalian muscle spindle, *J. Physiol., Lond.*, **253**, 175–206

Harvey, R. J. and Matthews, P. B. C. (1961). The response of de-efferented muscle spindle endings in the cat's soleus to slow extension of the muscle, *J. Physiol., Lond.*, **157**, 370–392

Hulliger, M., Matthews, P. B. C. and Noth, J. (1977). Static and dynamic fusimotor action on the response of Ia fibres to low frequency sinusoidal stretch of widely ranging amplitudes, *J. Physiol., Lond.*, **267**, 811–838

Jami, L., Lan-Couton, D. and Petit, J. (1980). A study with the glycogen-depletion method of intrafusal distribution of γ-axons that increase sensitivity of spindle secondary endings, *J. Neurophysiol.*, **43**, 16–26

Jami, L. and Petit, J. (1978). Fusimotor actions on sensitivity of spindle secondary endings to slow muscle stretch in cat peroneus tertius, *J. Neurophysiol.*, **41**, 860–869

Jami, L. and Petit, J. (1979). Dynamic and static responses of primary and secondary spindle endings of the cat peroneus tertius muscle, *J. Physiol., Lond.*, **296**, 109P

Lännergren, J. (1971). The effect of low-level activation on the mechanical properties of isolated frog muscle fibres, *J. gen. Physiol.*, **58**, 145–162

Lennerstrand, G. (1968). Position and velocity sensitivity of muscle spindles in the cat. I: Primary and secondary endings deprived of fusimotor activation, *Acta physiol. scand.*, **73**, 281–299

Lennerstrand, G. and Thoden, U. (1968). Position and velocity sensitivity of muscle spindles in the cat. III: Static fusimotor single fibre activation of primary and secondary endings, *Acta physiol. scand.*, **74**, 30–49

Matthews, P. B. C. and Stein, R. B. (1969). The sensitivity of muscle spindle afferents to small sinusoidal changes of length, *J. Physiol., Lond.*, **200**, 723–743

Poppele, R. E. (1973). System approach to the study of muscle spindles. In *Control of Posture and Locomotion* (eds R. B. Stein, K. G. Pearson, R. S. Smith and J. B. Redford), New York and London, Plenum Press, 127–146

Poppele, R. E., Kennedy, W. R. and Quick, D. C. (1979). A determination of static mechanical properties of intrafusal muscle in isolated cat muscle spindles, *Neurosci.*, **4**, 401–411

Renkin, B. Z. and Vallbo, A. B. (1964). Simultaneous responses of group I and II cat muscle spindle afferents to muscle position and movement, *J. Neurophysiol.*, **27**, 429–450

Muscle stretch as a way of detecting brief activation of bag_1 fibres by dynamic axons

F. EMONET-DÉNAND* and Y. LAPORTE*

SUMMARY

Stimulation of single dynamic α or β axons by very brief trains of shocks (2–3 shocks at 2–500/sec) or by single shocks, which at constant muscle length have a negligible effect on the discharge of primary endings, may strongly excite these endings when applied during dynamic stretch.

This effect was studied by the technique of frequencygrams (Bessou *et al.*, 1968) on de-efferented soleus and tenuissimus cat spindles. Muscles were submitted to periodic changes of length of triangular symmetrical form and the stimulation of the axons was applied at preset position during stretch. Stretches of 3–4 mm of amplitude with velocities ranging from 0.8 mm/sec to 32 mm/sec were used. Frequencygrams showed a fast rising phase followed by a slow decay, the whole response lasting about 0.4 sec; their amplitude increase with the velocity of stretch. Increases of 200 impulses/sec in the firing rate of the endings were observed.

The intrafusal distribution of a dynamic axon eliciting this effect in a tenuissimus spindle was determined by the glycogen-depletion method. This axon selectively supplied the bag_1 fibre of the spindle.

It is now well established that γ and β dynamic axons exert their action on primary endings through the activation of bag_1 fibres (Barker *et al.*, 1976b, 1977; Boyd, 1976; Boyd *et al.*, 1977, 1979; Banks *et al.*, 1978). Cinematographical studies of intrafusal muscle fibres have shown that even when a dynamic axon is repetitively stimulated, the contraction of the bag_1 fibre (the dynamic nuclear bag fibre of Boyd) is very weak. Shortening of sarcomeres is observed in a very limited region at some distance from the equatorial region and the deformation of the spirals of the primary ending is extremely small (Bessou and Pagès, 1975; Boyd *et al.*, 1977; Banks *et al.*, 1978).

*Laboratoire de Neurophysiologie, Collège de France, Paris

Frequencygrams of primary endings obtained during stimulation of dynamic axons reflect the weakness of bag_1 fibre contraction. In tenuissimus spindles, Bessou *et al.* (1968) reported that single shocks to dynamic axons elicited either no responses or very low amplitude responses and that repetitive stimulation of some duration was necessary to obtain frequencygram responses which were then strikingly different from the frequencygrams given by static axons in that they had a smooth contour for relatively low rates of stimulation.

Recently, Emonet-Dénand and Laporte (1978b) have studied in some hind-limb muscles (flexor hallucis longus, soleus and peroneus brevis) the frequencygrams of primary endings which on stimulation of γ dynamic axons gave type II responses of the classification of Emonet-Dénand *et al.* (1977). These responses differ from type I responses because when the muscle is held at a constant length, repetitive stimulation of the axon produces a relatively large increase in the firing rate of the ending and because the discharge is irregular. In these frequencygrams small increments of frequency could be detected after each stimulus, suggesting that the terminals of the endings were submitted to a stronger stress than in tenuissimus spindles activated by dynamic axons.

Primary ending frequencygrams obtained during stimulation of dynamic axons have so far been studied at constant muscle length. We report here that the stimulation of dynamic axons by single shocks or by very brief trains of shocks, which at constant muscle length would have had a negligible effect, is capable of eliciting very large increases in the firing rate of primary endings provided the stimulation is applied during stretch.

This observation was made on de-efferented soleus and tenuissimus muscle spindles in cats anaesthetised with Nembutal. Single Ia afferent fibres and single dynamic axons were prepared and identified as in previous studies on spindle motor innervation. Discharges of afferent fibres were recorded with an instantaneous frequencymeter and frequencygrams were obtained by superimposing on the screen of a storage oscilloscope several records obtained during the stimulation of a single dynamic axon while the muscle was being stretched by a servo-controlled electromagnetic puller. The shocks were synchronised with the sweep and hence with the puller because the output of the length transducer of the puller generated the sweep after appropriate amplification. The puller was driven by a low-frequency generator delivering pulses of triangular symmetrical shape. Stretches of 3–4 mm amplitude were applied to the muscle at velocities ranging from 0.8 mm/sec to 32 mm/sec. Single dynamic axons were stimulated either by single shocks or by very brief trains of shocks (2–3 shocks at 2/sec–500/sec) applied at preset positions during the dynamic stretch; in some instances they were continuously stimulated at low frequency, the stimulation also being synchronised with the stretch. Frequencygrams with a satisfactory definition were obtained by superimposing 5–20 traces.

Figure 1 illustrates the effect of dynamic stretch on the action exerted by a soleus dynamic axon repetitively stimulated at frequencies ranging from 2/sec to 7/sec.

After the effect exerted by the axon was identified as a type II dynamic effect (records 1–2), the axon was stimulated at 7/sec while the muscle was successively held at two lengths, L_0 and L_1, as indicated by the lowest trace of records 3–4. At length L_0 (record 3) the ending was silent and the stimulation of the axon did not activate it. At length L_1, i.e. after the muscle length had been increased by 4 mm, the ending displayed a steady discharge but the stimulation had virtually no effect on this discharge (record 4). The muscle was then submitted to periodical triangular stretches of 4 mm amplitude, between L_0 and L_1, at a velocity of 5 mm/sec. During the release phases (not illustrated) the ending fell silent but during the stretch phases (records 6–8) the frequencygrams displayed large increments (30–40 impulses/sec) which evolved from a progressively rising level, each one being related to a shock.

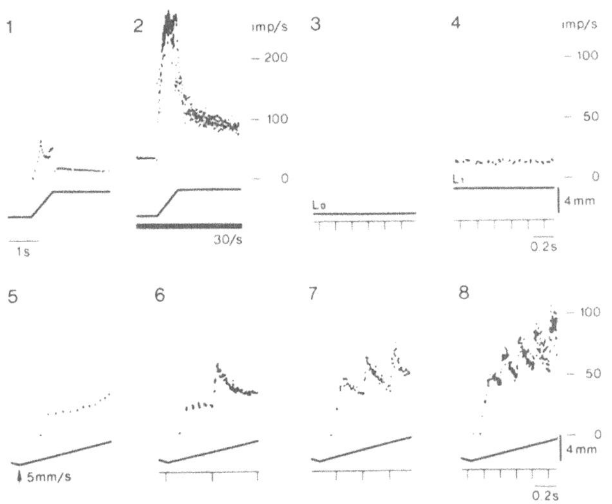

Figure 1. Frequencygrams of a primary ending elicited by the stimulation of a dynamic γ axon during muscle stretch (cat soleus muscle). 1: Passive response of the ending during a ramp-and-hold stretch of 4 mm applied at 6 mm/sec. 2: Strong dynamic effect due to the stimulation of the axon at 30/sec. 3: The stimulation of the axon at 7/sec has no effect on the ending discharge when the muscle is held at the initial length L_0, or a negligible one (4) when it is held at the final length L_1. 5: Discharge of the ending during stretch at 5 mm/sec. 6, 7 and 8: Frequencygrams obtained during similar stretches by stimulating the axon at different low frequencies (2.5/sec, 4/sec and 7/sec). The shocks were synchronised with the sweep.

Record 8 was obtained for a rate of stimulation (7/sec) similar to that used in record 4 but lower rates were used for records 6 (4/sec) and 7 (2.5/sec). In record 6, as the interval between successive shocks was sufficiently long, the complete time course of the response elicited by a shock applied in the middle of the stretch could be observed. This response, which lasted 0.4 sec, consisted of a fast rising phase followed by a much longer decay. Record 6 also shows that a shock applied at the very end of a release phase (first shock on this record) elicited no response, possibly because this ending (*see* record 5) did not resume firing at the onset of the stretch phase.

Figure 2 shows that the amplitude of the responses increases with the velocity of the stretch. In this experiment, a soleus dynamic γ axon was stimulated by three successive shocks at 500/sec. At constant muscle length (not illustrated) this brief stimulation had practically no effect on the firing rate of the primary ending. The muscle was then submitted to periodical triangular stretches of approximately 4 mm and a brief tetanic stimulation was applied during each stretching phase at a preset muscle length, regardless of

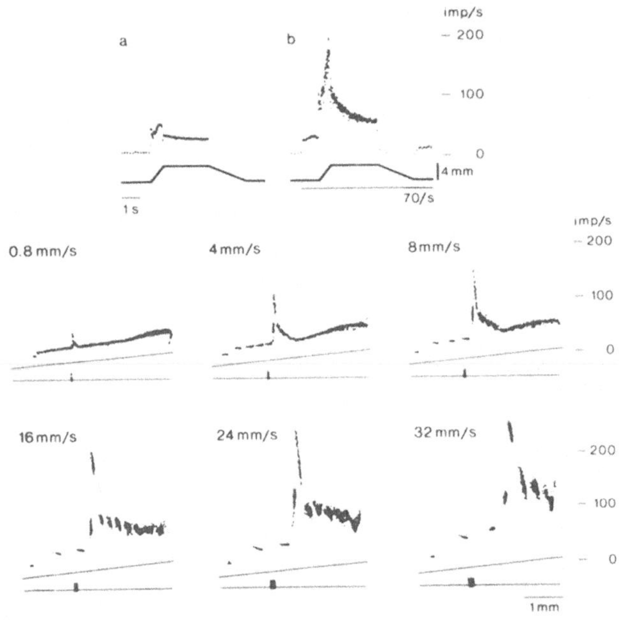

Figure 2. Effect of the velocity of stretch on the amplitude of frequencygrams of a primary ending elicited by the stimulation of a dynamic axon (cat soleus muscle). a and b: Identification of the dynamic action exerted by this fusimotor axon. a: Passive response of the ending during a ramp stretch of 4 mm applied at 6 mm/sec. b: Response observed during repetitive stimulation of the axon at 70/sec. The lower records show frequencygrams obtained by a brief stimulation of the axon (3 shocks at 500/sec) applied during muscle stretches of increasing velocities. Note that the abscissae are calibrated in length units.

the velocity of stretch. Various velocities of stretch were obtained by changing the frequency of the generator. Six velocities ranging from 0.8 mm/sec to 32 mm/sec were used. In each of the six records the upper trace shows the discharge of the primary ending, the middle trace the length of the muscle and the lower trace the stimulation of the axon, which took place when the muscle had been stretched to 1.33 mm. It should be noted that in order to avoid having different time scales for each record, the abscissae are calibrated in length units. When the muscle was slowly stretched (0.8 mm/sec), the increase of firing rate elicited by stimulation of the dynamic axon was small. For 4 mm/sec velocity, the amplitude of the frequencygram was already much larger and further increases were observed for faster stretches up to 24 mm/sec. At that velocity, the combined effect of stimulating the axon and stretch resulted in an increase of nearly 200 impulses/sec. Faster stretching did not increase the response. As in figure 1, all frequencygrams showed an abrupt rising phase but in the present series the peak frequency was followed by a quick fall before a slower decay was observed.

The effect of muscle stretch was also observed for a dynamic β axon as illustrated by figure 3. Single shocks applied to a soleus β axon in the middle of a stretch at 0.8 mm/sec had virtually no effect on the ending discharge (compare records 1 and 3) except for a very slight decrease of firing which possibly

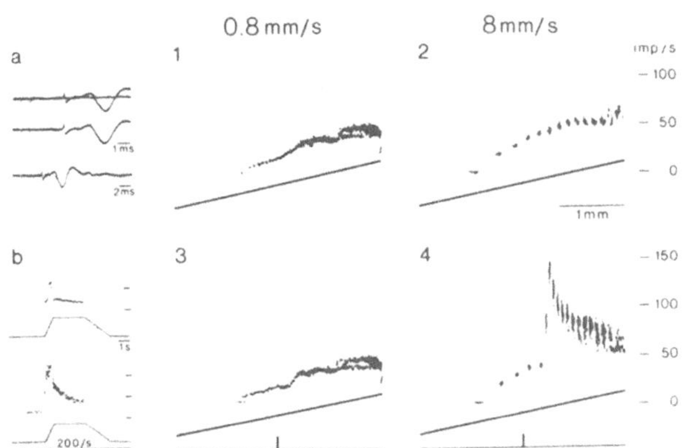

Figure 3. Frequencygram of a primary ending obtained by stimulating a dynamic β axon during muscle stretch at 8 mm/sec (cat soleus muscle).

a: Nerve and muscle action potentials recorded from a small nerve branch to the soleus muscle after stimulating an S1 ventral root filament. Superimposed sweeps. Upper trace: threshold stimulation of the axon. Middle trace: supra-threshold stimulation. Lower trace: supra-maximal stimulation. Note the slower sweep speed.

b: Identification of the dynamic action exerted by the skeletofusimotor axon.

1 and 2: Discharges of the ending during muscle stretch at two constant velocities—0.8 mm/sec and 8 mm/sec.

3 and 4: The β axon is stimulated (*see* lower trace) by a single stimulus applied at the middle part of the stretch.

was due to the contraction of extrafusal muscle fibres. However, when the soleus was stretched at 8 mm/sec (record 4) a single shock elicited a large increase in the firing rate of the ending. The frequencygram in this case also displayed a fast rising phase followed by a slower decay.

As all these observations were made on soleus spindles, in which the action of dynamic axons is generally stronger than in tenuissimus spindles, we thought it was necessary to verify whether stretch had a similar effect on spindles of this muscle. We also had in mind that if that were the case we would have the opportunity, by using the glycogen-depletion technique, to ascertain whether the bag_1 fibre was the only intrafusal fibre responsible for it. Such an experiment is illustrated in figure 4. The dynamic action of a γ axon (type I effect) is shown by comparing the passive response of a primary ending to a ramp-and-hold stretch (record 1) to the response during stimulation of the axon at 100/sec (record 2), and the lack of fatigability of this action is shown by records 3 and 4.

The muscle was submitted to periodical triangular stretches of 4 mm amplitude at various frequencies in order to change the velocity of stretch (records 5 to 10). Each record shows the action potentials led from a primary afferent fibre and their instantaneous frequency, both recorded during a single complete cycle. The discharges of the ending in the absence of stimulation are illustrated by records 5, 6 and 7 whereas the corresponding records 8, 9 and 10 show the alterations of the discharges elicited by a brief stimulation (3 shocks at 500/sec) applied after the first third of the stretch. Comparison of records 8, 9 and 10, respectively, to records 5, 6 and 7 clearly shows that stimulation of the axon elicited a marked increase in the firing rate of the ending and that this effect increased with the velocity of stretch.

Subsequently, the spindle on which these observations were made was located by the technique of Bessou and Laporte (1965) and the dynamic γ axon was repetitively stimulated in order to elicit glycogen depletion in the muscle fibres it supplied. The regime of stimulation used in this experiment to obtain glycogen depletion in intrafusal muscle fibres was different from that previously used in our laboratory. Instead of reducing the blood flow through the muscle during the periods of stimulation (5–20 periods of stimulation at 30–100/sec lasting 1 min and separated by one-minute intervals during which the circulation in the muscle was re-established), the circulation in the tenuissimus was kept normal throughout a single period of repetitive stimulation which lasted no less than 2 h. The rate of stimulation was never lower than 50/sec and approximately every 10 min it was progressively increased up to 500/sec for a few minutes. Continuous monitoring of the discharge of the primary ending showed that the axon kept activating the ending throughout this long period of stimulation and that signs of fatigue were seen only towards the end of this long period of stimulation.

The segment of tenuissimus muscle containing the spindle was processed for histochemical detection of glycogen as described in previous papers. In

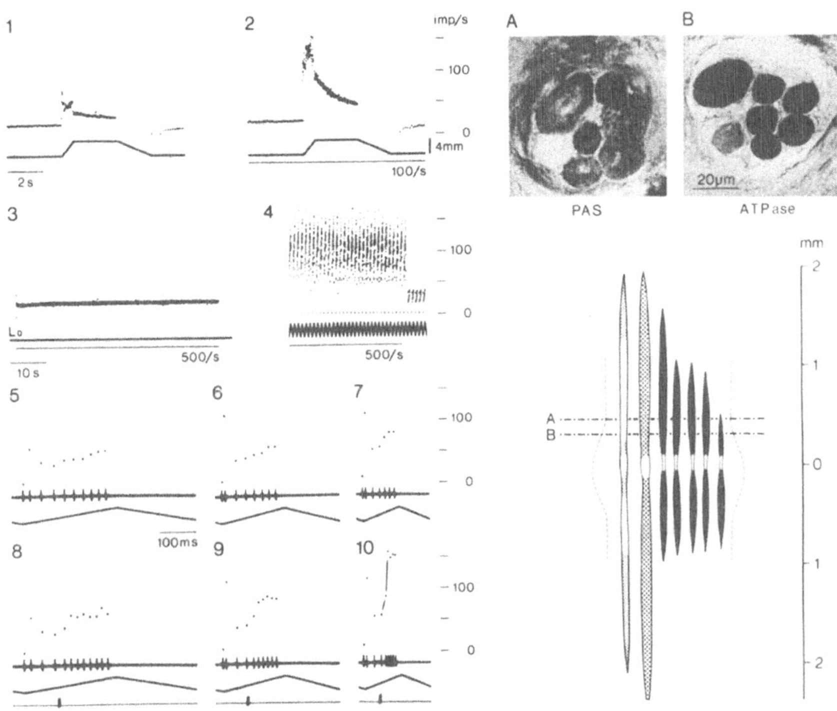

Figure 4. Alteration of the responses to stretch of a primary ending during the stimulation of a dynamic γ axon (cat tenuissimus spindle).

1 and 2: Identification of the dynamic action exerted by the axon.

3 and 4: Stimulation of the axon at 500/sec for nearly 1 min at constant muscle length (3) and during periodical triangular stretches at 1.5 Hz (4) showing the striking resistance to fatigue of the effects elicited by stimulating the dynamic axon (*see* Emonet-Dénand and Laporte, 1978a).

5, 6 and 7: Passive response of the ending during three complete cycles of different duration.

8, 9 and 10: The dynamic γ axon is stimulated (three shocks at 500/sec) after the first third of each stretch phase.

A: Transverse section treated with PAS showing that one intrafusal fibre was totally depleted of its glycogen content.

B: Transverse section treated for detecting ATPase activity after pre-incubation at pH 10.3 showing that the blanched fibre was the only intrafusal fibre that stained low (bag_1 fibre).

The diagram shows a reconstruction of a spindle made from serial transverse sections. An extensive zone of complete glycogen depletion was observed in each pole of the bag_1 fibre.

transverse sections of spindles stained for glycogen with the PAS method it is not always possible to distinguish between bag_1 and bag_2 fibres. The best criterion is the close association of bag_2 with chain fibres in the equatorial region, the bag_1 fibre being clearly separated from that group. Unfortunately this disposition is not a constant one and in previous studies we often had to eliminate spindles in which only one bag fibre was depleted because it was not possible to identify with certainty which of the two bag fibres was depleted (Emonet-Dénand *et al.*, 1980).

The ATPase staining reactions of intrafusal muscle fibres give the possibility of identifying bag_1 fibres in their intracapsular parts (regions A and B of Barker, Banks, Harker, Milburn and Stacey (1976a)) independently of their situation at the equator. After pre-incubation at pH 10.3, bag_1 fibres show a low ATPase activity whereas chain and bag_2 fibres show much higher activities. With this pre-incubation the contrast between bag_1 fibres and the other intrafusal fibres is more marked than previously described (*see* Ovalle and Smith, 1972; Burke and Tsairis, 1977; Kucera *et al.*, 1978). Consequently not all the transverse sections were treated for glycogen depletion but every 400–500 μm, two 10 μm sections were processed for ATPase after pre-incubation at pH 10.3. In the upper-right part of figure 4, microphotographs A and B are transverse sections of the spindle taken through the levels A and B indicated in the diagram. Section A, which was processed for glycogen, shows that one intrafusal fibre was completely depleted of its glycogen content. Section B, which was stained for ATPase, shows that the blanched fibre was the only one to stain light, i.e. that it was a bag_1 fibre. The reconstruction of the spindle made from serial transverse sections shows a rather extensive region of complete glycogen depletion in each pole of the bag_1 fibre, indicating that this dynamic axon supplied both poles of this fibre.

The powerful effect exerted on primary endings by brief activation of bag_1 fibres, when the muscle is being stretched, contrasts with the negligible effect of the same activation occurring when the muscle length does not change. Direct observations of living spindles held at a constant length have shown that prolonged repetitive stimulation of dynamic γ axons elicits a very weak contraction of bag_1 fibres during which only a few sarcomeres actually shorten. Brief stimulation of dynamic axons can be expected to give an even weaker effect. However, the brief stimulation of dynamic axons must elicit some significant alterations in the mechanical properties of bag_1 fibres since concomitant stretching produces a large increase in the firing rate of primary endings. Impulses in dynamic axons are known to generate through their synaptic junctions a partial depolarisation of the membrane of bag_1 fibres (Barker *et al.*, 1978; *see also* Bessou and Pagès, 1972). This depolarisation apparently does not trigger in the great majority of sarcomeres the usual sliding process between myofilaments which leads to sarcomere shortening, but it might lead to an alteration in the properties of the bonds between actin and myosin filaments and/or to an increase in their number. In this view, the large responses of the primary endings observed during stretch could be interpreted as resulting from transitory changes in stiffness and viscosity of the bag_I fibre induced by dynamic axons. The steep fall of frequency which may be observed after the peak value of large amplitude frequencygrams (*see* figure 2) could be due to the transducing properties of the bag_I terminal branch of the primary ending. Hunt *et al.* (1978) have shown that immediately after fast ramp stretches of large amplitude the generator potential goes through a fast repolarising phase due to an increase in K^+ conductance. The slow decay of the last phase of the frequencygram is probably related to the time taken by the sarcomeres to return to their preactivation state. Whichever may be the mechanism of the effect we report here, its functional implication seems clear. It shows that dynamic γ or β motoneurons can exert a powerful control of the dynamic sensitivity of primary endings for very low firing rates and that this effect depends on the velocity of muscle stretch (*see also* Emonet-Dénand and Laporte, 1979). Our observations also suggest that in experiments in which intrafusal muscle fibres are directly observed during the stimulation of single fusimotor axons, the lack of an obvious contraction in a bag_I fibre should be carefully interpreted; it may not necessarily mean that the fibre has not been activated.

ACKNOWLEDGEMENTS

The technical assistance of M. Ginapé, L. Decorte (histology), M. A. Thomas (illustrations and photography) and A. Tristant is gratefully acknowledged. This investigation was supported by a grant from the Fondation pour la Recherche Médicale Française.

Received on July 9th, 1980.

REFERENCES

Banks, R. W., Barker, D., Bessou, P., Pagès, B. and Stacey, M. J. (1978). Histological analysis of cat muscle spindles following direct observations of the effects of stimulating dynamic and static fusimotor axons, *J. Physiol., Lond.*, **283**, 605–619

Barker, D., Banks, R. W., Harker, D. W., Milburn, A. and Stacey, M. J. (1976a). Study of the histochemistry ultrastructure motor innervation and regeneration of mammalian intrafusal muscle fibres, *Prog. Brain Res.*, **44**, 67–87

Barker, D., Bessou, P., Jankowska, E., Pagès, B. and Stacey, M. J. (1978) Identification of intrafusal muscle fibres activated by single fusimotor axons and injected with fluorescent dye in cat tenuissimus spindles, *J. Physiol., Lond.*, **275**, 149–165

Barker, D., Emonet-Dénand, F., Harker, D. W., Jami, L. and Laporte, Y. (1976b). Distribution of fusimotor axons to intrafusal muscle fibres in cat tenuissimus spindles as determined by the glycogen depletion method, *J. Physiol., Lond.*, **261**, 49–69

Barker, E., Emonet-Dénand, F., Harker, D. W., Jami, L. and Laporte, Y. (1977). Types of intra and extrafusal muscle fibre innervated by dynamic skeleto fusimotor axons in cat peroneus brevis and tenuissimus muscles as determined by the glycogen depletion method, *J. Physiol., Lond.*, **266**, 713–726

Bessou, P. and Laporte, Y. (1965). Technique de préparation d'une fibre afférente I et d'une fibre afférente II innervant le même fuseau neuromusculaire chez le Chat, *J. Physiol., Paris,* **57**, 511–520

Bessou, P., Laporte, Y. and Pagès, B. (1968). Frequencygrams of spindle primary endings elicited by stimulation of static and dynamic fusimotor fibres, *J. Physiol., Lond.*, **196**, 47–63

Bessou, P. and Pagès, B. (1972). Intracellular potentials from intrafusal muscle fibres evoked by stimulation of static and dynamic fusimotor axons in the cat, *J. Physiol., Lond.*, **227**, 709–727

Bessou, P. and Pagès, B. (1975). Cinematographical analysis of contractile events produced in intrafusal muscle fibres by stimulation of static and dynamic fusimotor axons, *J. Physiol., Lond.*, **252**, 397–427

Boyd, I. A. (1976). The mechanical properties of dynamic nuclear bag fibres, static nuclear bag fibres and nuclear chain fibres in isolated cat muscle spindles, *Prog. Brain Res.*, **44**, 33–50

Boyd, I. A., Gladden, M. H., McWilliam, P. N. and Ward, J. (1977). Control of dynamic and static nuclear bag fibres and nuclear chain fibres by gamma and beta axons in isolated cat muscle spindles, *J. Physiol., Lond.*, **265**, 133–162

Boyd, I. A., Gladden, M. H. and Ward, J. (1979). The effect of contraction in the three types of intrafusal fibres in isolated cat muscle spindles on the Ia discharge during stretch, *J. Physiol., Lond.*, **296**, 41P

Burke, R. E. and Tsairis, P. (1977). Histochemical and physiological profile of a skeletofusimotor (beta) unit in cat soleus muscle, *Brain Res.*, **129**, 341–395

Emonet-Dénand, F., Jami, L., Laporte, Y. and Tankov, N. (1980). Glycogen depletion of bag_1 fibres elicited by stimulation of static γ axons in cat peroneus brevis muscle spindles, *J. Physiol., Lond.*, **302**, 311–321

Emonet-Dénand, F. and Laporte, Y. (1978a). Effects of prolonged stimulation at high frequency of static and dynamic γ axons on spindle primary ending, *Brain Res.*, **151**, 593–598

Emonet-Dénand, F. and Laporte, Y. (1978b). Fréquencegrammes dûs à la stimulation d'axones γ dynamiques exerçant des effects du type II, *C.R. Acad. Sci. Paris,* **287**, 531–534

Emonet-Dénand, F. and Laporte, Y. (1979). Effets sur les terminaisons primaires de la stimulation à fréquence basse d'axones β dynamiques, *J. Physiol., Paris,* **75**, 37A

Emonet-Dénand, F., Laporte, Y., Matthews, P. B. C. and Petit, J. (1977). On the subdivision of static and dynamic fusimotor action on the primary ending of the cat muscle spindle, *J. Physiol., Lond.*, **268**, 827–861

Hunt, C. C., Wilkinson, R. S. and Fukami, Y. (1978). Ionic basis of the receptor potential in primary endings of mammalian muscle spindles, *J. Gen. Physiol.*, **71**, 683–698

Kucera, J., Dorovini-Zis, K. and King, E. W. (1978). Histochemistry of rat intrafusal muscle fibres and their motor innervation, *J. Histochem. Cytochem.*, **26**, 973–988

Ovalle, W. K. and Smith, R. S. (1972). Histochemical identification of three types of intrafusal muscle fibres in the cat and monkey based on the myosin ATPase reaction, *Canad. J. Physiol. Pharmacol.*, **50**, 195–202

Security of driving of Ia afferents by vibration that simultaneously elicits a tonic vibration reflex

F. J. CLARK*, P. B. C. MATTHEWS† AND R. B. MUIR‡

SUMMARY

Experiments were performed in the decerebrate cat to test how well the soleus Ia fibres are driven 1:1 by vibration (150 Hz, 50μm pulses, combined with a 1 mm stretch) in the presence of a tonic vibration reflex, with particular reference as to whether any phasic Ia 'misbehaviour' could be responsible for the tremor seen in the reflex contraction. In order to keep the reflex arc intact the unitary recording was performed by means of a capillary glass microelectrode inserted into the soleus nerve. In general, driving proved to be remarkably secure in spite of the unloading and any other effects of the reflex contraction, and it is concluded that the tremor in the reflex tension, recorded isometrically, is not attributable to oscillatory stretch reflex activity.

In both animals and man high-frequency small-amplitude vibration has proved to be a useful tool for injecting an excess of Ia input into the CNS so as to study the central actions of the Ia fibres from the spindle primary endings. The particular value of vibration for analytical, as opposed to qualitative, studies hangs upon two features of its action. First, the spindle primary endings have proved to be much more sensitive than the spindle secondary endings or the Golgi tendon organs, so that under suitable conditions they can be activated rather specifically. Second, with appropriate amplitudes of vibration the Ia discharge can be locked in 1:1 synchrony with the vibration, so that the frequency of Ia firing can be set to a known constant value (frequency clamp). Both situations can be achieved readily on applying vibration longitudinally to the de-efferented cat soleus muscle using a peak-to-peak ampli-

*†‡ University Laboratory of Physiology, Oxford. Present addresses:
* Department of Physiology, University of Nebraska; ‡ Anatomy Department, Erasmus University, Rotterdam

tude of 25–50 μm of sinusoidal stretching at a frequency of 100–300 Hz, and with the muscle well stretched (Brown *et al.*, 1967). In man, however, where the application of vibration is inevitably somewhat crude, direct afferent recordings show that these two requirements for quantitative study are poorly satisfied thus limiting the precision of analysis (Burke *et al.*, 1976). Even in the cat, dealing with just a single muscle, there continues to be uncertainty as to how well the Ia afferents can be driven 1:1 when the muscle is contracting and so tending to unload the spindles. Contraction would also seem to increase the stiffness of the muscle so that relatively more of a given amplitude of vibration is taken up within the tendon and so fails to reach the spindles (Brown *et al.*, 1967). In addition, the Ib afferents from the tendon organs are much more readily influenced by vibration when the muscle is contracting so that complete selectivity is inevitably lost.

The tonic vibration reflex provides a situation where these difficulties in the use of vibration come to the fore, since the Ia activation takes place in the presence of a concomitant reflex activation of the muscle. However, secure Ia driving is usually believed to be still readily achievable, since with progressive increase of the amplitude of vibration at a fixed frequency the resulting reflex soon settles at a maximal value. This is most simply attributed to the occurrence of stable 1:1 driving of the majority of the Ia afferents of the vibrated muscle, when further increase in the stimulus should produce no further augmentation of the afferent input (Matthews, 1966). Kanda and Rymer (1977) confirmed that this was so for a sample of five Ia afferents, from three cats, studied in cut dorsal root filaments with the receptor arc otherwise intact, but Jack and Roberts (1978) observed a variable degree of Ia 'misbehaviour' (failure to follow 1:1) in about half of a slightly larger sample of Ia afferents during the tonic vibration reflex, and produced indirect evidence that these were not isolated occurrences. However, as Jack and Roberts noted, their own experimental arrangement may be presumed to have favoured the occurrence of misbehaviour through spindle unloading since in nearly all their preparations 'a large fraction of the tonically active gamma efferent supply must have been interrupted by the ventral root section', which was performed on about half the relevant motor outflow to provide a peripheral ventral root stump for stimulation.

The possible occurrence of Ia misbehaviour is of particular interest in relation to recent experiments on the tremor which is invariably shown by the tonic vibration reflex of the decerebrate cat, recorded isometrically. This is illustrated in figure 1 which shows the small oscillations of up to 0.5 N peak-to-peak amplitude superimposed upon a steady level of about 6 N produced by vibration at 140 Hz. The amplitude of vibration was 50 μm, peak-to-peak, which was amply sufficient to produce a maximal reflex so that the Ia fibres should have had their discharge clamped at the vibration frequency and so have been unable to change their firing in response to the small oscillatory variations in tension. If this be accepted, it follows that the tremor cannot be ascribed to the usual standby of 'oscillation in the Ia stretch reflex arc' and some other mechanism must be sought. Spectral analysis shows that the tremor has a well-defined peak falling in the range 4–11 Hz, with the precise value varying from preparation to preparation (Cussons *et al.*, 1979). The tonic vibration reflex thus offers an interesting animal model for the study of the genesis of tremor. Several indirect arguments have already been adduced in favour of the occurrence of secure driving in this situation (Cussons *et al.*, 1979), but it seemed desirable to check up on the matter experimentally.

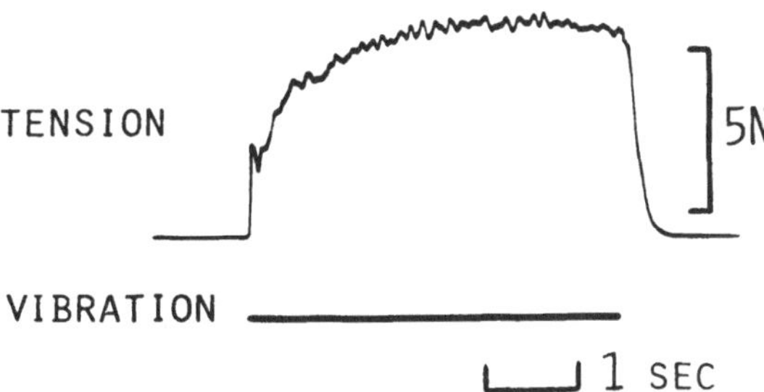

Figure 1. Myographic record of the tonic vibration reflex, elicited isometrically, showing well-marked irregularities in the force of the contraction. 50 μm amplitude vibration at 140 Hz applied to the soleus muscle of the decerebrate cat; the vibration consisted of discrete pulses and was applied with the muscle length held constant.

Unitary Ia discharges proved to be conveniently recorded with a conventional glass capillary micro-electrode (10–15 MΩ) inserted into the nerve to soleus 2–3 cm above its point of entry to the muscle, after incising the nerve sheath. The afferents were identified on the basis of their functional characteristics and on the measurement of their conduction velocity by spike-triggered averaging of the potential recorded with bipolar electrodes from the sciatic nerve. Fuller details are given elsewhere (Clark *et al.*, 1981a), but three points about the technique bear mention. First, in fortunate recordings a unit could be held for as long as was desired in spite of the mechanical interference produced by vibration and the resulting tonic vibration reflex, as well as that due to a 1 mm stretch. Second, when a unit was lost it could quite often be found again on making very small movements of the electrode; likewise, on moving the electrode slightly past a unit it could be found again on withdrawing the electrode. Thus the functional continuity of the afferents seems unlikely to have been interrupted by the recording, and up to 15 units were studied in a single preparation without any obvious deterioration of the reflex. Such preservation of the integrity of the reflex arc contrasts with the situation when units are obtained by splitting dorsal root filaments. Third, the micro-electrode picked up the activity of small, as well as of large, fibres and an appreciable number of spontaneously discharging gamma efferents were encountered; the identification was again provided by spike-triggered averaging. Such recordings thus offer an alternative to the isolation of gamma efferents by splitting of a peripheral nerve.

Every Ia afferent isolated was tested for its responsiveness to a vibration at 150 Hz, 50 μm amplitude, combined with a stretch of 1 mm applied after the beginning of the vibration to help counteract any tendency to spindle unloading. The vibration consisted of a series of discrete pulses, each lasting about 3.5 msec—these produce driving at a lower amplitude than does vibration consisting of a pure sinusoid (McGrath and Matthews, 1973). Table 1 shows the rather uniform findings for 63 afferents studied in six decerebrate cats.

Table 1. Numbers of Ia afferents observed with secure 1:1 driving in response to 50 μm amplitude vibration at 150 Hz. The units have been subdivided according to the size of the tonic vibration reflex recorded at the same time. (The vibration consisted of discrete pulses and was combined with a 1 mm steady stretch.)

	Secure	*Insecure*
Reasonable TVR (>1 N)	43	1
Poor TVR (<1 N)	14	4
Totals	57	5

When the tonic vibration reflex was reasonably developed (>1 N) then 1:1 driving was almost invariably completely secure throughout a 10 sec period of equilibrium when the reflex had reached its steady level; some endings showed a variable insecurity while the reflex was equilibrating. However, when the reflex was present but, for unknown reasons, was poorly developed, an appreciably greater degree of misbehaviour was observed (22% as opposed to 2%, during the steady reflex). This is slightly surprising since the unloading of the spindle by the contraction would then be less powerful, and the explanation seems to be that the smaller reflex was associated with a lower level of fusimotor drive so that any slack created in the spindles by the contraction failed to be taken up as efficiently. Support for this view was obtained by reflexly augmenting the level of fusimotor activity by gently manipulating the pinna or the base of the tail—this led to an increase in the resting level of Ia firing, and the three of the 'misbehaving' endings for which this was done then became securely driven by the 50 μm vibration, in spite of the fact that the tonic vibration reflex then also increased to above 1 N. All six preparations showed a typical tremor in the reflex tension and thus it may be concluded that the tremor in this particular situation with isometric recording must originate elsewhere than in an oscillation of the Ia mediated stretch reflex. Motor unit recordings suggest that the most likely mechanism is simply the chance synchronisation of the activity of a number of different units that happen to be firing at rather similar frequencies and well below the tetanic fusion frequency of the muscle (Clark *et al.*, 1981b). In a wider context the results show that secure Ia driving is readily achievable in spite of muscle contraction when the fusimotor system is active, though it should be noted that in the present experiments the matter has only been tested with the muscle within a few millimetres of its maximum length *in situ* and when the vibration was combined with a small tautening stretch.

Received on August 1st, 1980.

REFERENCES

Brown, M. C., Engberg, I. and Matthews, P. B. C. (1967). The relative sensitivity to vibration of muscle receptors of the cat, *J. Physiol., Lond.*, **192**, 773–800

Burke, D., Hagbarth, K.-E., Löfstedt, L. and Wallin, B. G. (1976). The responses of human muscle spindle endings to vibration during isometric contraction, *J. Physiol., Lond.*, **261**, 695–711

Clark, F. J., Matthews, P. B. C. and Muir, R. B. (1981a). Response of soleus Ia afferents to vibration in the presence of the tonic vibration reflex in the decerebrate cat, *J. Physiol., Lond.*, to be published

Clark, F. J., Matthews, P. B. C. and Muir, R. B. (1981b). Motor unit firing and its relation to tremor in the tonic vibration reflex of the decerebrate cat, *J. Physiol., Lond.*, to be published

Cussons, P. D., Matthews, P. B. C. and Muir, R. B. (1979). Tremor in the tension developed isometrically by soleus during the tonic vibration reflex in the decerebrate cat, *J. Physiol., Lond.*, **292**, 35–57

Jack, J. J. B. and Roberts, R. C. (1978). The role of muscle spindle afferents in stretch and vibration reflexes of the soleus muscle of the decerebrate cat, *Brain Res.*, **146**, 366–372

Kanda, K. and Rymer, W. Z. (1977). An estimate of the secondary spindle receptor afferent contribution to the stretch reflex in extensor muscles of the decerebrate cat, *J. Physiol., Lond.*, **264**, 63–87

McGrath, G. J. and Matthews, P. B. C. (1973). Evidence from the use of vibration during procaine nerve block that the spindle group II fibres contribute excitation to the tonic stretch reflex of the decerebrate cat, *J. Physiol., Lond.*, **235**, 371–408

Matthews, P. B. C. (1966). The reflex excitation of the soleus muscle of the decerebrate cat caused by vibration applied to its tendon, *J. Physiol., Lond.*, **184**, 450–472

SECTION 2
CNS CONTROL OF FUSIMOTOR NEURONS

Overview

A great deal is now known about the properties and specialisation of alpha motoneurons and their central connections. However, the equivalent information about fusimotor neurons is scanty and the contributions of this section are a welcome addition to it. The reason for the shortage of data on fusimotor neurons is largely the technical difficulty of intracellular study of such small cells. Westbury has worked patiently in the face of these difficulties, and has explored the electrical properties of fusimotor neurons and the relationship of discharge rate to injected current. So far no cellular properties seem to distinguish static from dynamic fusimotor neurons, but Appelberg describes one difference in presynaptic organisation as the basis for an operational test for dynamic fusimotor neurons. Another and elegant approach to the problem of distinguishing activity in the two parts of the fusimotor system is described by Gladden. Direct observations of contraction of intrafusal fibres in the exteriorised tenuissimus muscle still attached by its neuromuscular pedicle reveal a considerable potential independence of the control of the different intrafusal fibres. Video films illustrating this were shown at the symposium.

Ellaway's group have studied the reflex responses of fusimotor neurons to muscle afferent stimulation and extend the evidence for an organisation quite different from that of the alpha motoneurons. Further information on the reflex responses of the fusimotor system is provided by Hulliger's presentation in which selective activation of static or dynamic efferents was demonstrable by the responses of a Ia afferent to sinusoidal muscle stretch. A degree of independence of action was emphasised for skeletomotor and fusimotor units, especially for the dynamic group. Rymer and his colleagues show that the fusimotor system can exert marked effects on muscle spindles as a result of reflex activation before alpha motoneurons are involved. Indeed the fusimotor effects may saturate before skeletofusimotors are added. The theme of independence of action in fusimotor and alpha motor systems is further emphasised by Taylor and Appenteng, who make a case for two different patterns of activation for static and dynamic fusimotor neurons to masseter muscle during reflex jaw movements.

Altogether this session adds substantially to the evidence that static and dynamic fusimotor systems may be organised and used differently from the alpha system. Methods are beginning to be developed for recognising the actions of static and dynamic systems and we may look forward to a period when they can be exploited more widely in non-reduced preparations. We may soon have a clearer idea of the strategy actually adopted for the use of the fusimotor system in normal movements.

The papers in this section were read at the symposium, with the exception of those by Bessou *et al.* and Noth, which appeared as posters.

Electrophysiological characteristics of spinal gamma motoneurons in the cat

D. R. WESTBURY*

SUMMARY

Intracellular microelectrode recording methods have been used to study the electrical properties and discharge characteristics of spinal gamma motoneurons (axonal conduction velocity in the range 21–55 m/sec) of the anaesthetised cat. Small hyperpolarising current pulses of long duration applied through the recording electrode were used to measure the input resistances and the time constants of the neurons. Input resistances in the range 1.3–4.9 MΩ (mean, 2.65 MΩ) were found, with principal time constants of 4.6–11.4 msec (mean, 8.8 msec). Long-lasting depolarising current pulses evoked maintained discharges of action potentials at a high rate. These discharges began at a very high rate but then adapted over about 150 msec to reach a steady discharge. The relationship between steady discharge rate and injected current lay in the range 23–61 imp/sec/nA.

These properties are discussed in relation to the properties of alpha motoneurons, and also in relation to the functions of gamma motoneurons in activating intrafusal muscle fibres in muscle spindles.

INTRODUCTION

The electrical characteristics of neurons form an important background to studies of synaptic input and function. For alpha motoneurons, which innervate extrafusal muscle fibres, the morphology, electrical properties and the discharges of action potentials are well described (Eccles *et al.*, 1958; Kernell, 1965a, b and c; Burke, 1968; Baldissera and Gustafsson, 1971; Barrett and Crill, 1974; amongst many contributions). Almost all of these studies have

* Department of Physiology, University of Birmingham Medical School

depended upon the use of intracellular microelectrode methods to study single motoneurons.

By contrast, detailed measurements of the electrical characteristics of gamma motoneurons are few because of the technical difficulty of applying intracellular recording methods to these small neurons. Adequate records from spinal gamma motoneurons can be achieved, but much less frequently than has been the experience with the larger alpha motoneurons. Some results have however been forthcoming (Eccles *et al.*, 1960; Kemm and Westbury, 1978; Westbury and Kemm, 1980) and these are complemented by results obtained by more indirect means (Hunt and Paintal, 1958; Ellaway, 1972; Gustafsson and Lipski, 1979).

The results presented in this paper are an attempt to extend this background knowledge of the properties of gamma motoneurons by further intracellular recordings. Although a picture of the properties of gamma motoneurons is emerging, the number of adequate intracellular records made is still limited and so cannot be taken to represent the full range of the characteristics of these neurons.

METHODS

The experiments were performed upon adult cats of both sexes, weighing between 2.2 kg and 4.7 kg. Anaesthesia was induced by means of pentobarbitone sodium (Sagatal—May and Baker Ltd), and maintained by further injections as necessary. The lumbosacral spinal cord was exposed by laminectomy, and the dura mater was opened. The dorsal roots were cut (S3–L6) and the peripheral ends were mounted on bipolar electrodes for stimulation. The exposed tissues were covered with a pool of warmed liquid paraffin. The sciatic nerve above the origin of the hamstring nerve was stimulated electrically with bipolar electrodes. The animals were paralysed with gallamine triethiodide (Flaxedil—May and Baker Ltd), bilateral pneumothorax was performed and respiration was maintained artificially. End-tidal CO_2 was kept at 4–4.5%. The temperature of the animals was maintained at 37°C.

Intracellular records were made from motoneurons in the ventral horn of segments L7 and S1 using conventional single barrel glass micropipettes. These were filled with either 3M KCl or 2M potassium citrate solution, or a solution of horseradish peroxidase (Sigma type VI) in 0.1M tris-HCl buffer with 0.2M KCl added. The electrodes commonly had resistances in the range 15–40 MΩ. A conventional electrometer with high input impedance and negative capacitance feedback was employed. The microelectrode was arranged in a bridge circuit to permit current to be passed into the motoneurons whilst recording the responses of the cells through the single electrode. As far as possible, electrodes were chosen which did not change their resistance when current was passed but this was often difficult to achieve with small electrode-tip diameters. In about half of the recordings, the electrodes were coated with silicone (dimethyldichlorosilane—Hopkin and Williams Ltd) which improved the sealing of electrodes into neurons after penetration so that recorded membrane potentials were higher and more stable. Recorded membrane potentials of higher than 50 mV were considered adequate.

Motoneurons were identified by their antidromic response following electrical stimulation of the sciatic nerve. Further identification of the destination of the motoneuron axons was not usually achieved. Motoneurons with axonal conduction velocities in the range 55–21 m/sec were studied. None of the neurons presented in this paper received monosynaptic excitatory inputs from group Ia afferent axons following dorsal root stimulation, and so they may be presumed to be fusimotor in function (Hunt and Paintal, 1958; Kemm and Westbury, 1978).

RESULTS

Input resistance and time constant

The input resistance and the time constants of the motoneurons were measured by passing small hyperpolarising currents of 50–100 msec duration through the recording electrode.

Input resistance was calculated from the current passed and the final amplitude of the potential change caused by this. The measurements were taken to be adequate only when steady values of resistance were obtained, and when the membrane potential was stable. The results obtained had a considerable scatter, with values ranging from 1.3 to 4.9 MΩ, and are shown in table 1. These values are comparable with those obtained for small alpha motoneurons (type S; Burke, 1968). The main source of error in making these measurements lay in the behaviour of the microelectrodes when current was

Table 1 Values of input resistance and principal time constant for six gamma motoneurons

Motoneuron	*Conduction velocity* (m/sec)	*Input resistance* (MΩ)	*Time constant* (msec)
1	22	1.7	9.4
2	52	1.6	11.4
3	55	1.3	4.6
4	35	1.6	8.4
5	21	4.9	10.7
6	25	4.8	8.3

passed through them. Although electrodes were selected which did not change resistance when current was passed, and the bridge circuit was balanced as far as possible before impalement of the cell, electrode characteristics may have changed during penetration. Usually the electrodes could be shown to be unchanged after withdrawal from the neuron.

The principal time constant for the motoneurons was calculated from the shape of the voltage transient recorded following the onset of the pulse of hyperpolarising current using the method of Rall (1960). The values obtained ranged from 4.6 to 11.4 msec (table 1), and are comparable with those measured for small alpha motoneurons by Burke (1968) but longer than those measured by Barrett and Crill (1974).

The use of this method for determining time constant involves the acceptance of several assumptions about the morphology of the neurons, and of the way in which injected current is distributed within them. In particular, it is only appropriate if the ratio of dendritic to somatic conductance is ≥ 2. The method has been accepted for the study of alpha motoneurons, but it is difficult to be certain whether it is appropriate for other neurons.

Because of the difficulties inherent in the method, measurements such as these made by the single microelectrode method are open to considerable error. The errors can be minimised by the careful selection of microelectrodes, but the use of large-diameter microelectrodes which are more reliable (Burke and ten Bruggencate, 1971) was precluded here. The use of microelectrodes coated with silicone resulted in recordings that were more stable, and values of input resistance that were significantly higher than those made with untreated electrodes. This suggests that some of the lower values of input resistance obtained in these experiments may have been influenced by an increased 'leak' in the membrane as a result of damage by electrode penetration, even though adequate stable membrane potentials were obtained, and there was no injury discharge.

Neuronal input resistance will be reduced by synaptic activity, whether excitatory or inhibitory. Such background activity should have been reduced by the use of pentobarbitone anaesthesia. However, synaptic input will not have been eliminated entirely so that the values of input resistance and of time constant will be rather lower than the true values without synaptic conductance changes.

The discharge of action potentials

Gamma motoneurons proved to be very sensitive to the passage of depolarising current through the recording electrode, and responded with a discharge of action potentials at a high rate. Examples of this discharge are shown in figures 1 and 2. The rate of discharge increased with increasing depolarising current. However, increasing current also led to a progressive decrease in the size of the action potentials.

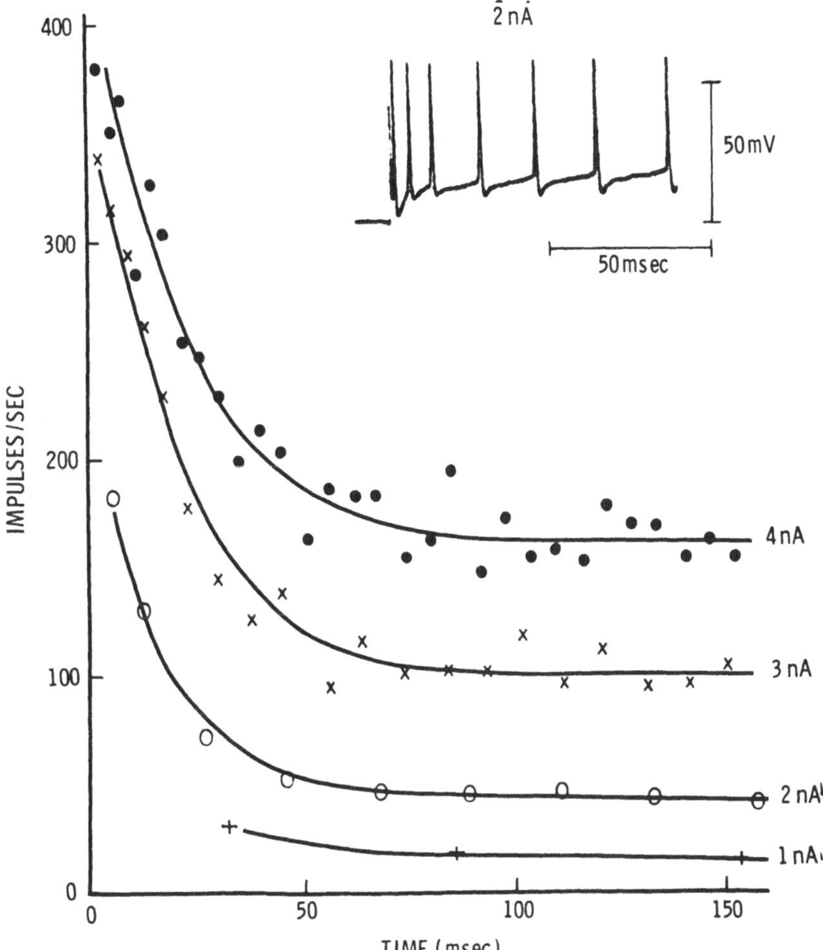

Figure 1. Adaptation of discharge during a depolarising current pulse. The discharge rate of a motoneuron, determined as the reciprocal of each interspike interval, is plotted against the time since the beginning of the injection of current for each of four different currents, 1–4 nA. The axonal conduction velocity of the neuron was 25 m/sec, and the resting membrane potential was −63 mV. A typical record of the early part of a maintained discharge of this motoneuron in response to a 2 nA current is shown as the inset. There was no discharge before current was passed.

The discharges began at a very high rate, but then adapted. This is illustrated in figure 1. The inset record shows the discharge of the motoneuron during a current injection of 2 nA. The initial high frequency slowed to reach a steady rate after about 120 to 150 msec. The graphs show in more detail the time course of adaptation of the discharge at different current levels.

Figure 2 shows a typical relationship between the rate of discharge and the

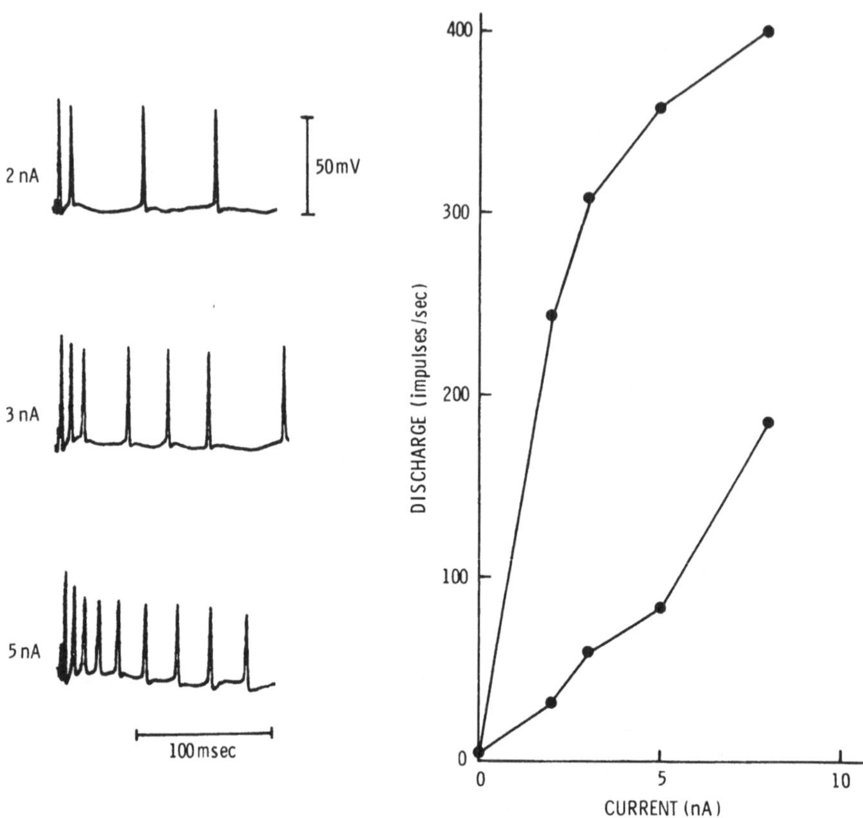

Figure 2. Relationship between discharge rate and injected current during depolarising current pulses. The records shown on the left of the figure illustrate the early part of the maintained discharge of the motoneuron in response to the injection of depolarising currents of different value. The current pulses start at the beginning of the traces. On the right is shown a graph relating rate of discharge of this motoneuron to the injected current. The upper line represents the initial discharge rate calculated as the reciprocal of the first interval; the lower shows the steady rate attained later in the discharge. The background discharge rate was at about 3 imp/sec. Resting membrane potential was −60 mV. Axonal conduction velocity was 22 m/sec.

injected current for these neurons. The rate of background discharge was low <10 imp/sec. The relationship was a simple one, with no sign of the primary and secondary discharge ranges demonstrated for alpha motoneurons by Kernell (1965b). The slope of the relationship between discharge rate after adaptation and injected current varied between 23 and 61 imp/sec/nA. Although detailed measurements of this relationship could only be made from three neurons, records from eight other neurons suggest that these are representative of the properties of many gamma motoneurons. This slope is very high when compared with the same index for alpha motoneurons, measured as 0.4–4.5 imp/sec/nA by Kernell (1965a). The rate of discharge calculated from the first interval was much higher than that found for the steady discharge reached after adaptation, as was expected from the adaptation relationship of figure 1. The slope of the curve relating initial discharge rate to injected current was steeper than that for steady discharge. A similar relationship was found by Kernell (1965a) in the discharge of alpha motoneurons but at much lower discharge rates.

DISCUSSION

The input resistance of a neuron is determined by the surface area of the cell membrane and so by its morphological features, and by the specific resistivity of its membrane. The values for input resistance found here for gamma motoneurons are lower than would be expected from extrapolation of measurements made on alpha motoneurons (Kernell, 1966; Burke, 1968). It is difficult to make stable penetrations of these small motoneurons, so it is possible that despite adequate stable membrane potentials, damage from impalement caused increased leakage conductance. This possibility is borne out by the finding that coating the microelectrode with silicone increased significantly the values obtained for input resistance. Against this possibility however is the finding that in only one case was a low input resistance associated with a short measured time constant, a consequence which should accompany increased leakage conductance. Even if the higher values obtained in these experiments are taken as representative of the properties of gamma motoneurons, the values are lower than would be expected from their morphology, which shows small cell bodies with few thin dendrites which branch relatively little (Cullheim and Ulfhake, 1979; Westbury, 1979; Westbury, in preparation). If true, this would imply that the specific membrane resistivity of the small gamma motoneurons is less than that of the large alpha motoneurons, but it will require careful matching of electrical and morphological measurements to establish this. A comparable situation was found in the study of abducens motoneurons by Grantyn and Grantyn (1978).

Gamma motoneurons have a low threshold for the generation of action potentials in response to injected current when compared with alpha motoneurons, and are capable of very high discharge rates. Sustained discharges at up to 200 imp/sec have been observed, which are beyond those seen in alpha motoneurons.

The main feature which controls the discharge of motoneurons is the after-hyperpolarisation conductance (Baldissera and Gustafsson, 1971). In alpha motoneurons, this is a dominant property. In gamma motoneurons, two hyperpolarisations follow the action potential. The first is a small brief hyperpolarisation lasting up to 4 msec. The second, apparently equivalent to the after-hyperpolarisation of alpha motoneurons, varies widely in its duration and amplitude—from neurons where it is almost non-existent, e.g. that of figure 1, through neurons where the amplitude is more significant but still small, e.g. in figure 2, to a minority of motoneurons in the conduction velocity range below 55 m/sec which resemble small alpha motoneurons in this respect and have prominent after-hyperpolarisations (Eccles *et al.*, 1960; Kemm and Westbury, 1978; Gustafsson and Lipski, 1979). In the majority of gamma motoneurons, however, the effect of the after-hyperpolarisation on discharge is much less than in alpha motoneurons and this contributes to the higher rate of discharge.

Adaptation of discharge in alpha motoneurons has been attributed to a summation of the after-hyperpolarisation conductance so enhancing its effect over the first few interspike intervals, and it seems likely that adaptation in gamma motoneurons has a similar cause. Comparing gamma motoneurons with alpha motoneurons, the repetitive discharges seem to be regulated by similar processes but with important quantitative differences between them.

A close parallel was found between the discharge characteristics of gamma motoneurons and those of abducens motoneurons (Grantyn and Grantyn, 1978). The slope of the relationship between discharge rate and injected current for gamma motoneurons was similar to those for abducens motoneurons (19–44 imp/sec/nA), and very much higher than the comparable values for alpha motoneurons.

These electrical properties underlie the great reflex responsiveness which seems to be a feature of the gamma motoneurons (Hunt, 1951; Hunt and Paintal, 1958). The discharge characteristics could to some extent offset the reduced surface area available for synaptic input to gamma motoneurons because a much smaller synaptic conductance change will be required to cause a given change in gamma motoneuron discharge than would be the case for alpha motoneurons. However, too little is known of the arrangement of synaptic input to gamma motoneurons.

The properties of gamma motoneurons seem well correlated with the requirements for activation of intrafusal muscle fibres. The effects of fusimotor stimulation on muscle spindles increases with rate up to more than 150 imp/sec (Matthews, 1972) and this is well matched to the range of discharge found in these experiments. Kernell (1979) found that for a range of alpha motoneurons, the after-hyperpolarisation (and so the frequency–current relationship) and the range of discharge rate were related quite closely to the frequency–tension relationship for the motor unit which they innervated. In a similar way, the properties of abducens motoneurons are related to the properties of their motor units. This suggests, because of the similarity between gamma and abducens motoneurons, that in some ways the properties of intrafusal muscle fibres may be similar to those of the fibres of lateral rectus muscle. The measurement of electrical properties has not revealed any subgrouping of gamma motoneurons which might correspond to the subdivision into static and dynamic types, although the number of motoneurons studied so far is small. This finding is perhaps not surprising, however, as the static and dynamic fusimotor fibres seem to require rather similar ranges of discharge.

ACKNOWLEDGEMENT

This research was supported by a grant from The Wellcome Trust.

Received on June 23rd, 1980.

REFERENCES

Baldissera, F. and Gustafsson, B. (1971). Regulation of repetitive firing in motoneurons by the after-hyperpolarisation conductance, *Brain Res.*, **30**, 431–434

Barrett, J. N. and Crill, W. E. (1974). Specific membrane properties of cat motoneurones, *J. Physiol., Lond.*, **239**, 301–324

Burke, R. E. (1968). Group Ia synaptic input to fast and slow twitch motor units of cat triceps surae, *J. Physiol.*, **196**, 605–630

Burke, R. E. and ten Bruggencate, G. (1971). Electrotonic characteristics of alpha motoneurones of varying size, *J. Physiol., Lond.*, **212**, 1–20

Cullheim, S. and Ulfhake, B. (1979). Observations on the morphology of intracellularly stained γ-motoneurons in relation to their axon conduction velocity, *Neurosci. Lett.*, **13**, 47–50

Eccles, J. C., Eccles, R. M., Iggo, A. and Lundberg, A. (1960). Electrophysiological studies on gamma motoneurones, *Acta physiol. scand.*, **50**, 32–40

Eccles, J. C., Eccles, R. M. and Lundberg, A. (1958). The action potentials of the motoneurones supplying fast and slow muscles, *J. Physiol., Lond.*, **142**, 275–291

Ellaway, P. H. (1972). The variability in discharge of fusimotor neurones in the decerebrate cat, *Expl Brain Res.*, **14**, 105–117

Grantyn, R. and Grantyn, A. (1978). Morphological and electrophysiological properties of cat abducens motoneurons, *Expl Brain Res.*, **31**, 249–274

Gustafsson, B. and Lipski, J. (1979). Do γ-motoneurones lack a long-lasting afterhyperpolarisation? *Brain Res.*, **172**, 349–353

Hunt, C. C. (1951). The reflex activity of mammalian small-nerve fibres, *J. Physiol., Lond.*, **115**, 456–469

Hunt, C. C. and Paintal, A. S. (1958). Spinal reflex regulation of fusimotor neurones, *J. Physiol., Lond.*, **143**, 195–212

Kemm, R. E. and Westbury, D. R. (1978). Some properties of spinal γ-motoneurones in the cat, determined by micro-electrode recording, *J. Physiol., Lond.*, **282**, 59–71

Kernell, D. (1965a). The adaptation and the relation between discharge frequency and current strength of cat lumbosacral motoneurones stimulated by long-lasting injected currents, *Acta physiol. scand.*, **65**, 65–73

Kernell, D. (1965b). High-frequency repetitive firing of cat lumbosacral motoneurones stimulated by long-lasting injected currents, *Acta physiol. scand.*, **65**, 74–86

Kernell, D. (1965c). The limits of firing frequency in cat lumbosacral motoneurones possessing different time course of afterhyperpolarization, *Acta physiol. scand.*, **65**, 87–100

Kernell, D. (1966). Input resistance, electrical excitability and size of ventral horn cells in cat spinal cord, *Science, N.Y.*, **152**, 1637–1640

Kernell, D. (1979). Rhythmic properties of motoneurones innervating muscle fibres of different speed in m. gastrocnemius medialis of the cat, *Brain Res.*, **160**, 159–162

Matthews, P. B. C. (1972). *Mammalian Muscle Receptors and Their Central Actions,* London, Edward Arnold

Rall, W. (1960). Membrane potential transients and membrane time constant of motoneurons, *Expl Neurol.*, **2**, 503–532

Westbury, D. R. (1979). The morphology of four gamma motoneurones examined by horseradish peroxidase histochemistry, *J. Physiol., Lond.*, **292**, 25–26P

Westbury, D. R. and Kemm, R. E. (1980). Some implications of the properties of spinal gamma motoneurones. *Progress in Clinical Neurophysiology,* Vol. 8 (ed. J. E. Desmedt), Basel, Karger, 28–32

Selective central control of dynamic gamma motoneurons utilised for the functional classification of gamma cells

B. APPELBERG*

SUMMARY

This report deals with supraspinal control of gamma motoneurons. Knowledge about a system originating in the rubral region in the mesencephalon, possibly from the rostral part of the red nucleus itself, is reviewed in particular. The pathway from this region (denoted the mesencephalic area for dynamic control (MesADC-region)) descends ipsilaterally down to the medullary level, sends collaterals to the inferior olivary nuclear complex and then crosses the midline to proceed in the dorsolateral funiculus of the contralateral spinal cord.

The MesADC-system acts on dynamic gamma motoneurons without any simultaneous effects on static gamma or on alpha motoneurons. A technique is described for the selective activation of the MesADC-system. It is argued that such selective activation can be used experimentally for the functional classification of gamma motoneurons. The MesADC-system furnishes an example of a motor system organised in a way not easily compatible with current ideas of alpha–gamma linkage.

In 1961 a short note by Jansen and Matthews came as the first indication of independent control of static and dynamic properties of muscle spindles. The findings were reported in full the following year (Jansen and Matthews, 1962) and gained immediate support by observations made by Appelberg (1962a, b) and also by Granit and van der Meulen (1962). In addition, two functional populations of gamma efferents were demonstrated by Matthews (1962).

Knowledge about muscle spindle structure and innervation was accumulating rapidly at that time (cf. Barker, 1962; Boyd, 1962). The physiological findings could therefore be interpreted on a morphological basis. It was supposed that spontaneous fluctuations or stimulation-induced changes in gamma cell drive could quite independently affect the one or the other of two

* Department of Physiology, University of Umeå

different populations of gamma motoneurons. Thereby different types of intrafusal muscle fibres were influenced and the spindles consequently changed either their tonic length-measuring capability or their sensitivity to dynamic changes in muscle length. Later work on the organisation of the fusimotor system has proved these original interpretations to be essentially correct. Static and dynamic gamma cells and efferent fibres were shown to have different intrafusal distributions, and when active, to produce fundamentally different effects (cf. Matthews, 1972; Barker *et al.*, 1976).

Appelberg (1962a, b) interpreted his findings of inhibition of gamma efferent activity and extensor spindle afferent activity from the red nucleus as a selective effect on static gamma cells. The simultaneously observed increase of spindle dynamic sensitivity was ascribed to an increased load being imposed passively on intrafusal nuclear bag fibres. Somewhat later (Appelberg, 1963) it became evident, however, that electrical stimulation in the region of the red nucleus could increase spindle dynamic sensitivity without indications of a simultaneous influence on the static gamma system. The red nucleus itself seemed to give reciprocal effects on static gamma motoneurons, inhibiting those to extensors while exciting those to flexors (Appelberg and Kosary, 1963).

METHODOLOGICAL ASPECTS

Much work in the author's laboratory has since been devoted to the task of studying in more detail the control of fusimotor neurons from the rubral region of the mesencephalic brain stem in cats. Mostly, in these experiments, changes in muscle spindle afferent activity caused by central stimulation during ramp extension or twitch-contraction tests were taken as an indicator of changes in dynamic and/or static fusimotor cell drive. A lack of electromyographic or tension changes in the small muscles usually used in these studies was taken as evidence that alpha or beta motoneurons were not involved. Instead the effects obtained were ascribed to influences on gamma motoneurons. When studying the effect on muscle spindle afferent activity of repetitive electrical stimulation in the brain stem, the nature of the effect could be appreciated by utilising the experience gained by others who stimulated isolated dynamic or static gamma efferents. Gamma-induced changes in the dynamic index, in the period of silence upon release of stretch, in the regularity of firing and in the pause of discharge during a twitch contraction are well known and have been elucidated in detail (cf. Matthews, 1972; Emonet-Dénand *et al.*, 1977). Evidently there are means available which allow one to judge whether centrally evoked effects on muscle spindles are predominantly static, dynamic or clearly mixed.

WORK ON MESENCEPHALIC CONTROL OF GAMMA CELLS REVIEWED

In the following description of various experimental series carried out in attempts to prove the existence of a specific descending system controlling dynamic gamma motoneurons, reference will be made to figure 1. In 1965,

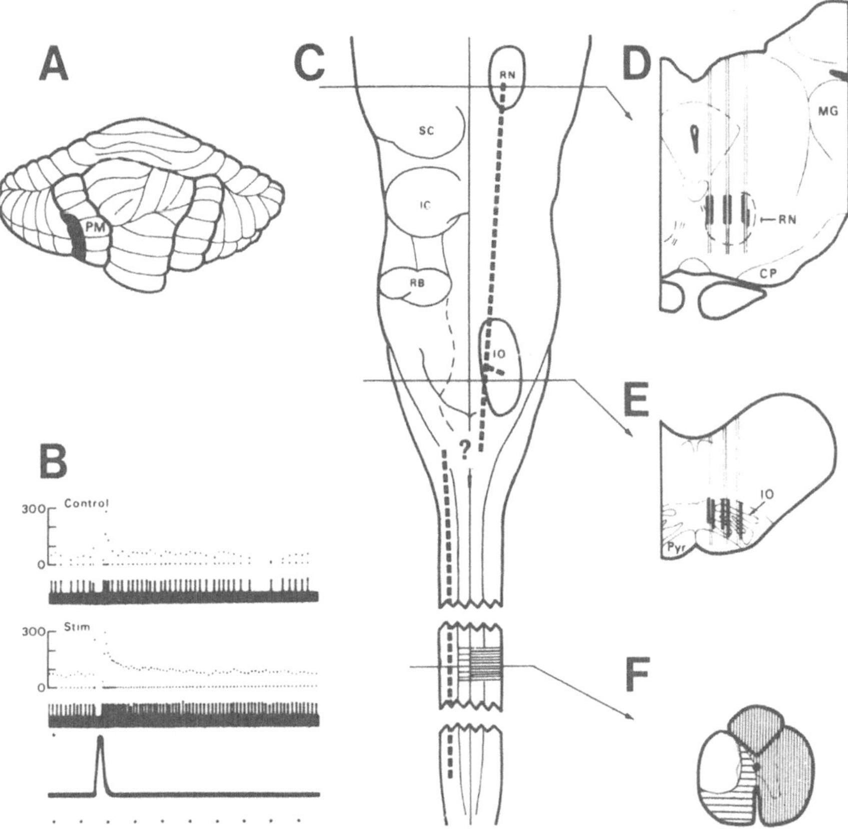

Figure 1. Composite figure illustrating various aspects of the MesADC-system. A, cerebellar posterior lobe viewed from behind; area in black is the d_1-zone. B, recordings from a muscle spindle primary afferent during twitch-test under control conditions and during MesADC-stimulation. C, dorsal view of mesencephalic, pontine and medullary brain stem and spinal cord. D, transverse section through mesencephalon at level indicated. E, transverse section through medullary brain stem at level indicated. F, transverse section through the spinal cord with lesions of dorsal funiculi, right spinal half and left ventral funiculus indicated. Abbreviations: CP, cerebral peduncle; IO, inferior olivary nucleus; MG, medial geniculate body; PM, paramedian lobule; Pyr, pyramid; RB, restiform body; RN, red nucleus; SC, superior colliculus. (B and D modified from Jeneskog, 1974a.)

Appelberg and Emonet-Dénand showed that effects of the type illustrated with twitch tests in figure 1B were obtained in extensor as well as flexor primary spindle afferents on stimulation in one and the same locus in the rubral region. The effects were exerted without simultaneous signs of extrafusal activation. The region in the mesencephalon from which such dynamic effects were evoked was found to comprise dorsal parts of the red nucleus but also to extend dorsally and caudally to that structure (Appelberg and Molander, 1967). These authors and also Appelberg (1967) reported that stimulation in sites causing typical spindle effects of the dynamic type also regularly evoked climbing fibre responses in the posterior lobe of the cerebellar cortex. A medullary region within the inferior olivary nuclear complex was found to affect muscle spindles similarly. It was also pointed out, however, that the descending effect on the muscle spindles was not abolished by cerebellectomy. The term MesADC (the mesencephalic area for dynamic control) which was introduced at that time will be used in the following text.

In a series of investigations by Appelberg and Jeneskog (1968, 1969, 1972) it was first of all demonstrated that clearly separable pathways exist in the spinal cord which in a selective way influence static or dynamic gamma motoneurons. Furthermore, the pathway transmitting the effects from the MesADC-region to dynamic gamma cells could be shown to proceed in the contralateral dorsolateral funiculus (DLF) of the spinal cord, i.e. in the same region as rubrospinal and corticospinal tracts. Such information was achieved by making cord lesions abolishing or saving the central influence on the spindles (cf. figures 1C and 1F). Particular efforts were made to investigate whether the pathway conveying the effects to dynamic gamma cells from the MesADC-region was either the rubrospinal or the corticospinal tract. This was all the more important, as the red nucleus and the sensory-motor cortex were previously indicated to affect gamma motoneurons (Appelberg, 1962b; Vedel, 1966). However, lesions of the rubrospinal tract at mesencephalic or medullary levels, and the destruction of the pyramid or degeneration of the corticospinal tract caused by sensory-motor cortex ablation, did not remove the effect on dynamic spindle sensitivity caused by MesADC-stimulation. *The existence of a particular MesADC-pathway to dynamic gamma motoneurons which was separable from other major motor systems passing in the dorsolateral funiculus was thereby established.* Appelberg and Jeneskog (1972) also reported an area ventral to the red nucleus from which spindles could be influenced. The effects from this region were mainly static and were transmitted via a ventral or ventrolateral spinal pathway. This ventral area has also been observed to act on alpha motoneurons (Appelberg and Jeneskog, unpublished observations).

The observations reported by Appelberg (1967) that MesADC-stimulation also regularly caused climbing fibre responses in the posterior lobe of the cerebellum were confirmed by Appelberg and Jeneskog (1972) and by Jeneskog (1974a, b). It could be shown that such responses were in fact localised to a very restricted part of the contralateral paramedian (PM) lobule, the so-called d_1-zone (figure 1A). From single-stimulating electrode tracks through the rubral region, climbing fibre responses and dynamic spindle effects were found to be evoked in remarkable parallelism. An example of this is illustrated in figure 1D where black bars to the left of the tracks indicate evoked dynamic spindle effects and bars to the right of the tracks indicate evoked climbing fibre responses (in both cases at a stimulating intensity of 70 μA). Also at a medullary level such a coincidence between the two stimulating effects could be demonstrated (figure 1E).

Interestingly enough the climbing fibre projection zone in the paramedian lobule for the MesADC-system was found to be the same as the projection area for an ascending system, a spino-olivo-cerebellar path conveying information from low threshold cutaneous receptors in distal parts of the hind limb (the DLF-SOCP described by Larsson *et al.*, 1969). It was suggested by Appelberg and Jeneskog (1973) and by Jeneskog (1974a) that information about on-going activity in a descending system (the MesADC-system) and about peripheral events (cutaneous receptors) would be integrated in the inferior olive and used by the cerebellum in its task of updating continuously the outflow in the motor system.

The MesADC-system may have its origin in the rostral part of the red nucleus. This statement is based upon indirect evidence, namely the observation of a latency difference of about 1 msec between climbing fibre responses evoked from caudal and rostral positions within the MesADC (Jeneskog, 1974a). The shift in latency was observed to occur when the stimulating electrode was moved rostrally into the rostral pole of the red nucleus. Further, in figure 1C the question-mark just caudal to the inferior olivary nucleus indicates the possible existence of a synaptic relay in the MesADC-pathway at this level. The strongest evidence suggesting such a relay seems to be the inability to produce distinct descending volleys in the pathway by single shock stimulation within the MesADC. A train of impulses, on the other hand, leads to a gradually growing descending activity in the DLF (cf. Appelberg *et al.*, 1975, and figure 2B). The MesADC-system thus seems to be a rubro-bulbo-spinal system sending collaterals to certain parts of the inferior olivary nucleus (Jeneskog, 1974c; Jeneskog and Johansson, 1977). These authors also pointed out the remarkable similarities in several respects between this system and the dorsal reticulospinal system (Engberg *et al.*, 1968a,b; Baldissera *et al.*, 1972b) and suggested the possibility that the two were in fact one and the same (cf. also Jeneskog, 1979).

SELECTIVE ACTIVATION OF THE MesADC-SYSTEM

The rubral region in the mesencephalon is probably one of the most well-studied areas in the entire brain stem. From work by Lundberg and collaborators (cf. Hongo *et al.*, 1969a,b; Hongo *et al.*, 1972a,b; Baldissera *et al.*, 1972a,b) and the work reviewed above, it is clear that electrical stimulation in the vicinity of the red nucleus may excite a vast number of ascending and descending pathways including ones with known influence on alpha motoneurons as well as gamma motoneurons. It seems equally clear, however, that when the necessary precautions are taken, the MesADC-system can be activated in isolation. The following procedure should be employed to achieve this.

1. Two stereotactically guided stimulating electrodes should be used.
2. One of these should be placed in the 'ventral border zone' ventrocaudally in the red nucleus (cf. Baldissera *et al.*, 1972b).
3. The other stimulating electrode should be placed within the MesADC dorsally and caudally to the red nucleus.
4. Spinal lesions sparing only the DLF on the contralateral side should be performed to avoid gamma and alpha effects from the ventral mesencephalic area (see above) and effects possibly transmitted ipsilaterally.

The correct placement of the rubral electrode should be guided by recording the descending rubrospinal volley on the dorsolateral aspect of the spinal cord. On penetrating the caudal pole of the red nucleus, a distinct volley with a latency of about 2.5 msec in the lower thoracic region, will be recorded. When proceeding further ventrally the volley will gradually change into one with a latency about 0.5 msec longer (cf. Baldissera *et al.*, 1972b, figure 1). Presynaptic interposito-rubral fibres are now being stimulated and no direct

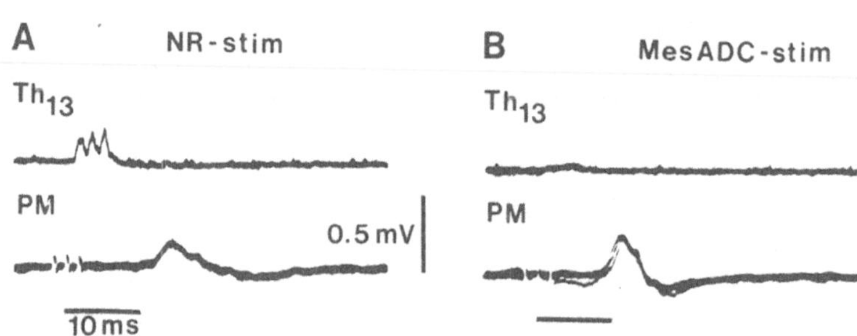

Figure 2. Descending volleys and climbing fibre responses caused by rubral and MesADC-stimulation. A, stimulation in nucleus ruber (three pulses at 600 Hz, 60 μA), recording on dorsolateral aspect of spinal cord in Th_{13} segment and in d_1-zone of the paramedian lobule. B, stimulation in the MesADC (three pulses at 600 Hz, 60 μA), same recording sites. In both cases the climbing fibre responses were evoked by the second stimulating pulse with a latency of 12.5 msec in A and 8.0 msec in B. Note three distinct descending volleys from NR but weak, diffuse and slowly rising volley from MesADC. Negativity is down in all records.

stimulation of rubral cellular elements takes place. From this electrode position a pulse-to-pulse descending volley will thus be evoked at a threshold not higher than 20 μA. In addition to this a d_l-zone climbing fibre response in the contralateral PM with a latency of 9–14 msec will frequently be observed, particularly if a train of stimuli is delivered. Such a combination of responses evoked from a rubral electrode is shown in figure 2A.

When placing the MesADC-electrode, a short train of pulses should be delivered via the stimulating electrode, and the appearance of a d_l-zone climbing fibre response with a latency not longer than 8–8.5 msec should be looked for. When such a response is evoked at 20–30 μA, either the same or a slightly stronger stimulus also usually gives rise to an indistinct, slowly rising descending volley in the spinal cord. Figure 2B illustrates these responses. With the two electrodes placed to fulfil these requirements the inter-tip distance between them will be only 1–2 mm. This means that at stimulating strengths above 100 μA, current may spread from the one region to the other. Usually intensities above 70 μA should not be used and the lack of rubrospinal descending volleys on MesADC-stimulation should be carefully ascertained.

A GAMMA CELL CLASSIFICATION METHOD

The experience accumulated over many years in the present author's laboratory seems to allow the conclusion that *the MesADC-system influences dynamic gamma motoneurons but neither static gamma nor alpha moto neurons*. The existence of a descending system which exerts a very specific segmental effect and which can be selectively activated, led to the idea of utilising this sytem as a methodological tool allowing a functional classification of gamma motoneurons. In a series of experiments by Appelberg *et al.* (1975) this method was first tested in an attempt to clarify rubrospinal control of static and dynamic gamma motoneurons. Cells, mainly extracellularly recorded, which were influenced from the MesADC were regarded as dynamic, all others as static. The provisional conclusion was reached that the rubrospinal system influences static and dynamic gamma cells and, according to previous knowledge, alpha motoneurons in a way allowing a linked control of all three cell categories from the red nucleus. Clearly, these experiments served as a test of the feasibility of the method, but yielded no information concerning its relevance.

Recently, however, it was disclosed that intracellularly or extracellularly recorded gamma cells, classified as dynamic with the aid of the MesADC-method, were regularly and strongly influenced by electrical stimuli applied to muscle nerves at group II strength, thus including secondary afferents from muscle spindles (Appelberg *et al.*, 1977). Figure 3 shows the identification of a triceps gamma cell (A) classified as dynamic by being excited from the

MesADC (B). Typically, this cell received a clear, short latency excitatory (EPSPs) influence of homonymous (C) as well as heteronymous (D) muscle nerves stimulated at group II strength. Frequently such cells were, on the other hand, inhibited from higher threshold afferents (E).

These findings led to the development of an experimental strategy aimed at disclosing whether the indicated reflex influence on dynamic gamma motoneurons emanated from secondary muscle spindle afferents. Appelberg *et al.* (1978a,b; Appelberg *et al.*, to be published) obtained strong evidence to the effect that such a reflex connection exists. A clear increase in triceps muscle spindle dynamic sensitivity was frequently caused by stretch of the ipsilateral posterior biceps/semitendinosus muscles. This also occurred in animals spinalised at a low thoracic or high lumbar level. It seems that the technique of classifying gamma motoneurons by MesADC-stimulation has thereby proved its worth, since a reflex connection indicated by its use could be confirmed with a different experimental procedure.

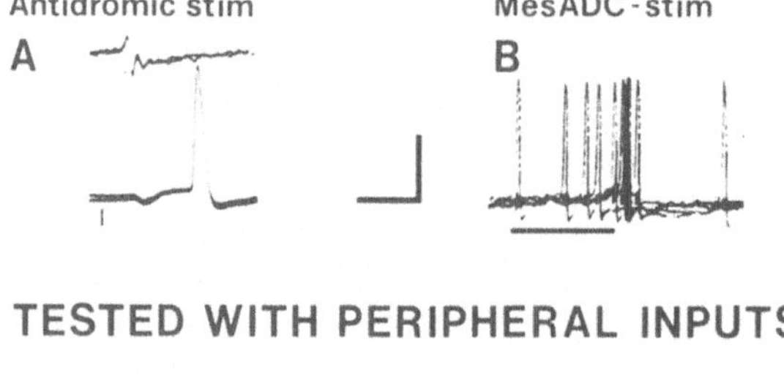

TESTED WITH PERIPHERAL INPUTS

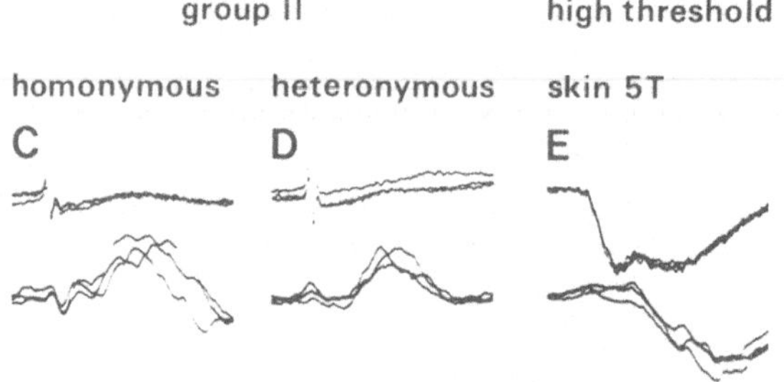

Figure 3. Intracellular recording from triceps gamma motoneuron with an axonal conduction velocity of 31 m/sec. A, antidromic identification. B, superimposed traces showing spike firing caused by MesADC-stimulation (long train at 600 Hz, 70 μA). C, EPSPs evoked from the triceps nerve stimulated at 5T. D, EPSPs evoked from the posterior biceps/semitendinosus nerve stimulated at 5T. E, inhibition evoked from high threshold afferents in superficial peroneal nerve. Calibration bars between A and B are 4 msec for A, C, D and E, 20 msec for B, and 20 mV for A, 10 mV for B and 4 mV for C, D and E. Extracellular fields were negligible and therefore omitted. (Rearranged and simplified from Appelberg *et al.*, 1975.)

CONCLUSION

Our understanding of how the fusimotor system is utilised in motor control is still utterly fragmented. Much emphasis has, in recent years, been put on the idea of α–γ linkage. In lower animals only skeletofusimotor fibres exist and a total synchronisation of the nervous actions on extrafusal and intrafusal muscle fibres thus seems to be, in a way, a basic principle for the control of skeletal muscle. A prerequisite for rigid linkage is available in higher animals in the form of beta motoneurons and their efferents directed to extrafusal as well as intrafusal muscle fibres. On the other hand, a specific fusimotor system has also developed, being functionally divided into static and dynamic gamma cells and their efferent fibres. Whether this type of organisation prevails in man does not seem to be fully clear (cf. Vallbo *et al.*, 1979).

It is tempting to assume that, as in evolution several different functional groups of motor cells—alphas, static and dynamic betas and static as well as dynamic gammas—have developed, these groups should also be available for independent reflex and supraspinal control. It is of considerable importance that in cats, α–γ linkage can be experimentally broken (Granit *et al.*, 1955) and that in intact cats studied during free falls, alpha and gamma cell drive does not seem to be linked (Prochazka *et al.*, 1976, 1977).

The major motor systems (corticospinal, rubrospinal and vestibulospinal as well as the MLF-system) in cats seem to be organised in a way which would allow them to work on static gamma motoneurons and alpha motoneurons in a linked fashion. Whether they really do this or not is not known. In contrast to this, the MesADC-system seems to act exclusively on one functional group of motor cells, i.e. the dynamic gammas. It was demonstrated by Jeneskog and Johansson (1977) that there is a complete correspondence, at the mesencephalic level, between the MesADC-area and the area from which the dorsal reticulospinal system can be activated. The dorsal reticulospinal system at this level was studied by Baldissera *et al.* (1972a) and is known, on the basis of the careful studies of the same system at a medullary level by Engberg *et al.* (1968a,b), to give inhibition to interneuronal pathways but not to act on alpha motoneurons. The reasonable conclusion must be that the MesADC-system does not act on alpha motoneurons. Observations supporting this view were made by Appelberg *et al.* (1975).

The MesADC-system is, therefore, the only known example of a motor system not being able to work according to the linkage idea. If it is ever active under natural circumstances it will, in an exclusive way, change spindle sensitivity via an action on dynamic gamma cells without causing any direct effects on other categories of motoneurons.

Received on May 27th, 1980.

REFERENCES

Appelberg, B. (1962a). The effect of electrical stimulation of nucleus ruber on the gamma motor system, *Acta physiol. scand.*, **55**, 150–159

Appelberg, B. (1962b). The effect of electrical stimulation in nucleus ruber on the response to stretch in primary and secondary muscle spindle afferents, *Acta physiol. scand.*, **56**, 140–151

Appelberg, B. (1963). Central control of extensor muscle spindle dynamic sensitivity, *Life Sci.*, **9**, 706–708

Appelberg, B. (1967). A rubro-olivary pathway. II: Simultaneous action on dynamic fusimotor neurones and the activity of the posterior lobe cerebellar cortex, *Expl Brain Res.*, **3**, 382–390

Appelberg, B. and Emonet-Dénand, F. (1965). Central control of static and dynamic sensitivities of muscle spindle primary endings, *Acta physiol. scand.*, **63**, 487–494

Appelberg, B. Hulliger, M., Johansson, H. and Sojka, P. (1978a). The action of joint and secondary muscle-spindle afferents on dynamic γ-motoneurones of the cat triceps surae muscle, *J. Physiol., Lond.*, **284**, 176–177

Appelberg, B., Hulliger, M., Johansson, H. and Sojka, P. (1978b). The influence of secondary muscle spindle afferents on dynamic γ-motoneurones to the triceps surae of the cat, *Neurosci. Lett.*, Suppl. No. 1

Appelberg, B. and Jeneskog, T. (1968). Extension and twitch of muscle as tests of muscle spindle static and dynamic sensitivities, *Life Sci.*, **7**, 1277–1282

Appelberg, B. and Jeneskog, T. (1969). A dorso-lateral spinal pathway mediating information from the mesencephalon to dynamic fusimotor neurones, *Acta physiol. scand.*, **77**, 159–171

Appelberg, B. and Jeneskog, T. (1972). Mesencephalic fusimotor control, *Expl Brain Res.*, **15**, 97–112

Appelberg, B. and Jeneskog, T. (1973). Parallel activation from the cat brain stem of hind limb dynamic fusimotor neurones and climbing fibres to the cerebellar paramedian lobule, *Brain Res.*, **58**, 229–233

Appelberg, B., Jeneskog, T. and Johansson, H. (1975). Rubrospinal control of static and dynamic fusimotor neurones, *Acta physiol. scand.*, **95**, 413–440

Appelberg, B., Johannson, H. and Kalistratov, G. (1977). The influence of group II muscle afferents and low threshold skin afferents on dynamic fusimotor neurones to the triceps surae of the cat, *Brain Res.*, **132**, 153–158

Appelberg, B. and Kosary, I. Z. (1963). Excitation of flexor fusimotor neurones by electrical stimulation in the red nucleus, *Acta physiol. scand.*, **59**, 445–453

Appelberg, B. and Molander, C. (1967). A rubro-olivary pathway. I: Identification of a descending system for control of the dynamic sensitivity of muscle spindles, *Expl Brain Res.*, **3**, 372–381

Baldissera, F., Lundberg, A. and Udo, M. (1972a). Activity evoked from the mesencephalic tegmentum in descending pathways other than the rubrospinal tract, *Expl Brain Res.*, **15**, 133–150

Baldissera, F., Lundberg, A. and Udo, M. (1972b). Stimulation of pre- and post-synaptic elements in the red nucleus, *Expl Brain Res.*, **15**, 151–167

Barker, D. (1962). The structure and distribution of muscle receptors. In *Symposium on Muscle Receptors* (ed. D. Barker), Hong Kong, Hong Kong University Press, 227–240

Barker, D., Emonet-Dénand, F., Harker, D. W., Jami, L. and Laporte, Y. (1976). Distribution of fusimotor axons to intrafusal muscle fibres in cat tenuissimus spindles as determined by the glycogen-depletion method, *J. Physiol., Lond.*, **261**, 49–69

Boyd, I. (1962). The structure and innervation of the nuclear bag muscle fibre system and the nuclear chain muscle fibre system in mammalian muscle spindles, *Phil. Trans. R. Soc. B*, **245**, 81–136

Emonet-Dénand, F., Laporte, Y., Matthews, P. B. C. and Petit, J. (1977). On the subdivision of static and dynamic fusimotor actions on the primary ending of the cat muscle spindle, *J. Physiol., Lond.*, **268**, 827–861

Engberg, I., Lundberg, A. and Ryall, R. W. (1968a). Reticulospinal inhibition of transmission in reflex pathways, *J. Physiol., Lond.*, **194**, 201–223

Engberg, I., Lundberg, A. and Ryall, R. W. (1968b). Reticulospinal inhibition of interneurones, *J. Physiol., Lond.*, **194**, 225–236

Granit, R., Holmgren, B. and Merton, P. A. (1955). The two routes for excitation of muscle and their subservience to the cerebellum, *J. Physiol., Lond.*, **130**, 213–224

Granit, R. and van der Meulen, J. P. (1962). The pause during contraction in the discharge of the spindle afferents from primary end organs in cat extensor muscles, *Acta physiol. scand.*, **55**, 231–244

Hongo, T., Jankowska, E. and Lundberg, A. (1969a). The rubrospinal tract. I: Effects on alpha-motoneurones innervating hindlimb muscles in the cat, *Expl Brain Res.*, **7**, 344–364

Hongo, T., Jankowska, E. and Lundberg, A. (1969b). The rubrospinal tract. II: Facilitation of interneuronal transmission in reflex paths to motoneurones, *Expl Brain Res.*, **7**, 365–391

Hongo, T., Jankowska, E. and Lundberg, A. (1972a). The rubrospinal tract. III: Effects on primary afferent terminals, *Expl Brain Res.*, **15**, 39–53

Hongo, T., Jankowska, E. and Lundberg, A. (1972b). The rubrospinal tract. IV: Effects on interneurones, *Expl Brain Res.*, **15**, 54–78

Jansen, J. K. S. and Matthews, P. B. C. (1961). The dynamic responses to slow stretch of muscle spindles in the decerebrate cat, *J. Physiol., Lond.*, **159**, 20–22

Jansen, J. K. S. and Matthews, P. B. C. (1962). The central control of the dynamic response of muscle spindle receptors, *J. Physiol., Lond.*, **161**, 357–378

Jeneskog, T. (1974a). Parallel activation of dynamic fusimotor neurones and a climbing fibre system from the cat brain stem. I: Effects from the rubral region, *Acta physiol. scand.*, **91**, 223–242

Jeneskog, T. (1974b). Parallel activation of dynamic fusimotor neurones and a climbing fibre system from the cat brain stem. II: Effects from the inferior olivary region, *Acta physiol. scand.*, **92**, 66–83

Jeneskog, T. (1974c). A descending pathway to dynamic fusimotor neurones and its possible relation to a climbing fibre system, *Umeå University Dissertations*, No. 14, Centraltryckeriet Umeå

Jeneskog, T. (1979). Inhibitory actions from low and high threshold cutaneous afferents on group II and III muscle afferent pathways in the spinal cat, *Acta physiol. scand.*, **107**, 297–708

Jeneskog, T. and Johansson, H. (1977). The rubro-bulbospinal path. A descending system known to influence dynamic fusimotor neurones and its interaction with distal cutaneous afferents in the control of flexor reflex afferent pathways, *Expl Brain Res.*, **27**, 161–179

Larsson, B., Miller, S. and Oscarsson, O. (1969). Termination and functional organization of the dorsolateral spino-olivocerebellar path, *J. Physiol., Lond.*, **203**, 611–640

Matthews, P. B. C. (1962). The differentiation of two types of fusimotor fibre by their effects on the dynamic response of muscle spindle primary endings, *Q. Jl exp. Physiol.*, **47**, 324–333

Matthews, P. B. C. (1972). *Mammalian Muscle Receptors and Their Central Actions*, London, Edward Arnold

Prochazka, A., Westerman, R. A. and Ziccone, S. P. (1976). Discharges of single hindlimb afferents in the freely moving cat, *J. Neurophysiol.*, **39**, 1090–1104

Prochazka, A., Westerman, R. A. and Ziccone, S. P. (1977). Ia afferent activity during a variety of voluntary movements of the cat, *J. Physiol., Lond.*, **268**, 423–448

Vallbo, Å. B., Hagbarth, K.-E., Torebjörk, H. E. and Wallin, B. G. (1979). Somatosensory, proprioceptive and sympathetic activity in human peripheral nerves, *Physiol. Rev.*, **59**, 919–957

Vedel, J. P. (1966). *Etude du Controle Supraspinal des Sensibilités Dynamique et Statique des Terminaisons Primaires du Fuseau Neuromusculaire Chez le Chat,* Thèse doctorat, Université d'Aix-Marseille

The activity of intrafusal muscle fibres during central stimulation in the cat

M. H. GLADDEN*

SUMMARY

Results from a new method of studying fusimotor neuron activity are reviewed. The method involves observation of the movement of intrafusal muscle fibres in muscle spindles isolated from the tenuissimus muscle. The muscle is exteriorised with a pedicle containing its nerve and blood supply and the spinal roots are left intact. Observations were made in cats under barbiturate anaesthesia, during stimulation of the sensorimotor cortex and following decerebration and spinalisation. The results suggested that the central control of γ_d motoneurons influences muscle spindle afferents through dynamic bag_1 fibre activity, and that the central control of γ_s motoneurons influences spindle afferents through static bag_2 and nuclear chain activity. Unexpectedly the evidence further suggested a division of γ_s motoneurons into two groups under separate central control, exerting effects through predominantly static bag_2 fibre contraction or predominantly nuclear chain fibre contraction. The development of the method to give quantitative information on γ motoneuron firing rates by intracellular recording from intrafusal muscle fibres is described. Preliminary accounts are given of experiments on the spread of junctional potentials into the poles of dynamic bag_1 fibres and static bag_2 fibres, and also experiments employing intracellular recording from dynamic bag_1 fibres during γ_s axon stimulation to clarify whether the innervation of these fibres by γ_d motoneurons, and not γ_s motoneurons, is as exclusive as observation of the contraction of dynamic bag_1 fibres suggests.

INTRODUCTION

Gladden and McWilliam (1977a and b) described a new method of studying

* Institute of Physiology, University of Glasgow

the activity of fusimotor neurons. Rather than recording directly from γ efferents (Hunt, 1951) or γ motoneurons (Eccles *et al.*, 1960), we observed the movements of intrafusal muscle fibres during spontaneous fusimotor discharge and during central or reflex activation. Muscle spindles were isolated in the tenuissimus muscle attached to the cat by a pedicle containing the muscle nerve and blood supply, as in the experiments of Boyd *et al.* (1975, 1977), but the spinal roots were left intact. Our results, which will be reviewed here, suggest that the method will be useful not only in augmenting information on γ motoneuron activity obtained by other means, but also in interpreting afferent responses from muscle spindles under different central conditions. This is because the form of afferent response depends on the type of intrafusal muscle fibre activated, whether dynamic bag_1 fibre, static bag_2 fibre or nuclear chain fibres (*see* Emonet-Dénand *et al.*, 1977; Boyd *et al.*, 1979).

The chief limitation of our method was that it did not produce quantitative information on γ motoneuron firing rates, and this paper describes the development of the method to resolve this limitation by intracellular recording from individual intrafusal muscle fibres. Junctional potentials can be recorded from both types of nuclear bag fibres and action potentials from nuclear chain fibres. In the case of junctional potentials it was necessary to discover whether their detection depended critically on the distance between the recording electrodes and γ motor end plates. A preliminary report on this subject is given and also on the validity of reasoning that intracellular potentials recorded from dynamic bag_1 fibres result solely from dynamic γ motoneuron activity.

THE MOVEMENTS OF INTRAFUSAL MUSCLE FIBRES IN LIGHT ANAESTHESIA, DURING CORTICAL STIMULATION AND IN DECEREBRATE AND SPINAL ANIMALS

The preparation

Sixteen muscle spindles were isolated from the tenuissimus muscle in 13 cats. The same muscle spindles were observed for up to 24 h after isolation, usually first in light barbiturate anaesthesia, then after decerebration, and finally following spinalisation at T_{12}–T_{13}. Four cats were anaesthetised with halothane and nitrous oxide prior to decerebration.

Immediately after muscle spindles were isolated in lightly anaesthetised or decerebrate animals small-amplitude movements of intrafusal muscle fibres were observed. These were due to small variations in spontaneous activity of γ motoneurons. Nuclear chain fibres were recognised from their smaller diameters than nuclear bag fibres. Dynamic bag_1 fibres were distinguished from static bag_2 fibres by their greater sensitivity to topically applied acetylcholine (Gladden, 1976). Increased γ efferent activity, either reflexly induced or during central stimulation, caused an increase in the frequency of the small-amplitude movements, a movement of the intrafusal fibres towards their contraction foci and an opening of the primary sensory spirals in the equatorial region of the contracting muscle fibres. Inhibition caused a reduction or cessation of the small-amplitude movements and a relaxation, seen as a movement of the intrafusal fibres away from the contraction foci. Extrafusal muscle fibre contraction was detected by direct observation, and usually put the muscle spindle out of focus.

The activity of three types of intrafusal fibre

These results are summarised in table 1. Table 1 includes a few more recent results additional to those reported (Gladden and McWilliam, 1977a and b),

Table 1. The activity of the three types of intrafusal muscle fibre in muscle spindles isolated in the tenuissimus muscle under different central conditions. Each muscle spindle had nuclear chain fibres, a dynamic bag_1 fibre and one or more static bag_2 fibres. (After Gladden and McWilliam, 1977a and b)

	Nuclear chain fibres	*Static bag_2 fibres*	*Dynamic bag_1 fibres*
Spontaneous activity—light anaesthesia or decerebration (16 spindles)	11	16	3 (2)*
Recruited on cortical stimulation (8 spindles)	4	8	6
Activity after spinalisation (8 spindles)	1†	2 (3)‡*	3
	Presumed activity of γ_s motoneurons		Presumed activity of γ_d motoneurons

* The dynamic bag_1 fibre was spontaneously active in three spindles and in two additional spindles could be activated reflexly.
† Only two nuclear chain fibres became active in one spindle.
‡ The static bag_2 fibre became active in two spindles and in three was spontaneously active both before and after spinalisation.

but these results agree with our initial findings. Our analysis involved the assumption that movement of dynamic bag_1 fibres signified that dynamic γ motoneurons were active and movement of static bag_2 fibres or nuclear chain fibres signified activity of static γ motoneurons. This assumption followed from the microscopic observations of living spindles during stimulation of single static γ axons, and dynamic γ or β axons (Boyd *et al.*, 1975, 1977; Bessou and Pagès, 1975). Both groups agreed that movement of dynamic bag_1 fibres occurred only during stimulation of dynamic γ and β axons, and stimulation of static γ axons caused movement of static bag_2 fibres and/or nuclear chain fibres only.

It was remarkable that the static bag_2 fibres were spontaneously active in all 16 muscle spindles studied in light anaesthesia or in decerebrate cats, either at one or both poles. Static bag_2 fibres could be recruited in all eight muscle spindles observed during stimulation of the sensorimotor cortex (*see* below). When anaesthetised cats were decerebrated, spontaneous activity tended to continue in the same poles of the same intrafusal fibres, so presumably the same γ motoneurons were involved in the two conditions. However, spontaneous activity tended to decline in decerebrate animals with time and after many hours some intrafusal fibres that were not previously active became active while activity in others was depressed. Six cats were spinalised following decerebration. Before spinalisation in five of the eight muscle spindles from these cats, static bag_2 fibres were spontaneously active. After spinalisation the spontaneous activity stopped in two cases and continued in three, but at a lower frequency than before. In three of the eight muscle spindles the static bag_2 fibres were inactive before spinalisation. After spinalisation one fibre remained inactive, but the other two became active although at low frequency.

Nuclear chain fibres were not so likely to be spontaneously active as static bag_2 fibres in light anaesthesia or decerebrate animals (table 1). In 11 of the 16 muscle spindles nuclear chain fibres were active, and in one of the eleven the activity involved only one chain fibre. Nuclear chain fibres could be recruited in four of the eight muscle spindles observed during cortical stimulation (*see* below). After spinalisation nuclear chain fibres became active in only one of the eight muscle spindles studied, and only two nuclear chain fibres at one pole were involved.

Dynamic bag_1 fibres were seen to be spontaneously active in light anaesthesia in only three of 16 cases, and never in decerebrate cats. Of these three cases two muscle spindles were in the same muscle, and so the same γ motoneuron may have been responsible. In all three the activity was more susceptible to deepening anaesthesia than spontaneous activity in other intrafusal fibres of the same spindle. In addition, in two spindles in which the dynamic bag_1 fibres were inactive, activity could be induced by noxious stimulation of the skin covering the ipsilateral hip. On cortical stimulation dynamic bag_1 fibres were recruited in six of eight cases (*see* below). After spinalisation the dynamic bag_1 fibres became active in three of eight muscle spindles observed. In two of these three the same dynamic bag_1 fibres had been active in light anaesthesia.

Cortical stimulation

The exact area from which fusimotor effects were obtained differed from cat to cat, but involved an area around the postcruciate dimple and a smaller area just anterior to the cruciate salcus. The size of the area where stimulation produced contraction of intrafusal muscle fibres diminished with deepening anaesthesia. Anodal and cathodal stimulation was used at 100–700 Hz.

As anaesthesia lightened, the first intrafusal fibre recruited by cortical stimulation was the static bag_2 fibre, usually alone. The area where stimulation was effective then increased, and stimulation of smaller areas within this area then resulted in contraction of nuclear chain fibres, dynamic bag_1 fibres or extrafusal fibres, or some combination of these. In only two instances contraction of the static bag_2 fibres could not be separated from contraction of nuclear chain fibres. The most likely explanation was that stimulation over a wide area recruited γ_s motoneurons supplying static bag_2 fibres selectively (six cases) and γ_s motoneurons supplying static bag_2 fibres plus nuclear chain fibres (two cases). In two other muscle spindles nuclear chain fibres could be recruited by cortical stimulation, and the static bag_2 fibres were recruited simultaneously. However, in one case nuclear chain recruitment required a higher stimulus strength. In the other case we recruited nuclear chain fibres by pushing the stimulating electrodes about 1 mm below the cortical surface. Clearly on these two occasions a γ_s motoneuron supplying nuclear chain fibres alone was activated in addition to a γ_s supplying the static bag_2 fibre alone.

In all six muscle spindles in which the dynamic bag_1 fibres were recruited by cortical stimulation the static bag_2 fibres were recruited simultaneously. Although in some cases it was possible to find cortical areas where stimulation produced greater movement in the dynamic bag_1 than in the static bag_2 fibre, in no instance was the movement unequivocally confined to the dynamic bag_1 fibre alone. However in every case it was possible to recruit the static bag_2 fibre alone by altering the electrode position. This suggests that a pure excitation of dynamic γ motoneurons is difficult to obtain by cortical stimulation.

Because extrafusal contraction put the muscle spindle out of focus concomitant intrafusal contraction was difficult to observe. In one case static bag_2 fibre contraction slightly preceded extrafusal contraction and in another both static bag_2 and dynamic bag_1 contraction accompanied extrafusal contraction.

Significance of results

Gladden and McWilliam (1977a and b) suggested firstly that their results were compatible with the conclusion that the central nervous system can control the dynamic bag_1 fibres of muscle spindles entirely separately from the static bag_2 fibres and nuclear chain fibres. Secondly they suggested that there might be a separate population of γ_s motoneurons innervating predominantly static

bag_2 fibres which were under separate central nervous control. This second proposal will be discussed later.

The reasons why these results support the view that dynamic bag_1 fibres are under separate control are as follows. Contraction of dynamic bag_1 fibres often occurred sequentially with that of other intrafusal muscle fibres. For example, they were the last to become spontaneously active in light anaesthesia and the first to stop during deepening anaesthesia. These fibres were usually judged to be inactive if their sarcomeres did not move, but in some instances static bag_2 fibre contraction unloaded and kinked the dynamic bag_1 fibres. Had there been contraction in the dynamic bag_1 fibres the kinks would have straightened out. Also, if dynamic bag_1 and static bag_2 fibres were simultaneously contracting, their movements were never in phase.

It has been suggested (*see* Laporte, 1979) that visual observation alone might fail to detect dynamic bag_1 fibre contraction, if there is simultaneous contraction of static bag_2 fibres, which is usually very much more powerful. In fact we observed separate movement of dynamic bag_1 and static bag_2 fibres in the same poles in 13 instances, involving eight muscle spindles. Nevertheless activation of γ motoneurons innervating solely nuclear bag fibres at $<$10–15 impulses/sec would not cause visible contraction (*see* Boyd, 1976) and hence would be undetected by observation alone. In addition our conclusions could not be reconciled with evidence from glycogen-depletion studies (Barker *et al.*, 1976) that static γ axons innervate dynamic bag_1 fibres in about half the tenuissimus muscle spindles they studied. For these reasons it was necessary to develop the technique of intracellular recording from intrafusal muscle fibres under visual control, especially as this would also provide quantitative information on γ motoneuron firing rates.

THE PROPAGATION OF POTENTIALS INTO THE POLES OF NUCLEAR BAG FIBRES

Impalement of nuclear bag fibres with microelectrodes is easier in the poles because they are not covered by layers of capsule cells. It is also easier to distinguish individual fibres where they project beyond the capsule, because the capsule gathers them closely together in a bundle. However, it was necessary to establish that single potentials set up at motor end plates in the capsular sleeve region would be detected by intracellular recording electrodes some distance away in the poles. Potentials in nuclear bag fibres are almost always junctional. Barker *et al.* (1978) recorded junctional potentials in nine dynamic bag_1 fibres and seven static bag_2 fibres. Action potentials occurred in only one static bag_2 fibre, but in all undamaged nuclear chain fibres. In the experiments described in this paper no action potentials were encountered in nuclear bag fibres. Junctional potentials were recorded from seven dynamic bag_1 fibres and nine static bag_2 fibres.

The spread of junctional potentials into the poles of nuclear bag fibres was investigated in experiments on muscle spindles isolated in tenuissimus muscles completely removed from the cat as in the experiments described by Boyd (*see* Boyd, 1976). γ axons were recruited by stimulating the muscle nerve, and extrafusal fibres were denervated.

Between two and six serial impalements were made along the poles of five dynamic bag_1 fibres and four static bag_2 fibres (table 2). The nuclear bag fibres

Table 2. The sites of multiple serial impalements in poles of nuclear bag fibres. In each experiment one or both nuclear bag fibres were impaled at one muscle spindle pole. Each number indicates an impalement and is the distance in millimetres of the recording electrode from the termination of the Ia axon. At all these sites junctional potentials were recorded when the γ axon innervating the nuclear bag fibre was stimulated in the muscle nerve at 1 Hz

Experiment number	*Dynamic bag_1 fibre*	*Static bag_2 fibre*
B1.2	1.6, 2.2, 2.6, 3.1	1.1, 1.6, 2.1, 2.6, 3.1
B1.4	—	2.6, 3.0
B1.6	2.4, 1.8	—
B1.7	3.0, 2.5, 1.9, 3.0, 3.3	3.4, 2.2
B1.8	3.0, 2.4, 1.7, 1.2, 2.5, 1.3	2.5, 3.0
B1.9	1.6, 2.0	—

were distinguished by their different mechanical properties during the experiment. At the end of the experiment one nuclear bag fibre was deliberately damaged to mark it and then identified as bag_1 or bag_2 histologically from the distribution of elastic fibres (*see* Gladden, 1976).

If the stimulus strength was sufficient to cause visible contraction of the nuclear bag fibres at higher frequencies, intracellular potentials could always be recorded when the frequency was reduced to 1 Hz, indicating that summation was not necessary for potentials to be recorded. In one bag fibre of each type potentials were recorded 2 mm away from the likely position of the motor end plate. Since these muscle spindles had no blood supply, their resting membrane potentials deteriorated with time. Despite this it was always possible to record potentials. It therefore seems fairly safe to assume that if no potentials are recorded at any position in the poles of nuclear bag fibres, the γ motoneurons innervating them are silent.

THE INNERVATION OF DYNAMIC BAG_1 FIBRES

Visual observations of isolated muscle spindles with intact innervation indicate that γ-static innervation of dynamic bag_1 fibres rarely occurs, whereas glycogen-depletion studies suggest a much greater incidence (*see* Laporte,

1979). Three types of experiment utilising intracellular recording have been employed to resolve the difficulty.

In the first type of experiment single static or dynamic γ axons were stimulated in ventral root filaments during intracellular recordings from dynamic bag_1 fibres. The type of bag fibres was confirmed histologically after the experiment. Table 3 summarises the findings. Although junctional potentials were recorded from dynamic bag_1 fibres when two dynamic γ axons were stimulated, no potentials were recorded when three static γ axons were stimulated. Stimulation of these three static γ axons caused visible contraction in the static bag_2 fibres of the same spindle pole, and nuclear chain fibres in addition in one case, and junctional potentials were recorded from the static bag_2 fibres.

Table 3. The occurrence of junctional potentials in dynamic bag_1 fibres (Db_1) and static bag_2 fibres (Sb_2) during stimulation of known static and dynamic γ axons in spinal root filaments

Experiment no.	*Type of axon stimulated*	*Type of fibre contracting*	*Intracellular potentials*	
			Dynamic bag_1 fibre	*Static bag_2 fibre*
011	γ_s	Sb_2	None	+
023	γ_s	Sb_2 + NC*	None	+
030	γ_s	Sb_2	None	+
017	γ_d	Db_1	+	Not impaled
026	γ_d	Db_1	+	Not impaled

* Nuclear chain fibres.

The second type of experiment involved simultaneous intracellular recording from dynamic bag_1 fibres and static bag_2 fibres in the same pole during recruitment of γ axons in the muscle nerve. At a stimulus strength sufficient to cause junctional potentials in the static bag_2 fibre (presumably a static γ axon was recruited) no potentials were seen in the dynamic bag_1 fibre. This was found on two occasions.

In the third experimental situation recordings were made from a dynamic bag_1 fibre in an isolated muscle spindle of a decerebrate cat with intact spinal roots. Two separate γ motoneurons were separately active because one innervated a nuclear chain fibre and the other innervated two static bag_2 fibres, and intracellular recording from these fibres showed that the activity occurred at different rates. Presumably these were two static γ motoneurons yet no potentials could be recorded from the dynamic bag_1 fibre at this spindle pole. This experiment will be referred to again in the next section.

Because of their technical difficulty the numbers of successful experiments of these three types are small compared to the numbers of muscle spindles studied by glycogen depletion. In each of the present experiments single muscle spindle poles were tested. Although in the experiments of Barker *et al.* (1976) γ_s stimulation caused depletion of bag$_1$ fibres in roughly half of tenuissimus muscle spindles, only about one third of spindle poles were affected. Nevertheless the present results do not support the glycogen-depletion evidence.

INTRACELLULAR RECORDING FROM INTRAFUSAL MUSCLE FIBRES DURING MIDBRAIN STIMULATION

Relatively few experiments of this type have been done but one experiment will be described to illustrate the potentiality of the method.

The preparation

The tenuissimus muscle was reflected from a cat into a bath continuously irrigated with Krebs solution at 37°C under barbituate anaesthesia. The blood pressure and end-tidal p_{CO_2} of the cat were monitored continuously for the duration of the experiment, and the blood pressure maintained with dextran given intravenously. The cat was decerebrated (pre-collicular) 6 h after the experiment began. The muscle received its nerve and blood supply from the animal via a pedicle. The fluid space and one pole of a muscle spindle, up to 3–4 mm from the equator, was exposed. Intrafusal muscle fibres were impaled under direct visual guidance with KCl-filled glass microelectrodes. The midbrain of the decerebrate cat were stimulated with a steel microelectrode. The stimulated position was marked by iontophoretic deposition of iron, and the mark identified after the experiment in serial sections of the midbrain stained with Perl's Prussian Blue method. Two nuclear bag fibres were also marked at the end of the experiment by deliberately damaging them at measured distances from the equator. These were then identified as bag$_1$ or bag$_2$ in serial sections of the muscle spindle stained for elastic fibres (Gladden, 1976).

Results

The muscle spindle in this experiment (figure 1) had an unusually large number of nuclear bag fibres which were shown to be three static bag$_2$ fibres and one dynamic bag$_1$ fibre by subsequent light microscopy (Gladden, 1976). There were six nuclear chain fibres. The dynamic bag$_1$ fibre (Fibre A, figure 1) and two of the static bag$_2$ fibres (Fibres B and C, figure 1) were visible fairly clearly for over 4 mm of one pole. The third static bag$_2$ fibre (Fibre D, figure 1) was only partially visible. In light anaesthesia, the static bag$_2$ and chain fibres were active, whilst the dynamic bag$_1$ fibre was inactive. Stimulation in

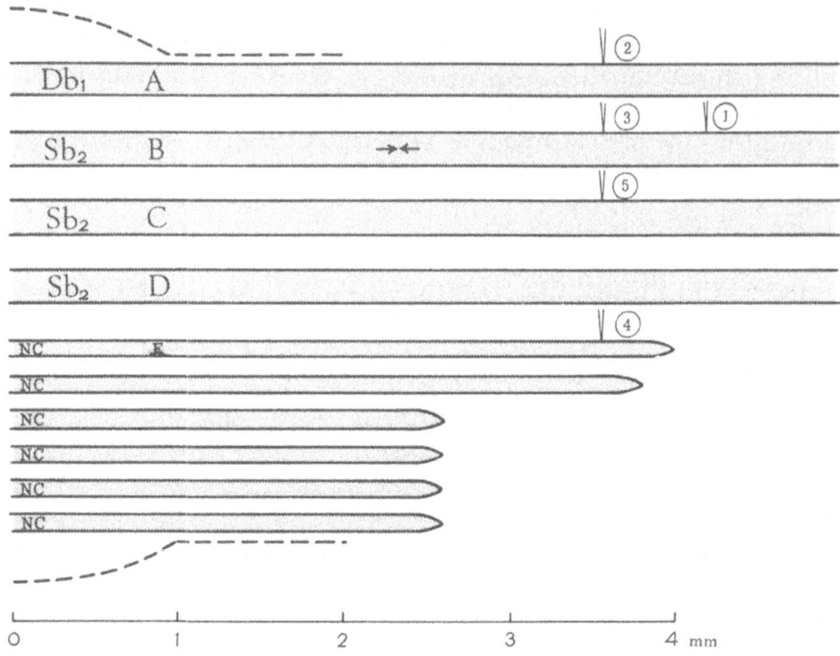

Figure 1. Schematic reconstruction of one pole of muscle spindle, which contained one dynamic bag$_1$ fibre (Db$_1$), three static bag$_2$ fibres (Sb$_2$) and six nuclear chain fibres (NC). The V's show the position of the impalements and the circled numbers show the order in which the impalements were made. The scale is in millimetres from the primary sensory ending. The fibres A–E are referred to in the text. The arrows indicate the position of a contraction site in Fibre B.

the region of the caudal part of the red nucleus caused inhibition of all spontaneous activity.

The first impalement of one static bag$_2$ fibre (Fibre B, figure 1) was made 14 h after a pre-collicular decerebration, 4.2 mm from the equator. Although a long nuclear chain fibre (Fibre E, figure 1), which was later impaled, was clearly spontaneously active, there was little spontaneous activity in the impaled fibre, and only occasional junctional potentials were recorded. Stimulation at 500 Hz just caudal to the red nucleus caused γ excitation initially at 10–15 impulses/sec (recorded from Fibre B, figure 1), quickly adapting to 4–7 impulses/sec 4 sec later. During the stimulation, another static bag$_2$ fibre (Fibre C, figure 1) contracted and there was also a weak contraction of extrafusal muscle fibres. No contraction was visible in the dynamic bag$_1$ fibre (Fibre A, figure 1).

Six hours later, there was still spontaneous activity in the long chain fibre (Fibre E, figure 1) and increased spontaneous activity in the static bag_2 fibre previously impaled. It was now clear that the same γ motoneuron innervated both this and another static bag_2 fibre (Fibres B and C, figure 1) as they contracted simultaneously. Impalement of both fibres 3.5 mm from the equator showed that the frequency of the spontaneous activity was about 5 impulses/sec. The microelectrode in the midbrain had not been moved since the previous impalement, but stimulation with the same parameters now caused inhibition which was complete for 1–2 sec and then incomplete until stimulation stopped. The resting membrane potentials of these fibres was only 20–25 mV because the muscle spindle had by now been isolated for 35 h and the blood supply of this particular muscle was poor. Despite this, and despite the likely position of the motor end plates being over 1 mm closer to the equator (a contraction site in one of these fibres (Fibre B, figure 1) was observed 2.3 mm from the equator), junctional potentials of about 4 mV were recorded. Each potential coincided with a visible contraction of the impaled fibre, though nearer to the equator than the impalement site.

Intracellular recording from the long chain fibre (Fibre E, figure 1) 3.6 mm from the equator showed that it had a different γ innervation from the two impaled static bag_2 fibres (Fibres B and C, figure 1) because the frequency of the spontaneous activity was higher, 20–25 impulses/sec, and identical central stimulation only partially inhibited this activity. Action potentials were recorded. No potentials were recorded from the dynamic bag_1 fibre (Fibre A, figure 1). The impalement site was 3.5 mm from the equator.

DISCUSSION

The experiment described above demonstrates that quantitative information on γ motoneuron firing rates can be obtained by intracellular recording from intrafusal fibres. Recordings can be obtained even during extrafusal contraction and several recordings obtained from the same intrafusal fibre during the course of an experiment. The microelectrodes are not usually dislodged by intrafusal contraction even if the rate of γ stimulation is much higher than those recorded in this particular experiment.

Intracellular recordings combined with direct visual observation of the intrafusal muscle fibres impaled do support the validity of our visual observations (Gladden and McWilliam, 1977a and b) reviewed in the first part of this paper. Such recordings have been obtained in a total of 14 experiments during spontaneous activity of γ motoneurons, during stimulation of known γ_s or γ_d axons in ventral spinal roots and during recruitment of γ axons in the muscle nerve. Junctional or action potentials were recorded only when the intrafusal fibre impaled was seen to contract, except in two circumstances—if the impalement was of a nuclear bag fibre and the stimulation frequency was less than 10 Hz, and if the impalement was close to a contraction site, in which case miniature end plate potentials were recorded. In fact stimulation of a nuclear bag fibre at less than 10 Hz will have little effect on the spindle afferents, so that if we observed a nuclear bag fibre to be quiescent there would be no concealed effect of physiological importance.

It can be important when interpreting spindle afferent responses in different central conditions to remember that intrafusal fibres may be active at one pole and inactive at the other. For example, in interpreting the results of Hongo and Shimazu (1965), who recorded from both primary and secondary sensory endings of the same tenuissimus muscle spindles in barbiturate anaesthesia, Granit (1970) concluded that dynamic γ motoneurons are more resistant to the anaesthetic than static γ motoneurons. Our experiments do not suggest this. The confusing observation was that facilitation of the primary sensory discharge was more resistant to deepening barbiturate anaesthesia than facilitation of the secondary sensory discharge in about half their experiments. The afferent response to stretching was not recorded and therefore the dynamic sensitivity not tested. One explanation would be that static γ motoneurons innervating the pole where the secondary lay were silenced, but static γ motoneurons innervating the other pole were still providing a facilitatory input to the primary sensory ending alone. Contraction of intrafusal fibres at one pole will not affect the discharge of a secondary ending at the other pole. Alternatively, in these muscle spindles deepening anaesthesia first silenced the static γ motoneurons innervating nuclear chain fibres leaving active static γ motoneurons innervating static bag_2 fibres. The input from the static bag_2 fibres to secondaries is maximally only 50% of that of the nuclear chain fibres (Boyd, 1980) and would be especially small at the low frequencies likely in deeper anaesthesia.

The prevalence of static bag_2 fibre activity compared with nuclear chain fibre activity (table 1), especially during cortical stimulation, is perplexing in the absence of any other clear experimental evidence of a separate group of static γ motoneurons innervating predominantly static bag_2 fibres. Static γ axons do not appear to innervate selectively either nuclear chain or static bag_2 fibres in the different muscle spindles which they supply (Barker *et al.*, 1976; Boyd *et al.*, 1977; Jami *et al.*, 1980). Nevertheless it seems surprising if the central nervous system cannot control static bag_2 and nuclear chain fibre contraction separately to some extent, even if not completely. Static bag_2 fibre contraction decreases or increases slightly the position sensitivity of the secondary sensory ending (Jami *et al.*, 1980; Boyd, 1980) and the primary sensory ending (Boyd, 1980). Nuclear chain fibre contraction markedly increases the length sensitivity of the secondary sensory ending and drives the primary sensory ending (Boyd, 1980). Static bag_2 fibre contraction would provide a rather non-specific excitation input, such as would be appropriate in arousal, whereas nuclear chain fibre contraction would be more appropriate during movement. Our suggestion (Gladden and McWilliam, 1977a) that there are γ_s motoneurons innervating predominantly static bag_2 fibres under separate central nervous control implies that the static fusimotor input to muscle spindles might switch between one increasing the length sensitivity of secondary sensory endings (nuclear chain) and a non-specific excitatory one to primary and secondary endings (static bag_2). This rather tentative suggestion may have some truth since Prochazka and Wand (1980) found that a predominant increase in length sensitivity of spindle afferents in unobstructed voluntary movements of cats converted to a predominant decrease in strongly resisted movements.

ACKNOWLEDGEMENTS

This investigation was supported by a grant from The Wellcome Foundation, and from the Medical Research Funds of Glasgow University. I am grateful to Professor I. A. Boyd and Dr P. N. McWilliam for their support and helpful discussion, and to Miss J. Wilson and Mr J. Ward for technical assistance.

Received on July 4th, 1980

REFERENCES

Barker, D., Bessou, P., Jankowska, E., Pagès, B. and Stacey, M. J. (1978). Identification of intrafusal muscle fibres activated by single fusimotor axons and injected with fluorescent dye in cat tenuissimus spindles, *J. Physiol., Lond.*, **275**, 149–165

Barker, D., Emonet-Dénand, F., Harker, D. W., Jami, L. and Laporte, Y. (1976). Distribution of fusimotor axons to intrafusal muscle fibres in cat tenuissimus spindles as determined by the glycogen-depletion method, *J. Physiol., Lond.*, **261**, 49–69

Bessou, P. and Pagès, B. (1975). Cinematographic analysis of contractile events produced in intrafusal muscle fibres by stimulation of static and dynamic fusimotor axons, *J. Physiol., Lond.*, **252**, 397–427

Boyd, I. A. (1976). The response of fast and slow nuclear bag fibres and nuclear chain fibres in isolated cat muscle spindles to fusimotor stimulation, and the effect of intrafusal contraction on the sensory endings, *Q. Jl exp. Physiol.*, **61**, 203–254

Boyd, I. A. (1980). The action of the three types of intrafusal fibre in isolated cat muscle spindles on the dynamic and length sensitivities of primary and secondary sensory endings, this publication

Boyd, I. A., Gladden, M. H., McWilliam, P. M. and Ward, J. (1975). 'Static' and 'dynamic' nuclear bag fibres in isolated cat muscle spindles, *J. Physiol., Lond.*, **250**, 11–12P

Boyd, I. A., Gladden, M. H., McWilliam, P. N. and Ward, J. (1977). Control of dynamic and static nuclear bag fibres and nuclear chain fibres by gamma and beta axons in isolated cat muscle spindles, *J. Physiol., Lond.*, **265**, 133–162

Boyd, I. A., Gladden, M. H. and Ward, J. (1979). The effect of contraction in the three types of intrafusal fibre in isolated cat muscle spindles on the Ia discharge during stretch, *J. Physiol., Lond.*, **296**, 41P

Eccles, J. C., Eccles, R. M., Iggo, A. and Lundberg, A. (1960). Electrophysiological studies on gamma motoneurons, *Acta physiol. scand.*, **50**, 32–40

Emonet-Dénand, F., Laporte, Y., Matthews, P. B. C. and Petit, J. (1977). On the subdivision of static and dynamic fusimotor actions on the primary endings of the cat muscle spindle, *J. Physiol., Lond.*, **268**, 827–861

Gladden, M. H. (1976). Structural features relative to the function of intrafusal muscle fibres in the cat, *Prog. Brain Res.*, **44**, 51–59

Gladden, M. H. and McWilliam, P. N. (1977a). The activity of intrafusal muscle fibres during cortical stimulation in the cat, *J. Physiol., Lond.*, **273**, 28–29P

Gladden, M. H. and McWilliam, P. N. (1977b). The activity of intrafusal muscle fibres in anaesthetized, decerebrate and spinal cats, *J. Physiol., Lond.*, **273**, 49–50P

Granit, R. (1970). *The Basis of Motor Control*, London and New York, Academic Press, 232

Hongo, T. and Shimazu, H. (1965). Centrifugal modifications of discharge rates of primary and secondary endings of muscle spindles in the hind limb of the cat, *J. Neurophysiol.*, **28**, 724–741

Hunt, C. C. (1951). The reflex activity of mammalian small-nerve fibres, *J. Physiol., Lond.*, **115**, 456–469

Jami, L., Lan-Couton, D. and Petit, J. (1980). Glycogen-depletion method of intrafusal distribution of γ-axons that increase sensitivity of spindle secondary endings, *J. Neurophysiol.*, **43**, 16–26

Laporte, Y. (1979). On the intrafusal distribution of dynamic and static fusimotor axons in cat muscle spindles, *Prog. Brain Res.*, **50**, 3–10

Prochazka, A. and Wand, P. (1980). Fusimotor action during normal movements, deduced from variations in muscle spindle sensitivity, *J. Physiol., Lond.*, to be published

A critique of the papers by Westbury, Appelberg and Gladden

Identification and Differentiation of Central Control of Fusimotor Effects

K. S. K. MURTHY*

POSSIBLE EFFECTS DUE TO COACTIVATION OF β AXONS

Several techniques have been used for identifying static and dynamic fusimotor effects resulting from a stimulation of central structures or from an activation of spinal reflex pathways. Matthews (1972) has reviewed some of the techniques introduced earlier and discussed some of the problems in interpreting the results as purely due to either static or dynamic fusimotor action. While a predominantly static fusimotor action may with confidence be attributed to the static γ efferents, dynamic fusimotor action, either found concurrently or selectively, is open for debate as to whether it is restricted to the activation of γ fusimotor axons or whether the dynamic β axons are also recruited. It is difficult to rule out a contribution by the dynamic β axons in experiments where the whole muscle tension is monitored with a low gain of the myograph. This problem is present when fusimotor effects are deduced either from spindle responses to sinusoidal and ramp stretches of the muscle (Vedel and Mouillac-Baudevin, 1969) or from the changes in the dynamic sensitivity of spindle afferents during a muscle twitch (Appelberg, 1980). It is important to bear in mind that the β efferents producing dynamic fusimotor actions are exclusively among the slower twitch type S motor units innervating the oxidative muscle fibres (Barker *et al.*, 1977) which contribute to tetanic tensions of less than 10 g in the smaller muscles of the hind limb (Jami and Petit, 1975) and up to 30 g in some larger muscles. Even if a few of these are activated by central stimulation, they may contribute to fusimotor actions without causing a significant increase in muscle tension when measured with a low gain of the transducer.

*Department of Surgery, The University of Texas Medical School at Houston

PROBLEMS DUE TO THE EXISTENCE OF MIXED FUSIMOTOR EFFECTS

In recent years it has become clear that a fusimotor axon may be shared by intrafusal muscle fibres of different accepted types within a muscle spindle, although those identified as being dynamic fusimotor in action have been found to innervate the dynamic bag_1 fibres almost exclusively (Barker *et al.*, 1976). Exceptional instances of dynamic γ innervation of the static bag_2 or chain fibres have indeed been observed (Barker *et al.*, 1978; Banks *et al.*, 1978). In a detailed study of the variations in the qualitative and quantitative features of fusimotor effects produced by individual γ efferents, Emonet-Dénand *et al.* (1977) showed that a mixture of dynamic and static fusimotor effects may be observed with purely static or purely dynamic fusimotor actions considered as among six classes of fusimotor effects.

COMMENTS ON THE TECHNIQUE OF MAKING DIRECT OBSERVATIONS OF INTRAFUSAL MUSCLE CONTRACTIONS

It is thus of interest to turn to the experiments in which direct visual observations of intrafusal muscle contractions have been made during central or peripheral stimulation (Gladden, 1980). Questions may be raised regarding conclusions of selectivity in the activation of various types of intrafusal fibres during either spontaneous or induced contractions. Banks *et al.* (1978) make a distinction in intrafusal contractions as producing either convergent movements of the sarcomeres or translational movements. It is not established to what extent a translational movement observed in an intrafusal muscle fibre may be attributed to direct fusimotor activation as opposed to it resulting from a passive mechanical action due to contractions in an adjacent intrafusal muscle fibre. However, these questions will doubtless be answered in due course as more data accumulate. On an optimistic note, Dr Gladden should be congratulated on making an effective initial contribution in what promises to be a valuable technique for assessing central regulation of fusimotor actions.

ON THE IMPORTANCE OF CONSIDERING THE DISCHARGE RATES OF FUSIMOTOR AXONS IN ASSESSING FUSIMOTOR EFFECTS

It has been pointed out earlier in a review (Murthy, 1978) that the magnitude of observed peripheral fusimotor effects is dependent upon the resting level of fusimotor activity. It is thus of some concern that Gladden (1980) has made her observations essentially under very low levels of spontaneous fusimotor activity (< 5 imp/sec) with maximum rates of only up to 15 imp/sec under conditions of central stimulation. It is perhaps not justifiable to draw strong conclusions about the order of recruitment of different types of intrafusal fibres in what may prove to be sub-threshold fusimotor activity as far as the sensory receptor is concerned (cf. also figure 12 in Boyd, 1976). It has been established that the majority of γ efferents discharge spontaneously at rates of 20/sec and above, both under light barbiturate anaesthesia (Gildenberg and Murthy, 1977) and in decerebrate preparations (Ellaway, 1972). Even those γ efferents which fire at lower rates may be reflexly activated to fire briefly at very high rates (figure 1). It is thus necessary to extend Gladden's observations to preparations displaying higher mean rates of γ activity before definite conclusions on the recruitment patterns of fusimotor effects may be attempted.

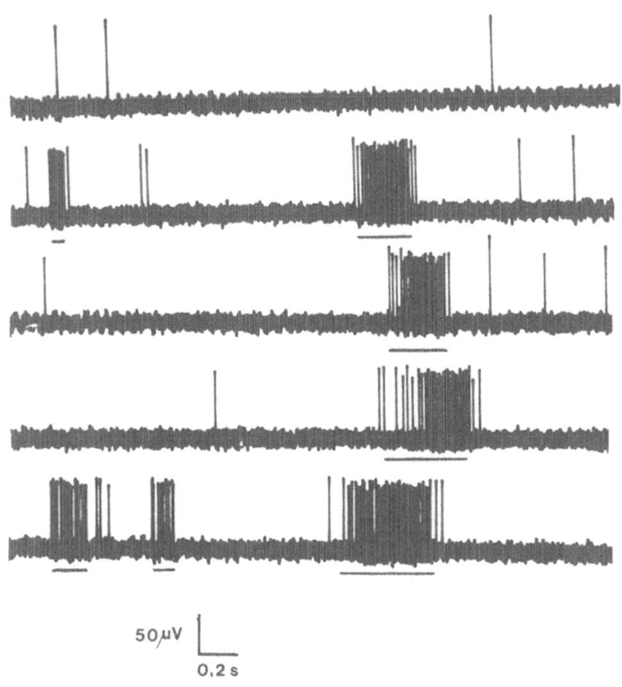

Figure 1. Gastrocnemius γ efferent with a low spontaneous rate of discharge (< 5 imp/sec) is seen to be excited by natural cutaneous stimulation with gentle brushing of the ipsilateral hind paw for brief periods denoted by the horizontal bars under the traces. Maximum rates of up to 150 imp/sec were observed during such excitation. Records continuous from top to bottom. Light barbiturate anaesthesia.

DO STATIC AND DYNAMIC γ EFFERENTS DISPLAY DIFFERENCES IN DISCHARGE CHARACTERISTICS?

Although differences in the discharge rates and in the regularity of impulses of spontaneously discharging γ efferents have been observed (cf. review by Murthy, 1978), it is not known whether such differences represent genuine functional characteristics of the static and dynamic γ efferents. The small sample of Westbury's data (Westbury, 1980) does not help to answer this question. Again, an extension of Gladden's experiments may provide the required answers.

THE PARADOXICAL NATURE OF THE DYNAMIC FUSIMOTOR EFFECTS

Understanding the effective use of the dynamic fusimotor system in the control of movements is made difficult due to the paradoxical characteristics of its peripheral effects (Goodwin *et al.*, 1975). Although stimulation of dynamic γ efferents produces an impressive increase of the dynamic index of the spindle primary afferent as observed during a ramp stretch (Crowe and Matthews, 1964), in a majority of cases this is observed only with fairly large amplitudes of muscle stretch performed at fairly high velocities and with the fusimotor stimulation at above 100 imp/sec. On the other hand, for small amplitude muscle stretch dynamic fusimotor stimulation has hardly any effect on the sensitivity of the primary afferent to passive stretch (Goodwin *et al.*, 1975). Further, in studying the problem of occlusion between simultaneously occurring static and dynamic fusimotor effects in the periphery, Hulliger *et al.* (1977) point out that if the dynamic fusimotor system is to be used to maintain the spindle sensitivity at a high level to detect small perturbations in movement, it is necessary for the static fusimotor system to be quiescent. This would then require the static fusimotor system to be switched off during maintained contractions in order to afford the dynamic fusimotor system the capability of boosting the dynamic sensitivity of the spindle afferent. This is, however, contradictory to the observations in man and in animals that there is a coactivation of the static γ system during maintained voluntary contractions (Hagbarth, 1980; Vallbo, 1980).

It is hence difficult to appreciate the need for a selective control of the dynamic γ system as argued for by Appelberg (1980). Although the dynamic fusimotor effects obtained by mesencephalic stimulation are indisputable, the extent to which this is selective to dynamic γ motoneurons to the exclusion of both static γ motoneurons and the α motoneurons is not clearly established (cf. Appelberg *et al.*, 1975).

There should also be some concern as to the significance of the small proportion of dynamic fusimotor efferents that are routinely observed in each experiment. In a thorough investigation of fusimotor innervation in peroneus tertius (Petit *et al.*, to be published) only 13 dynamic γ efferents were encountered as against 115 static γ's. There is no significant difference in either the mean or the range of conduction velocities of the two types of γ efferents. Earlier estimates of a proportion of 1:2–3 between dynamic and static γ's (Crowe and Matthews, 1964) may have been exaggerated due to the bias introduced by the technique of selection in favour of stronger fusimotor effects (cf. Brown *et al.*, 1965).

The frequencygrams of spindle primary afferents show rather feeble effects due to single stimuli applied to the dynamic γ axons with the muscle at constant length (Bessou *et al.*, 1968). On the other hand, the dramatic effect of such single stimuli applied to the dynamic γ axons during a slow stretch of the muscle has now been demonstrated by Emonet-Dénand and Laporte (1980). These findings reiterate the pitfalls of making conclusions on fusimotor effects from observations in rather restricted situations as found when a muscle is held at constant length.

Nevertheless, one is tempted to ask whether the dynamic sensitivity of the passive spindle primary ending is not sufficient to make it suitable for its presumed role in reflex regulation of movements (*see also* Vallbo *et al.*, 1979). The central nervous system has then only either to depress or switch off the γ activity to unmask this dynamic sensitivity whenever it is necessary to do so. For other situations the dynamic sensitivity is perhaps best kept low with the appropriate amount of γ drive.

Received on August 13th, 1980.

REFERENCES

Appelberg, B. (1980). Selective mid-brain control of dynamic gamma motoneurons utilised for the functional classification of gamma cells, this publication

Appelberg, B., Jeneskog, T. and Johansson, H. (1975). Rubrospinal control of static and dynamic fusimotor neurones, *Acta physiol. scand.*, **95**, 431–440

Banks, R. W., Barker, D., Bessou, P., Pagès, B. and Stacey, M. J. (1978). Histological analysis of cat muscle spindles following direct observation of the effects of stimulating dynamic and static motor axons, *J. Physiol., Lond.*, **283**, 605–619

Barker, D., Bessou, P., Jankowska, E., Pagès, B. and Stacey, M. J. (1978). Identification of intrafusal muscle fibres activated by single fusimotor axons and injected with fluorescent dye in cat tenuissimus spindles, *J. Physiol., Lond.*, **275**, 149–165

Barker, D., Emonet-Dénand, F., Harker, D. W., Jami, L. and Laporte, Y. (1976). Distribution of fusimotor axons to intrafusal muscle fibres in cat tenuissimus spindles as determined by the glycogen-depletion method, *J. Physiol., Lond.*, **261**, 49–69

Barker, D., Emonet-Dénand, F., Harker, D. W., Jami, L. and Laporte, Y. (1977). Types of intra- and extrafusal muscle fibre innervated by dynamic skeleto-fusimotor axons in cat peroneus brevis and tenuissimus muscles, as determined by the glycogen-depletion method, *J. Physiol., Lond.*, **266**, 713–726

Bessou, P., Laporte, Y. and Pagès, B. (1968). Frequencygrams of spindle primary endings elicited by stimulation of static and dynamic fusimotor fibres, *J. Physiol., Lond.*, **196**, 47–63

Boyd, I. A. (1976). The response of fast and slow nuclear bag fibres and nuclear chain fibres in isolated cat muscle spindles to fusimotor stimulation, and the effect of intrafusal contraction on the sensory endings, *Q. Jl exp. Physiol.*, **61**, 203–254

Brown, M. C., Crowe, A. and Matthews, P. B. C. (1965). Observations on the fusimotor fibres of the tibialis posterior muscle of the cat, *J. Physiol., Lond.*, **177**, 140–159

Crowe, A. and Matthews, P. B. C. (1964). The effects of stimulating static and dynamic fusimotor fibres on the response to stretching of the primary endings of muscle spindles, *J. Physiol., Lond.*, **174**, 109–131

Ellaway, P. H. (1972). The variability in discharge of fusimotor neurones in the decerebrate cat, *Expl Brain Res.*, **14**, 105–117

Emonet-Dénand, F. and Laporte, Y. (1980). Muscle stretch as a way of detecting brief activation of bag_1 fibres by dynamic axons, this publication

Emonet-Dénand, F., Laporte, Y., Matthews, P. B. C. and Petit, J. (1977). On the subdivision of static and dynamic fusimotor actions on the primary endings of the cat muscle spindle, *J. Physiol., Lond.*, **268**, 827–861

Gildenberg, P. L. and Murthy, K. S. K. (1977). Fusimotor discharge patterns in the lightly anesthetized cat, *J. Physiol., Lond.*, **272**, 68–69P

Gladden, M. H. (1980). The activity of intrafusal fibres during central stimulation in the cat, this publication

Goodwin, G. M., Hulliger, M. and Matthews, P. B. C. (1975). The effects of fusimotor stimulation during small-amplitude stretching on the frequency-response of the primary ending of the mammalian muscle spindle, *J. Physiol., Lond.*, **253**, 175–206

Hagbarth, K.-E. (1980). Fusimotor and stretch reflex functions studied in recordings from muscle spindle afferents in man, this publication

Hulliger, M., Matthews, P. B. C. and Noth, J. (1977). Effects of combining static and dynamic fusimotor stimulation on the response of the muscle spindle primary ending to sinusoidal stretching, *J. Physiol., Lond.*, **267**, 839–856

Jami, L. and Petit, J. (1975). Correlation between axonal conduction velocity and tetanic tension of motor units in four muscles of the cat hindlimb, *Brain Res.*, **96**, 114–118

Matthews, P. B. C. (1972). *Mammalian Muscle Receptors and Their Central Actions*, London, Edward Arnold

Murthy, K. S. K. (1978). Vertebrate fusimotor neurones and their influences on motor behaviour, *Prog. Neurobiol.*, **11**, 249–307

Petit, J. E., Barone, L. A., Cameron, W. E. and Murthy, K. S. K., Patterns of fusimotor innervation by gamma efferents in cat peroneus tertius, to be published

Vallbo, A. B. (1980). Basic patterns of muscle spindle discharge in man, this publication

Vallbo, A. B., Hagbarth, K. E., Torebjörk, H. E. and Wallin, B. G. (1979). Somatosensory, proprioceptive and sympathetic activity in human peripheral nerves, *Physiol. Rev.*, **59**, 919–957

Vedel, J. P. and Mouillac-Baudevin, J. (1969). Étude fonctionnelle du contrôle de l'activité des fibres fusimotrices dynamiques et statiques par les formations réticulées mésencéphaliques, pontique et bulbaire chez le chat, *Expl Brain Res.*, **9**, 325–345

Westbury, D. R. (1980). Electrophysiological characteristics of spinal gamma motoneurons in the cat, this publication

A critique of the papers by Westbury and by Appelberg

P. H. ELLAWAY*

ELECTROPHYSIOLOGICAL CHARACTERISTICS OF SPINAL GAMMA MOTONEURONS IN THE CAT: D. R. WESTBURY

The first intra-cellular recordings from cell bodies of gamma motoneurons in the cat were made by Eccles *et al.* (1960). The limited number of successful penetrations compared with those of alpha motoneurons indicated that the gamma cell bodies might be relatively small. The authors reported that for a given muscle, gamma motoneurons lay in the same motor nucleus as alpha motoneurons and exhibited similar spike potentials and associated after-hyperpolarisation (AHP). Gamma motoneurons, unlike alpha motoneurons, lacked a group Ia monosynaptic excitatory connection. Recently both Cullheim and Ulfhake (1979) and Westbury (1979) have studied the morphology of gamma motoneurons by impaling cells with micro-electrodes and injecting horse radish peroxidase. Their essential findings are that gamma motoneurons do indeed have smaller cell bodies (*see also* Bryan *et al.*, 1972; Burke *et al.*, 1977) and that they possess fewer dendrites which branch less extensively than those of alpha motoneurons. Only one of four gamma motoneurons in Westbury's (1979) study had axon collaterals and none of the six cells studied by Cullheim and Ulfhake (1979) showed this feature. A lack of axon collaterals is in agreement with recent work showing that antidromic impulses travelling in gamma motoneuron axons do not produce recurrent inhibition of alpha motoneurons (Westbury, 1980) or of other gamma motoneurons (Ellaway and Murphy, 1979, 1980a).

The study presented by Westbury in this book refers to results obtained with intracellular recording from six motoneurons. Axonal conduction was determined by noting the latency of invasion of an antidromic spike following electrical stimulation of the whole sciatic nerve. The muscle of origin was not determined. Two of the motoneurons had axonal conduction velocities of 52

*Department of Physiology, University College London

m/sec and 55 m/sec which makes it unlikely that they can be considered as gamma motoneurons. Boyd and Davey (1968) placed the upper limit for conduction velocities of gamma axons in cat hind limb muscles at 45 m/sec (e.g. popliteus, gastrocnemius, tibialis posterior and anterior). Other studies, where the fusimotor function of the gamma motoneuron was ascertained by recording the effect of motor axon stimulation upon the discharge of spindle afferents, have not indicated gamma motoneuron axons conducting at velocities above 50 m/sec (Kuffler *et al.*, 1951; Brown *et al.*, 1965; Ellaway *et al.*, 1972). It is known however that alpha motoneuron axons, particularly of small limb muscles such as the lumbricals (Bessou *et al.*, 1965), do conduct at velocities as low as, and lower than, 50 m/sec.

The other criterion used by Westbury to identify gamma motoneurons was a lack of monosynaptic excitation from group I axons in dorsal roots. Group I threshold of dorsal root axons will refer to the large limb muscles supplied by axons up to 20 μm in diameter. The axons supplying primary endings of spindles in smaller muscles have smaller diameters and hence higher thresholds to electrical stimulation (Lloyd and Chang, 1948). Such a negative finding should thus be used with caution.

Kemm and Westbury (1978) recorded intracellularly from 12 gamma motoneurons of which six had a long-lasting after-hyperpolarisation (AHP) of 30–100 msec following an antidromically conducted action potential. Using extracellular recording and a second test spike they claimed that 84 of 89 gamma cells behaved as if they had a short AHP of approximately 3 msec similar to another six cells recorded intracellularly.

Other workers' results, however, point to the conclusion that all gamma motoneurons generate soma spikes with long-duration AHP.

Gustafsson and Lipski (1979) also used the antidromic invasion technique but they noted the latency shifts that occur between the initial segment (IS) and the soma dendritic (SD) spike of a second test impulse. Extracellular recordings from gamma motoneurons showed latency shifts which were comparable to those seen for alpha motoneurons and indicated long (50–150 msec) AHP for all 35 gamma motoneurons studied.

In a study of the variability in discharge patterns of tonically firing gamma motoneurons in the decerebrated cat, it was found that the timing of a spike in a train of impulses was reset by an antidromically conducted impulse (Ellaway, 1972). Figure 1 from that study shows the phenomenon and indicates that the duration of the interspike interval following the antidromically conducted spike is dependent upon the length of the curtailed interval. At the shortest interspike interval the subsequent interval approaches 1.5 times the mean background interval. This ratio was independent of the actual frequency of discharge and the relationship was found for all 20 gamma motoneurons studied. Figure 2 shows that the line relating subsequent to curtailed interval could be higher for some gamma motoneurons and these neurons were found to be subject to recurrent inhibition (Ellaway, 1971). The most

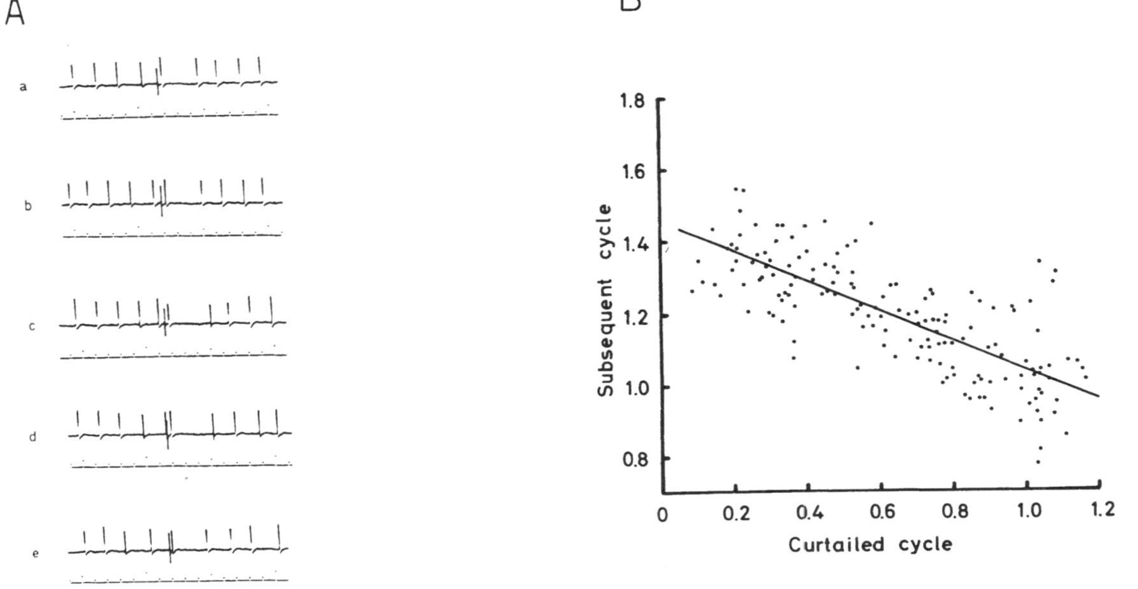

Figure 1. The effect of single antidromic motor volleys on the discharge of GM gamma motoneurons. A (a)–(e), the single diphasic potential in each record is the direct response of an alpha axon to stimulation of an intact ventral root and is followed after 3 msec by a direct response of the gamma motoneuron (36 m/sec) except in (e) where the axon is refractory to an orthodromic impulse. Rephasing of the rhythm of discharge and a negative correlation between curtailed and subsequent intervals is evident in (a)–(d). The increased interval duration in (e) shows that the neuron was also subject to recurrent inhibition. B, a plot of the duration of subsequent intervals against curtailed intervals created by antidromic impulses, as in A above, but for another gamma motoneuron (26 m/sec) which did not receive recurrent inhibition. The mean interval of the background discharge was 35 msec (± 3 msec S.D.) and is normalised on the graph as 1.0 cycle. The superimposed regression line has a slope of −0.44.

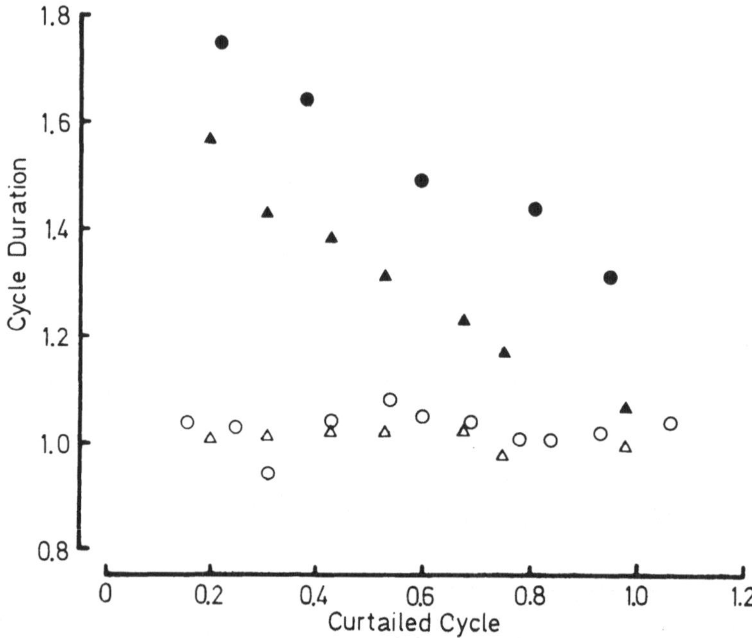

Figure 2. The action of single antidromic impulses on the discharges of two GL/Sol gamma motoneurons. One neuron (○, ● 25 m/sec) had a mean interval of 26 msec and the other (△, ▲ 29 m/sec) 15 msec. In this graph both the interval following the curtailed cycle (filled symbols) and the next interval in the train of impulses (open symbols) have been plotted against the interval curtailed by an antidromic impulse. The points represent average values for different durations of the curtailed cycle expressed in multiples of their current mean interval.

probable mechanism underlying the relation shown in figure 1B is summation of AHP as is known to occur for alpha motoneurons (Eccles and Hoff, 1932; Eccles, 1953) and has recently been demonstrated also for gamma motoneurons (Gustafsson and Lipski, 1979). Thus I cannot agree with the emphasis placed by Westbury in this book that most gamma motoneurons have a short duration AHP which is of limited effectiveness.

SELECTIVE CENTRAL CONTROL OF DYNAMIC GAMMA MOTONEURONS UTILISED FOR THE FUNCTIONAL CLASSIFICATION OF GAMMA CELLS: B. APPELBERG

This report by Appelberg reviews the evidence accumulated by his laboratory that electrical stimulation of an area of the mesencephalon (Mes ADC) can give excitation of dynamic gamma motoneurons without any simultaneous action on either static gamma motoneurons or alpha motoneurons. The technique was originally established by noting selective increases in dynamic sensitivity of primary endings of muscle spindles. Subsequently, intracellular recordings were made from gamma motoneurons and these were classified as dynamic or static according to the effect of Mes ADC stimulation. Some of the dynamic gamma motoneurons received a facilitatory input from group II muscle afferents (Appelberg *et al.*, 1977) and such an action has again been suggested after the employment of a different technique (Appelberg *et al.*, 1979). These latter authors found that stretch of ipsilateral hamstring muscles (PBST) could give enhanced dynamic sensitivity of gastrocnemius-soleus (GS) spindles and this effect was attributed to reflex action by PBST secondary spindle afferents. Although the methods of activation of secondary spindle afferents were not selective, either with electrical stimulation of the muscle nerve or with muscle stretch, the results provide circumstantial support that the Mes ADC stimulus is a selective means of activating dynamic gamma motoneurons.

The possible participation in the reflex control of gamma motoneurons of group II afferents receives support from other sources.

Noth and Thilmann (1980) showed that the discharge of 11 out of 22 GS gamma motoneurons recorded in decerebrated cats was facilitated by electrical stimulation of the homonymous muscle nerve at group II strength. Ellaway and Murphy (1980b) have observed late facilitation of gamma motoneurons during the relaxation of a muscle twitch at a time when secondary endings would have been excited. The facilitation was more evident in spinal than in decerebrate cats. In the case of the homonymous muscle the facilitation could terminate a period of inhibition caused by tendon organ discharge. If the inhibition was lacking, as is more likely with a heteronymous muscle, the facilitation appears alone at latencies ranging from 30 to 60 msec after the ventral root shock used to elicit the twitch contraction (Ellaway *et al.*, 1980, figure 6). Experiments are in hand to examine whether the rebound discharge of secondary endings during muscle relaxation is responsible for this facilitation. A group Ia action is discounted since such action on gamma motoneurons is weak (Ellaway and Trott, 1978) even in the spinal cat where late facilitation to contraction is a strong effect. An action by group III muscle afferents has, however, not been ruled out, since in a recent experiment late facilitation of a GL/Sol gamma motoneuron was caused by contraction of the GM muscle (spinal cat), but it needed shocks greater than ten times threshold for group I axons to elicit facilitation when the cut central end of the GM nerve was stimulated directly.

Appelberg *et al.* (1977) found that four out of 12 dynamic gamma motoneurons were excited by group II muscle nerve stimulation, but these four had been selected as neurons receiving group III inhibition. In that study it was more common to see static gamma motoneurons excited by group III axons, which supports the notion that in our studies in the spinal cat (Ellaway and

Murphy, 1980b) we may be selecting static as well as, or in preference to, dynamic gamma motoneurons. Clearly we await a more detailed analysis expected from Appelberg (personal communication) as to the relative actions of group II and III actions on static and dynamic gamma motoneurons.

ACKNOWLEDGEMENT

Permission to reproduce figures 1 and 2 was granted by the publishers of *Experimental Brain Research.*

Received on July 25th, 1980.

REFERENCES

Appelberg, B., Hulliger, M., Johansson, H. and Sojka, P. (1979). Excitation of dynamic fusimotor neurones of the cat triceps surae by contralateral joint afferents, *Brain Res.*, **160**, 529–532

Appelberg, B., Johansson, H. and Kalistratov, G. (1977). The influence of Group II muscle afferents and low threshold skin afferents on dynamic fusimotor neurones to the triceps surae of the cat, *Brain Res.*, **132**, 153–158

Bessou, P., Emonet-Dénand, F. and Laporte, Y. (1965). Motor fibres innervating extrafusal and intrafusal muscle fibres in the cat, *J. Physiol., Lond.*, **180**, 649–672

Boyd, I. A. and Davey, M. R. (1968). *Composition of Peripheral Nerves*, Edinburgh, Livingstone

Brown, M. C., Crowe, A. and Matthews, P. B. C. (1965). Observations on the fusimotor fibres of the tibialis posterior muscle of the cat, *J. Physiol., Lond.*, **177**, 140–159

Bryan, R. N., Trevino, D. L. and Willis, W. D. (1972). Evidence for a common location of alpha and gamma motoneurones, *Brain Res.*, **38**, 193–196

Burke, R. E., Strick, P. L., Kanda, K., Kim, C. C. and Walmsley, B. (1977). Anatomy of medial gastrocnemius and soleus motor nuclei in cat spinal cord, *J. Neurophysiol., Lond.*, **40**, 667–680

Cullheim, S. and Ulfhake, B. (1979). Observations on the morphology of intracellularly stained gamma motoneurones in relation to their axon conduction velocity, *Neurosci. Lett.*, **13**, 47–50

Eccles, J. C. (1953). *Neurophysiological Basis of Mind: The Principles of Neurophysiology*, Oxford, Clarendon Press, 174–178

Eccles, J. C., Eccles, R. M., Iggo, A. and Lundberg, A. (1960). Electrophysiological studies on gamma motoneurones, *Acta physiol. scand.*, **50**, 32–40

Eccles, J. C. and Hoff, H. E. (1932). The rhythmic discharge of motoneurones, *Proc. R. Soc. B*, **110**, 483–514

Ellaway, P. H. (1971). Recurrent inhibition of fusimotor neurones exhibiting background discharges in the decerebrate and the spinal cat, *J. Physiol., Lond.*, **216**, 419–439

Ellaway, P. H. (1972). The variability in discharge of fusimotor neurones in the decerebrate cat, *Expl Brain Res.*, **14**, 105–117

Ellaway, P. H., Emonet-Dénand, F., Jami, L. and Joffroy, M. (1972). Proportion des

fibres fusimotrices statiques et dynamiques dans les muscles peroneus longus et flexor hallucis longus du chat, *Compt. Rend.*, **274**, 3597–3600
Ellaway, P. H. and Murphy, P. R. (1979). Recurrent inhibition of gamma motoneurones, *Neurosci. Lett.*, Suppl 3, S314
Ellaway, P. H. and Murphy, P. R. (1980a). A quantitative comparison of recurrent inhibition of alpha and gamma motoneurones in the cat, *J. Physiol., Lond.*, **301**, 55–56P
Ellaway, P. H. and Murphy, P. R. (1980b). Autogenetic effects of muscle contraction on extensor gamma motoneurones in the cat, *Expl Brain Res.*, **38**, 305–312
Ellaway, P. H., Murphy, P. R. and Trott, J. R. (1980). Autogenetic effects from spindle primary endings and tendon organs on the discharge of gamma motoneurons in the cat, this publication
Ellaway, P. H. and Trott, J. R. (1978). Autogenetic reflex action on to gamma motoneurones by stretch of triceps surae in the decerebrated cat, *J. Physiol., Lond.*, **276**, 49–66
Gustafsson, B. and Lipski, J. (1979). Do gamma motoneurones lack a long lasting after-hyperpolarisation?, *Brain Res.*, **172**, 349–353
Kemm, R. E. and Westbury, D. R. (1978). Some properties of spinal gamma motoneurones in the cat, determined by microelectrode recording, *J. Physiol., Lond.*, **282**, 59–71
Kuffler, S. W., Hunt, C. C. and Quilliam, J. P. (1951). Function of medullated small nerve fibres in mammalian ventral roots: efferent muscle spindle innervation, *J. Neurophysiol.*, **14**, 21–54
Lloyd, D. P. C. and Chang, H. T. (1948). Afferent fibres in muscle nerves, *J. Neurophysiol.*, **6**, 199–208
Noth, J. and Thilmann, A. (1980). Autogenetic excitation of extensor gamma motoneurones by Group II muscle afferents in the cat, *Neurosci. Lett.*, **17**, 23–26
Westbury, D. R. (1979). The morphology of four gamma motoneurones examined by horseradish peroxidase histochemistry, *J. Physiol., Lond.*, **292**, 25–26P
Westbury, D. R. (1980). Lack of a contribution from gamma motoneurone axons to Renshaw inhibition in the cat spinal cord, *Brain Res.*, **186**, 217–221

Autogenetic effects from spindle primary endings and tendon organs on the discharge of gamma motoneurons in the cat

P. H. ELLAWAY*, P. R. MURPHY* AND J. R. TROTT*†

SUMMARY

This article reviews experiments which have been carried out in decerebrated and in spinal cats to determine the action of muscle afferents on the discharge of gamma motoneurons.

Single tonically firing gamma motoneurons were isolated in teased filaments of muscle nerve fascicles leaving much of the afferent nerve supply to the muscle intact.

Selective activation of primary endings of muscle spindles by brief stretch or vibration gave autogenetic facilitation of approximately half of the triceps surae (GS) gamma motoneurons studied. The central delay of this facilitation (5–14 msec) precludes any monosynaptic connection. Reciprocal Ia inhibition of antagonist gamma motoneurons was not evident.

Inhibition of gamma motoneurons was elicited by twitch contractions of GS and flexor digitorum longus (FDL) muscles. The inhibition could be ascribed to discharges elicited in Ib afferents from tendon organs. It had a short central delay comparable with that of the Ib pathway to alpha motoneurons. The inhibition had a particularly strong autogenetic component and was less evident between synergists (GS and FDL). Spinal section produced a marked potentiation of the inhibition.

INTRODUCTION

Compared with the wealth of information concerning the reflex behaviour of alpha motoneurons our knowledge of the spinal connections from peripheral

* Department of Physiology, University College, London
† Present address: Department of Physiology, The Medical School, University Walk, Bristol, BS8 1TD, England

receptors to gamma or fusimotor neurons is sparse. This has resulted from the difficulty experienced in penetrating gamma motoneuron cell bodies with micro-electrodes (Eccles *et al.*, 1960). There is now little doubt that gamma motoneuron cell bodies lie in among alpha motoneurons of the same muscle nucleus but the cells are small and have fewer dendrites (Burke *et al.*, 1977; Cullheim and Ulfhake, 1979; Westbury, 1979). We have made extensive use of an alternative method for detecting synaptic actions which can be used with tonically firing cells. The technique, which will be referred to as the peri-stimulus time histogram (p.s.t.h.) notes the time of occurrence of spikes with reference to a stimulus. Using this technique both facilitation and inhibition of tonically firing gamma motoneurons can be detected.

A large proportion of gamma motoneurons to various muscles in the decerebrated and decerebrated spinal cat show a background tonic discharge. This discharge in gamma motoneurons is readily influenced both by activity descending in the spinal cord from central nuclei and by reflex segmental inputs from the skin (*see* review by Murthy, 1978). What has not been clear, since the pioneer investigations of Hunt (1951) and Hunt and Paintal (1958), is the extent to which muscle receptors may influence gamma motoneuron discharge. Clearly it is necessary to establish the degree to which spindle endings influence gamma motoneurons, and in particular, their own gamma motoneurons, since anatomically these structures form a closed loop. We decided initially to see whether an action by primary spindle endings could be detected (Trott, 1976; Ellaway and Trott, 1978). This was simultaneously attempted by Fromm and Noth (1976) and both laboratories described a hitherto unexpected facilitatory action by primary endings on their own fusimotor neurons. However, when the primary ending discharge elicited by stretch or vibration was intense enough to elicit alpha motoneuron discharge, gamma motoneurons could be inhibited. The inhibition was attributed to recurrent inhibition (Fromm *et al.*, 1974) which gamma motoneurons are known to receive (Ellaway, 1971).

We have also investigated the action of Golgi tendon organ discharges on gamma motoneurons and found a marked autogenetic inhibitory influence especially in the spinal cat (Ellaway *et al.*, 1979; Ellaway and Murphy, 1980a). Other workers have shown connections from Gp II afferents in muscle nerves which, in triceps surae muscles, appear to be facilitatory (Noth and Thilmann, 1980) and may be preferentially directed to dynamic gamma motoneurons (Appelberg *et al.*, 1977).

One of the problems met in assessing the contribution of muscle afferents to the reflex control of gamma motoneurons is that any alpha motoneuron discharge induced by muscle or nerve stimulation can potentially inhibit gamma motoneuron discharge via the recurrent Renshaw cell loop (Ellaway, 1971). The effectiveness of this pathway in its own right should be considered in any assessment of spinal segmental control of gamma motoneurons. Despite recent work on anaesthetised cats suggesting that most gamma motoneurons lacked both recurrent i.p.s.p.s and the after-hyperpolarisation associated with a spike (Kemm and Westbury, 1978), further studies have indicated the presence of after-hyperpolarisation (Gustafsson and Lipski, 1979) and a relatively potent degree of recurrent inhibition (Ellaway and Murphy, 1980b).

This article reviews the contribution that we have made in the field of segmental reflex control of gamma motoneurons by muscle afferents. It has an emphasis on recent findings concerning the inhibition of gamma motoneurons by tendon organ afferents. The work is taken from articles by Ellaway and co-workers quoted in this introduction.

AUTOGENETIC FACILITATION BY PRIMARY SPINDLE ENDINGS

Figure 1 illustrates the effect of both vibration and brief stretch of triceps surae on homonymous gamma motoneurons. Approximately 50% of tonically firing GS gamma motoneurons in decerebrated cats could be facilitated with such stimuli. The threshold for facilitation was always at an amplitude of movement sufficient to excite primary endings alone among muscle receptors. As can be seen in figure 1B′, stretch of a muscle could also elicit inhibition but, with repetitive vibration (figure 1B), net facilitation only was seen. Furthermore, when inhibition was elicited by stretch of the muscle it always had a higher threshold than facilitation.

Latency measurements of the facilitation gave central delays to the incoming Gp Ia volley ranging from 5 to 14 msec. This confirms earlier reports that gamma motoneurons lack a monosynaptic Gp Ia connection (Eccles *et al.*, 1960; Grillner *et al.*, 1969) and raises the interesting possibility that this polysynaptic pathway is the same as that which contributes tonic excitation of alpha motoneurons during the stretch or vibration reflex.

It should be stressed that the facilitation of gamma motoneurons was weak. Nothing like the potency of the Ia monosynaptic excitation of alpha motoneurons was ever seen and quiet gamma motoneurons could not be excited to fire by stretch or vibration. Interestingly the facilitation was also evident in decerebrated cats with thoracic spinal cord section but could be either slightly stronger or weaker than before spinal section. The possibility remains that it is a reflex which could be potentiated by centres located higher than the position of the intercollicular decerebration (*see* Discussion).

We have looked to see whether reflex effects from Ia afferents to gamma motoneurons of antagonist muscles exist. In this study (Ellaway and Trott, unpublished observations) the effects of vibration and stretch of triceps surae on the discharges of tonically firing gamma motoneurons to hamstring (PBST) muscles were examined in decerebrated and in spinal cats. Among 21 gamma motoneurons no obvious pattern of reflex effect was seen. The only

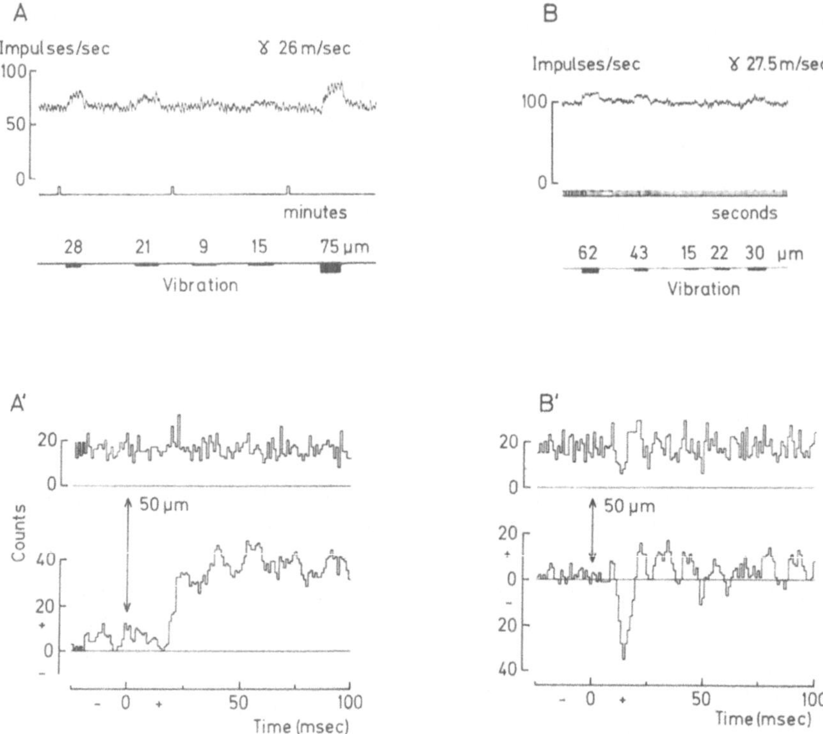

Figure 1. GS gamma motoneurons showing net facilitation in response to vibration of the homonymous muscle (A and B) but different responses to single brief stretches of 50 μm amplitude (A′ and B′). Decerebrated cat, spinal cord intact. A and B, integrated frequency of discharge. A′ and B′, p.s.t.h.s (above) and their cumulative sums (below) for the same two gamma motoneurons. Histogram bin widths 1 msec, number of trials 256. The ordinate for the cumulative sums (cusums) in this and subsequent figures represents the consecutively summed differences, of each point of the p.s.t.h., from a control mean bin count measured over a period of 250 msec before the stimulus. The cusum effectively integrates any departure from the count expected if no stimulus was presented (see Ellaway, 1978).

reproducible effect of stretch and vibration was inhibition in about half the sample. Generally, however, the threshold for inhibition was in excess of 50 to 75 μm. Since autogenetic facilitation regularly appeared with a threshold of between 10 and 35 μm we can conclude, at least, that no crossed facilitatory connection exists. Control studies showed that approximately 25% of both secondary spindle afferents and tendon organs are excited by stretch and vibration of an amplitude in the range 50–75 μm. The origin of the inhibition thus remains unclear.

In conclusion, neither a monosynaptic autogenetic facilitation nor a disynaptic reciprocal inhibition (which have been amply demonstrated to alpha motoneurons) appears to exist in the segmental control of gamma motoneurons.

INHIBITION FROM TENDON ORGANS

A study has been made of the effects of contraction in a muscle on the discharge of gamma motoneurons to both homonymous and synergist muscle groups in decerebrated and spinally transected cats. Twitch contractions were elicited by stimulating part of the ventral root supply (see figure 2) and discharges of gamma motoneurons were recorded in filaments of one of the nerve fascicles supplying a muscle. In the lower traces of figure 2 (B and C) the disturbance caused by a twitch to the firing of a GL/Sol gamma motoneuron can be seen as a pause in the discharge. When p.s.t.h.s of many trials were constructed a quantitative picture of the inhibition emerged (figure 3). Contraction can result in a brief silent period (figure 3A and C) or merely give a decreased probability of firing during the twitch (figure 3B and D).

Two lines of evidence point to discharges in Ib afferents from tendon

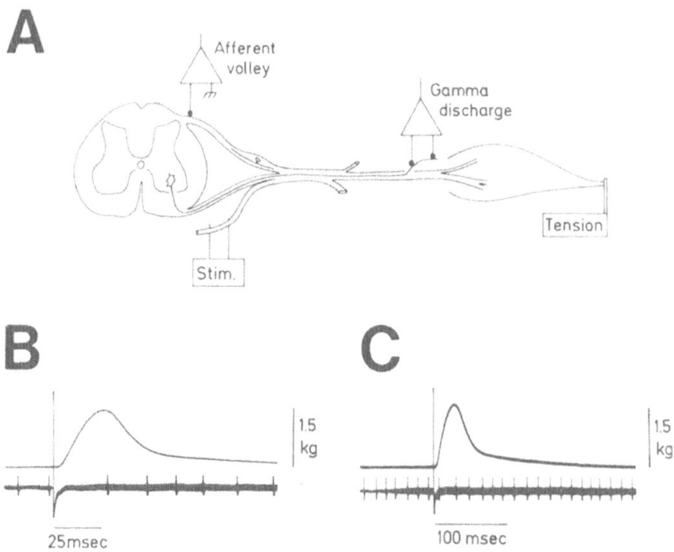

Figure 2. Method for studying the effect of contraction of a muscle on the discharge of homonymous gamma motoneurons in the decerebrated cat. A, the preparation. B and C, the response of a tonically firing GL/Sol gamma motoneuron (axonal conduction velocity 29 m/sec) to twitch contractions generated in triceps surae by stimuli applied to ventral root S1. Upper traces, tension developed in the muscle. Lower traces, gamma motoneuron discharge. The large spike is a direct alpha response recorded in the same filament.

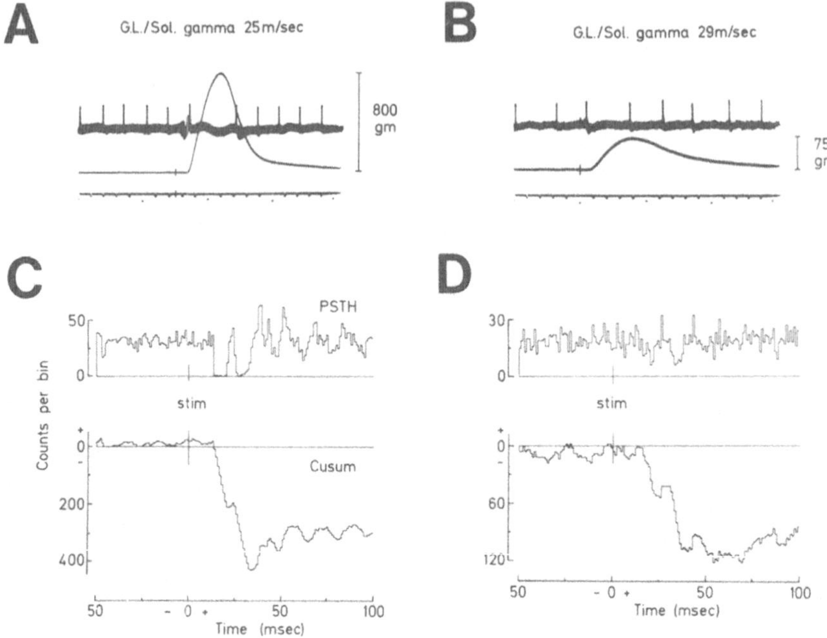

Figure 3. The effect of submaximal twitch contractions of triceps surae on the discharge of two homonymous gamma motoneurons. A and B, single sweeps showing strong (A) and weak (B) inhibition from twitch contractions of similar magnitude. Decerebrated cat, spinal cord intact. Traces, from above, gamma discharge, muscle tension, time (smallest units 10 msec). C and D, p.s.t.h.s (above) and their cumulative sums (below) for 512 sweeps of the form shown above. Note two periods of inhibition in C and a suggestion of a similar division in D. Oscillations in histogram C following the inhibition represent a partial resetting (or rephasing) of the regular gamma discharge.

organs causing the inhibition. In the first place the latency of the inhibition to a ventral root shock was short (10–20 msec) and included peripheral conduction delays in alpha, afferent and gamma axons. We calculated that, in many instances, the afferent volley responsible must have been an early discharge (Hunt and Kuffler, 1951) in Gp I afferents generated by the muscle contraction or action potential. Since selective Gp Ia excitation by muscle stretch facilitated gamma motoneurons in the same preparation (Ellaway and Trott, 1978), this result implicated the Ib axons. Furthermore, slackening the muscle so that no active tension could be recorded at the tendon of insertion invariably reduced the inhibition (figure 4). Often it was the later component that was reduced or abolished by this procedure (figure 4B). Control recordings from Ib afferents showed that the early discharges in these afferents were unaltered by shortening the muscle. In contrast the Ib discharge induced

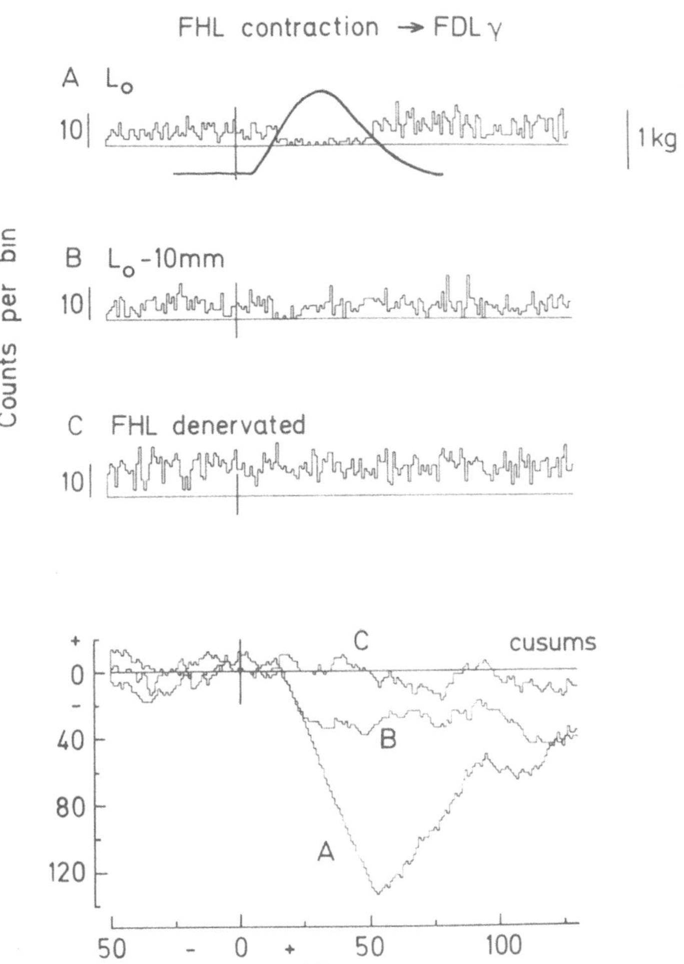

Figure 4. Autogenetic inhibition of an FDL gamma motoneuron, in the decerebrated spinal cat, to isometric contraction of the FHL muscle. A, B and C (above), p.s.t.h.s each of 512 trials with a single shock to part of ventral root L7. A, muscle length held at L_0, just short of the maximum physiological length. Time course of the muscle tension change is superimposed. B, the response at $L_0 - 10$ mm when there was no active tension developed at the FHL tendon. The inhibition is restricted to an initial component of 9 msec duration. C, absence of inhibition of L_0 following section of the FHL nerve. Below, the cumulative sums of the three p.s.t.h.s.

during the twitch is reduced by such a length change (Green and Kellerth, 1967).

On subtracting the various peripheral delays from the latency of inhibition, the central delay (1.2–3.2 msec) to the incoming Ib early discharge indicated, at most, a di- or tri-synaptic pathway. These central delays are similar to those found for the inhibitory Ib pathway to alpha motoneurons (Eccles *et al.*, 1957) and it thus seemed appropriate to test the system for other similarities.

The Ib inhibition of alpha motoneurons is depressed in the decerebrated cat but may be released by spinal section (Eccles and Lundberg, 1959). This proved also to be the case for gamma motoneurons although the study met the complication that few gamma motoneurons to extensor muscles exhibit a background discharge in the spinal state. In the decerebrated cat with intact spinal cord, 22 out of 47 gamma motoneurons could be inhibited by twitch contractions of the homonymous muscle (GS), and some only weakly. In cats with a spinal cord section, all nine gamma motoneurons studied were strongly inhibited by twitch contractions of the homonymous muscle (GS or FDL).

The effect of contraction in flexor digitorum longus or hallucis muscles was studied because of a report that Ib inhibition of alpha motoneurons from these muscles was particularly strong (Eccles *et al.*, 1957). Strong inhibition of FDL/FHL gamma motoneurons did occur in response to contraction of their own muscles (figure 4) but crossed effects between heteronymous synergists were not so evident. Thus contraction of FDL/FHL inhibited five of 11 GS gamma motoneurons and in only two instances was the inhibition as strong as on to the homonymous neurons. Likewise, contraction of GM inhibited only five of 12 FHL gamma motoneurons, and only two strongly. The results are summarised in figure 5. The degree of inhibition was measured both in terms

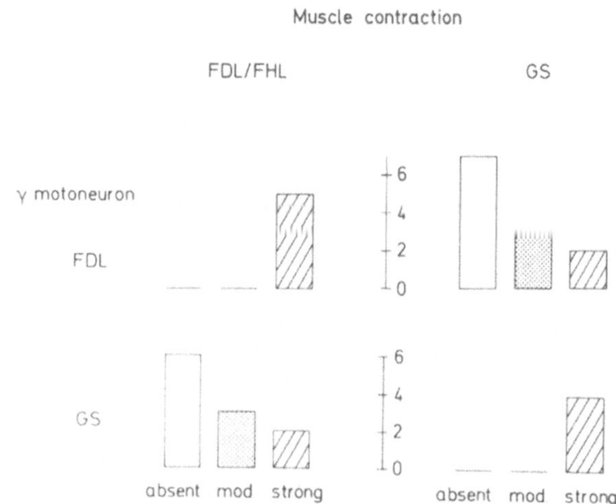

Figure 5. Summary of the incidence of inhibition caused by twitch contractions of FDL and GS muscles in the decerebrated spinal cat. The ordinate scales represent the numbers of gamma motoneurons studied. The terms moderate (mod.) and strong refer to arbitrary divisions based both on the duration and degree (decreased probability of firing) of inhibition. Note that autogenetic effects were invariably strong (upper left and lower right) and that inhibition was frequently absent between synergists (lower left and upper right).

of the average reduction in probability of firing and the duration of the response. Autogenetic inhibitory responses in the spinal cat showed average decreases in probability of firing in the range 35 to 90% and durations from 20 to 65 msec. Generally a short duration was coupled with a large decrease in probability of firing and vice versa.

Responses of gamma motoneurons to contraction could include periods of facilitation. In the spinal cat this facilitation could be as strong in terms of net excitation as the inhibition, but generally appeared at longer latency. The facilitation often appeared at the peak of contraction or during relaxation. It was also seen in gamma motoneurons lacking Ib inhibition (figure 6). We are

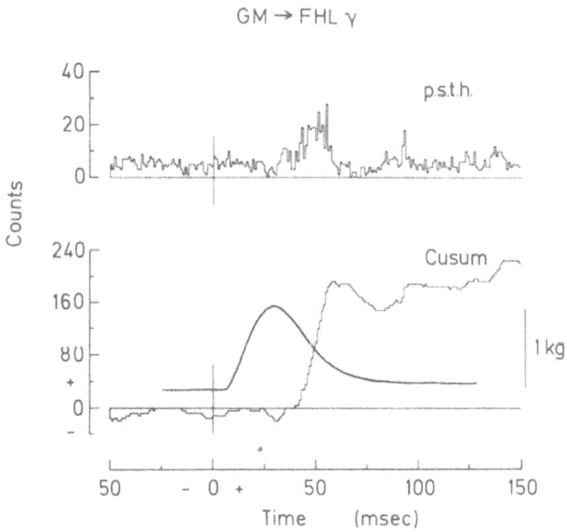

Figure 6. Facilitation of an FHL gamma motoneuron (axon conduction velocity 33 m/sec) by twitch contractions of GM in the decerebrated spinal cat. Top, p.s.t.h. of 512 trials. Stimulus, maximal for GM alpha motoneuron axons, applied to part of ventral root L7 at time zero. Bottom, cumulative sum of the p.s.t.h. Superimposed is a tracing of the tension recorded in the GM muscle in response to a single stimulus.

uncertain as to the origin of the facilitation but are now doubtful whether it could be due to Ia discharges. Discharges from primary spindle endings (Ia) are elicited both as early discharges and by re-extension during relaxation of the muscle. In our experiments vibration of the tendon did not evoke any increase in discharge of those gamma motoneurons which had responded to contraction with particularly strong facilitation. Firm tapping of the belly of the muscle, however, did elicit such discharges. We intend to explore the possibility that either Gp II or III discharges are responsible for the facilitation.

CONCLUSIONS

There are difficulties in assessing the likely roles of the segmental pathways from muscle afferents to gamma motoneurons not the least of which stem from the fact that the results have been obtained in decerebrated and in spinal cats where, judging from recordings in intact animals (Prochazka *et al.*, 1977), patterns of gamma activity may be abnormal. One fact that emerges from using these preparations, however, is the importance which should be attached to control of these reflexes by activity in descending spinal tracts.

The Ia autogenetic facilitation of gamma motoneurons in the decerebrated cat is neither suppressed nor elevated by spinal section. This is a reflex with a clear potential for positive feedback. Although the gain of the stretch reflex is thought to be quite low it may be adjustable (Marsden *et al.*, 1972) and thus the Ia facilitation of gamma motoneurons should preferably be under close central control. It is interesting in this context that the reflex is weak or absent in both decerebrated and spinal cats. It may thus need positive potentiation from a more central site in order to become effective. Regions of the mesencephalon, which are destroyed in the classical decerebrated preparations used in our studies, are known to be capable of facilitating the discharge of gamma motoneurons, particularly the dynamic gamma motoneurons (Appelberg and Jeneskog, 1972).

The inhibition of gamma motoneurons by Ib afferents is quite clearly under supraspinal control. As with the inhibitory pathway from Ib afferents to alpha motoneurons (Eccles and Lundberg, 1959) the reflex is depressed in the decerebrated cat and may be potentiated by spinal section. This should be considered in conjunction with the fact that Ib inhibition of gamma motoneurons has a strong autogenetic component. In this sense it differs from the Ib inhibition of alpha motoneurons where one particular muscle (e.g. FDL) may exert a strong inhibition over a variety of other hind limb muscles (Eccles *et al.*, 1957). Thus the intensity of the reflex inhibition of gamma motoneurons may be controlled and, presumably, restricted to specific muscles. Considering the variety of movements to which individual muscles contribute it is conceivable that such individual control may be required. As has been proposed by Houk *et al.* (1970) and elaborated upon by Lundberg *et al.* (1977), fine exploratory movements may require the suppression of muscle length control if, for example, tension in a contracting muscle rises as a result of a limb encountering an obstacle. Inhibition of gamma motoneurons would achieve this suppression. The advantage of an autogenetic system over a more widespread feedback is that it could specifically suppress activity in muscles disturbed by the obstacle while leaving muscles controlling the position of other joints unaffected.

ACKNOWLEDGEMENTS

We thank Mr J. E. Pascoe for programming a LINC-8 computer to form p.s.t.h.s, Maria Winder for her technical assistance and Dr Françoise Emonet-Dénand for advice on surgical techniques. We acknowledge the support of the M.R.C. (post-doctoral position for J.R.T.) and the Wellcome Trust (post-doctoral position for P.R.M.). Permission for the reproduction of figures 1, 2 and 3 was given by the *Journal of Physiology* and of figure 4 by *Experimental Brain Research*.

Received on June 2nd, 1980.

REFERENCES

Appelberg, B. and Jeneskog, T. (1972). Mesencephalic fusimotor control, *Expl Brain Res.*, **15**, 97–112

Appelberg, B., Johansson, H. and Kalistratov, G. (1977). The influence of Gp II muscle afferents and low threshold skin afferents on dynamic fusimotor neurones to the triceps surae of the cat, *Brain Res.*, **132**, 153–158

Burke, R. E., Strick, P. L., Kanda, K., Kim, C. C. and Walmsley, B. (1977). Anatomy of medial gastrocnemius and soleus motor nuclei in cat spinal cord, *J. Neurophysiol.*, **40**, 667–680

Cullheim, S. and Ulfhake, B. (1979). Observations on the morphology of intracellularly stained gamma motoneurones in relation to their axon conduction velocity, *Neurosci. Letts*, **13**, 47–50

Eccles, J. C., Eccles, R. M., Iggo, A. and Lundberg, A. (1960). Electrophysiological studies on gamma motoneurones, *Acta physiol. scand.*, **50**, 32–40

Eccles, J. C., Eccles, R. M. and Lundberg, A. (1957). Synaptic actions on motoneurones caused by impulses in Golgi tendon organ afferents, *J. Physiol., Lond.*, **138**, 227–252

Eccles, R. M. and Lundberg, A. (1959). Supraspinal control of interneurones mediating spinal reflexes, *J. Physiol., Lond.*, **147**, 565–584

Ellaway, P. H. (1971). Recurrent inhibition of fusimotor neurones exhibiting background discharges in the decerebrate and the spinal cat, *J. Physiol., Lond.*, **216**, 419–439

Ellaway, P. H. (1978). Cumulative sum technique and its application to the analysis of peri-stimulus time histograms, *Electroenceph. clin. Neurophysiol.*, **45**, 302–304

Ellaway, P. H. and Murphy, P. R. (1980a). Autogenetic effects of muscle contraction on extensor gamma motoneurones in the cat, *Expl Brain Res.*, **38**, 305–312

Ellaway, P. H. and Murphy, P. R. (1980b). A quantitative comparison of the recurrent inhibition of alpha and gamma motoneurones in the cat, *J. Physiol., Lond.*, to be published

Ellaway, P. H., Murphy, P. R. and Trott, J. R. (1979). Inhibition of gamma motoneurone discharge by contraction of the homonymous muscle in the decerebrated cat, *J. Physiol., Lond.*, **291**, 425–442

Ellaway. P. H. and Trott, J. R. (1978). Autogenetic reflex action on to gamma motoneurones by stretch of triceps surae in the decerebrated cat, *J. Physiol., Lond.*, **276**, 49–66

Fromm, C., Haase, J. and Noth, J. (1974). Length dependent autogenetic inhibition of extensor gamma motoneurones in the decerebrated cat, *Pflügers Arch.*, **346**, 251–262

Fromm, C. and Noth, J. (1976). Reflex responses of gamma motoneurones to vibration of the muscle they innervate, *J. Physiol., Lond.*, **256**, 117–136

Green, D. G. and Kellerth, J. O. (1967). Intracellular autogenetic and synergistic effects of muscular contraction on flexor motoneurones, *J. Physiol., Lond.*, **193**, 73–94

Grillner, S., Hongo, T. and Lund, S. (1969). Descending monosynaptic and reflex control of gamma motoneurones, *Acta physiol. scand.*, **75**, 592–613

Gustafsson, B. and Lipski, J. (1979). Do gamma motoneurones lack a long lasting afterhyperpolarisation?, *Brain Res.*, **172**, 349–353

Houk, J. C., Singer, J. J. and Goldman, M. R. (1970). An evaluation of length and force feedback to soleus muscles of decerebrate cats, *J. Neurophysiol.*, **33**, 784–811

Hunt, C. C. (1951). The reflex activity of mammalian small nerve fibres, *J. Physiol., Lond.*, **115**, 456–469

Hunt, C. C. and Kuffler, S. W. (1951). Stretch receptor discharges during muscle contraction, *J. Physiol., Lond.*, **113**, 298–315

Hunt, C. C. and Paintal, A. S. (1958). Spinal reflex regulation of fusimotor neurones, *J. Physiol., Lond.*, **143**, 195–212

Kemm, R. E. and Westbury, D. R. (1978). Some properties of spinal gamma motoneurones in the cat, determined by micro-electrode recording, *J. Physiol., Lond.*, **282**, 59–71

Lundberg, A., Malgrem, K. and Schomburg, E. D. (1977). Cutaneous facilitation of transmission in reflex pathways from Ib afferents to motoneurones, *J. Physiol., Lond.*, **265**, 763–780

Marsden, C. D., Merton, P. A. and Morton, H. B. (1972). Servo action in human voluntary movement, *Nature, Lond.*, **238**, 140–143

Murthy, K. S. (1978). Vertebrate fusimotor neurones and their influences on motor behaviour, *Progress in Neurobiology*, **11**, 249–307

Noth, J. and Thilmann, A. (1980). Autogenetic excitation of extensor gamma motoneurones by group II muscle afferents in the cat, *Neurosci. Letts.*, **17**, 23–26

Prochazka, A., Westerman, R. A. and Ziccone, S. (1977). Discharges of single hind limb afferents in the freely moving cat, *J. Neurophysiol.*, **39**, 1090–1104

Trott, J. R. (1976). The effect of low amplitude muscle vibration on the discharge of fusimotor neurones in the decerebrate cat, *J. Physiol., Lond.*, **255**, 635–649

Westbury, D. R. (1979). The morphology of four gamma motoneurones of the cat examined by horseradish peroxidase histochemistry, *J. Physiol., Lond.*, **292**, 25P

Reflex activation of dynamic fusimotor neurons by natural stimulation of muscle and joint receptor afferent units

B. APPELBERG*, M. HULLIGER†, H. JOHANSSON* and P. SOJKA*

SUMMARY

The reflex control of fusimotor neurons to triceps surae of the cat has been investigated by natural stimulation of ipsilateral posterior biceps-semitendinosus (PBSt) by tonic stretch, or of the contralateral hindlimb by full extension of its principal joints, or by a combination of these two forms of stimulation. The induced fusimotor activity was indirectly recorded by monitoring changes in responses, during sinusoidal stretching at 1 Hz 1 mm, of muscle spindle primary afferent units from triceps. A method is described which permits, first, the quantitative assessment of the size of the reflex and, second, the characterisation of the induced activity as involving dynamic, static or both types of fusimotor neurons (mixed).

Fusimotor reflexes could be induced in non-spinalised or spinalised preparations with both ipsilateral and contralateral stimulation. Stimulation of the contralateral hindlimb evoked fusimotor reflexes which usually were purely dynamic or mixed. Also, both joint and muscle afferents could elicit such reflex activity when they were selectively activated. Tonic stretch of ipsilateral PBSt provoked fusimotor reflexes to triceps which, however, were less potent and less uniform: in two thirds of the cases they were dynamic or mixed, and in one third they were static. Simultaneous ipsilateral and contralateral stimulation revealed two patterns of reflex interaction: tonic stretch of PBSt could enhance or inhibit the reflex excitation from the contralateral hindlimb. The reflex activation of fusimotor neurons occurred without concomitant activation of skeletomotor units. Thus the concept of rigid linkage of skeletomotor and fusimotor activity seems not to be a general rule as far as dynamic fusimotor neurons are concerned.

*Department of Physiology, University of Umeå
†Brain Research Institute, University of Zürich

INTRODUCTION

The activity of fusimotor neurons is controlled firstly by descending command signals, which may or may not be closely linked to the descending commands to skeletomotor neurons, and secondly by reflexes from peripheral receptors which may operate at a segmental or suprasegmental level.

This chapter considers two important questions concerning the reflex control of fusimotor activity. The first concerns the basic organisation of the peripheral reflex inputs: is the peripheral receptive area, from which individual static and dynamic fusimotor neurons can be excited or inhibited, a small territory consisting, for example, of only a single group of functionally linked muscles? Or, alternatively, are individual neurons influenced from a wide receptive area? Further, do these receptive areas contain only a single or a whole range of receptor types? The second question is: are the reflex pathways, which are known from studies with graded electrical stimulation of peripheral nerves, operative when natural stimulation of the afferent units is used? An analysis of this question was prompted by the recent finding (Appelberg *et al.*, 1977; Appelberg *et al.*, to be published) that gamma motoneurons which were classified as dynamic, using the method of mesencephalic stimulation (cf. Appelberg, 1980), received a strong excitatory input from electrically activated muscle group II (but only rarely group I) fibres from both homonymous and heteronymous muscles.

METHODOLOGICAL ASPECTS

The experiments were performed in cats lightly anaesthetised with Chloralose. A conventional nerve muscle preparation was performed for triceps (or soleus alone) and PBSt, and standard methods of single unit recording from afferents, mechanical stimulation and data analysis were used (for details see Hulliger *et al.*, 1977a; Appelberg *et al.*, 1979; Appelberg *et al.*, to be published). Here, attention is drawn to the particular method of detection and analysis of reflexly induced fusimotor activity. The activity of fusimotor neurons was not directly recorded. Instead the occurrence of such activity was inferred, and its amount estimated, from stimulation-induced changes in dynamic sensitivity of primary spindle afferents from triceps. These changes were then related to the characteristic alterations evoked in primary afferents by electrical stimulation of single static and dynamic fusimotor fibres (Hulliger *et al.*, 1977a,b). Thus the experimental arrangement was as follows: sinusoidal stretching (1 Hz, 1 mm p.t.p.) was applied to ipsilateral triceps at a mean length 2 mm below its physiological maximum (i.e. at the length where the reference data were also taken). The natural stimuli used to elicit fusimotor reflexes were either tonic stretch of the ipsilateral PBSt muscles or full extension of the three principal joints (hip, knee, ankle) of the contralateral hindlimb. The discharge of triceps spindle afferents was recorded in order to detect reflex changes in the activity of the pool of fusimotor neurons supplying these spindles.

Both in the present experiments and with the earlier studies, which provided the reference data, the sinusoidal responses of the spindle afferents were averaged to give cycle histograms to which a simple sinusoid was fitted. Thus, estimates of mean discharge rate (henceforth called *fitted mean*), depth of sinusoidal modulation and phase of the response were obtained (cf. figure 1A). For each of these parameters the difference was then calculated between a test response (during electrical stimulation of fusimotor fibres, as in figure 1, or during their reflex excitation) and a control response (in the absence of fusimotor stimulation, or when no reflex was evoked). For each pair of test and control responses the test-induced changes in modulation (figure 1B) and in phase (figure 1C) were then plotted against the change in fitted mean. In figures 1A and 1B the characteristic action of single static and dynamic fusimotor neurons is illustrated with a set of responses obtained during graded electrical stimulation (data from Hulliger *et al.*, 1977a, replotted). With increasing intensity of dynamic fusimotor stimulation there was a parallel increase in both fitted mean and modulation depth. Since the control values (passive units) varied very little, this gave rise to a progressive increase in both change in fitted mean and change in modulation (▲ in figure 1B). At the same time the test responses lagged increasingly behind the control responses; thus the change in phase became progressively more negative (▲ in figure 1C). During static fusimotor stimulation, large changes in fitted mean were typically accompanied by a sizeable reduction in modulation and thus by negative values for change in modulation (○ in figure 1B). Qualitatively the changes in phase during static stimulation did not differ from dynamic action, since with increasing size of static action the change in phase also became progressively more negative (○ in figure 1C).

It may be noted that the present scatter diagram display of sinusoidal responses is an extension of the scatter display for dynamic index introduced by Crowe and Matthews (1964). In order to obtain a more general measure of static, dynamic, and combined static and dynamic, action on sinusoidal responses, the results of Hulliger *et al.* (1977a,b) were recalculated to provide reference data. The changes in response induced by fusimotor stimulation were averaged for those primary afferent units for which data with both static and dynamic single fibre stimulation, at comparable and fixed intensities, were available. The slopes of the reference lines of figure 1B give the ratio of (the average change in modulation)/(the average change in fitted mean) for 12 primary spindle afferents, each studied under three conditions: with dynamic fusimotor stimulation alone (labelled γ_D), static stimulation alone (labelled γ_S), and combined static and dynamic stimulation at the same intensities (labelled $\gamma_S + \gamma_D$). Correspondingly, the slopes of the lines in figure 1C give the ratios of (average change in phase)/(average change in fitted mean) for the same data. It may be seen that there is a fair agreement between the responses during graded stimulation of individual fibres and the average data from the whole sample of units, the latter obtained with fixed rates of stimulation.

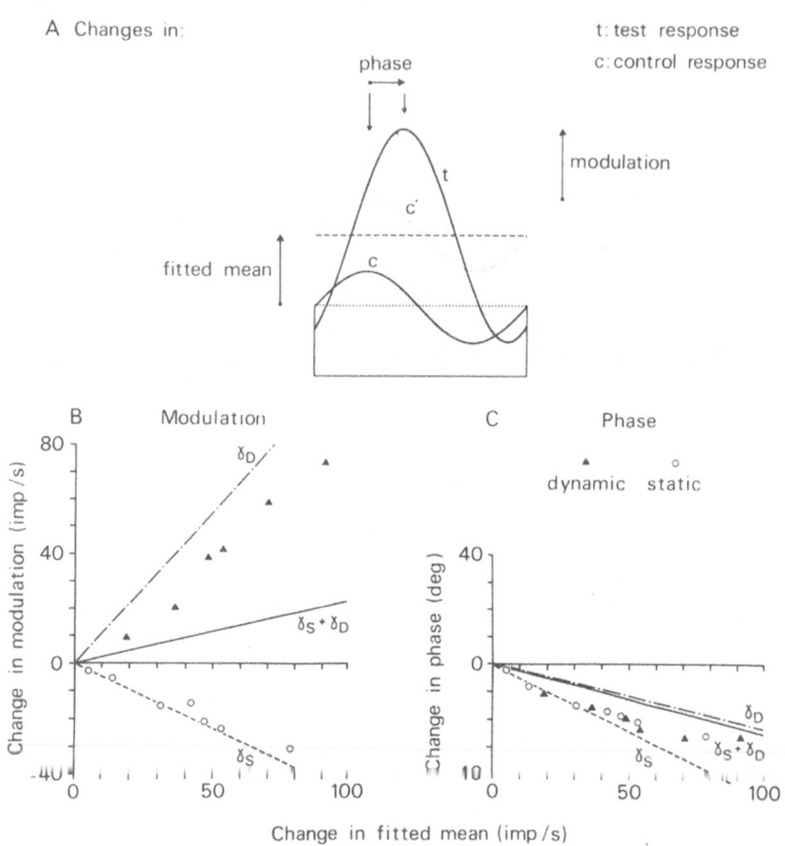

Figure 1. Scatter diagram display of fusimotor action on sinusoidal responses of primary spindle afferents.

In A, a control (c) response and test (t) response to sinusoidal stretching are drawn schematically as simple sinusoids (continuous lines); the dashed horizontal lines indicate the mean level of the response, the amplitude of the curves corresponds to the depth of modulation, and the phase values are 0° (c) and −60° (t). The differences in fitted mean, modulation and phase between test and control are indicated by arrows. For convenience the control response is redrawn (dotted line, c′) at the same mean level as the test response.

In B and C the changes in sinusoidal response (at 1 Hz, 1 mm) of two separate primary afferent units during graded electrical stimulation of a single dynamic (▲) and a single static (○) fusimotor fibre are illustrated in scatter diagrams, as explained in the text (recalculated and replotted data from Hulliger *et al.* 1977a).

For the present reflex studies, alterations in fusimotor activity were analysed using diagrams of the type shown in figure 1. Further support for the classification of reflexly induced fusimotor activity as predominantly dynamic was obtained by considering, in the original cycle histograms, the occurrence of any afferent silence during the release of stretching (cf. Crowe and Matthews, 1964; Hulliger *et al.*, 1977a,b). Finally, the occurrence of gamma rather than beta fusimotor activity was inferred, when alterations in fusimotor activity were not accompanied by any detectable manifestation or increase in surface e.m.g. activity. Normally, surface e.m.g. activity was completely absent during the present experiments (cf. below).

RESULTS

Fusimotor reflexes from the contralateral hindlimb

In figure 2 it is illustrated that mechanical stimulation of the contralateral hindlimb provoked drastic changes in sinusoidal response of primary spindle afferent units from ipsilateral triceps. In the following it is shown that in all likelihood this is attributable to the occurrence of crossed fusimotor reflexes, which predominantly involved dynamic fusimotor neurons and which were operative at the level of the lumbar spinal cord.

Sinusoidal responses of primary spindle afferents from triceps were analysed by the quantitative comparison of test and control responses as already explained for figure 1. Test responses were measured during full extension of the contralateral hindlimb, which was performed manually and maintained throughout the whole period of data collection (usually 10 sec). Control responses were taken when the contralateral hindlimb was in its resting position. For both units of figure 2, large increases in dynamic response—and thus sensitivity—were obtained during contralateral stimulation. Comparison with the reference lines labelled γ_D indicates that the changes induced clearly fall in the range which is characteristic of dynamic fusimotor action. In either case the increase in dynamic sensitivity was accompanied by afferent silence during the release of stretching. The question now arises as to whether the observed effects could have been due to some mechanical coupling between the contralateral and ipsilateral hindlimb. However, two kinds of observations strongly suggest that this was not the case, and that the effects indeed arose from reflex activation of dynamic fusimotor neurons to triceps. First, the size of the induced changes frequently showed considerable variability with successive pairs of test and control responses (as in figure 2A), although the type and intensity of contralateral stimulation were always the same. Second, it was repeatedly observed that the stimulation-induced increase in dynamic sensitivity disappeared as soon as the depth of anaesthesia was increased. This is illustrated in figure 2B by the cluster of points labelled 'deep anaesthesia'; its position around the origin indicates that consistent and large effects could no longer be induced.

Figure 2 also shows that reflex activation of dynamic fusimotor neurons from the contralateral hindlimb could be achieved both in anaesthetised animals with intact spinal cord (figure 2A) and after low spinalisation (in the present experiments always at L3, cf. figure 2B). So far, significant fusimotor reflexes have been found in 32 primary units. The significance was assessed using a paired *t*-test for modulation and/or fitted mean. In animals with intact spinal cords 12/15 units showed predominantly dynamic or mixed fusimotor reflexes. However, in 3/15 units contralateral stimulation evoked a predominantly static fusimotor reflex, as inferred from the scatter diagram analysis (cf. figure 1). In spinalised animals 14/17 units showed apparently pure dynamic fusimotor reflexes, one unit exhibited a mixed reflex and in two units signs of weak inhibition of spontaneous fusimotor activity were detected.

What types of receptors were mediating the crossed fusimotor reflexes described above? Since the innervation of the contralateral hindlimb was largely intact in the majority of preparations, several receptor types were presumably excited by full extension of the limb and might thus have contri-

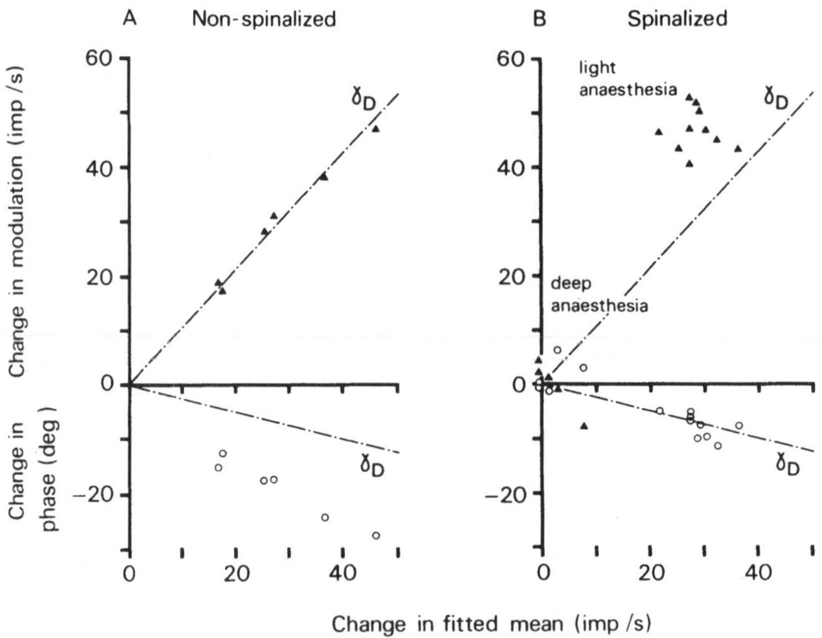

Figure 2. Changes in sinusoidal response (at 1 Hz, 1 mm) of two primary spindle afferents from triceps during full extension of contralateral hindlimb. Same scatter diagram display of change in modulation (▲) and change in phase (○) against change in fitted mean as in figure 1. A, preparation with intact spinal cord; B, animal spinalised at L3 (separate experiment). The measurements during deep anaesthesia were taken after an additional dosage of Mebumal (10 mg/kg).

buted to the fusimotor reflex. In short, evidence has been obtained that joint afferent units from the contralateral knee and ankle joints, and muscle afferents from contralateral extensors, could contribute to the reflex activation of ipsilateral fusimotor neurons. The type of experiments performed to demonstrate selective contribution from individual receptor types is illustrated in figure 3. Sometimes it was apparent that fusimotor reflexes could be

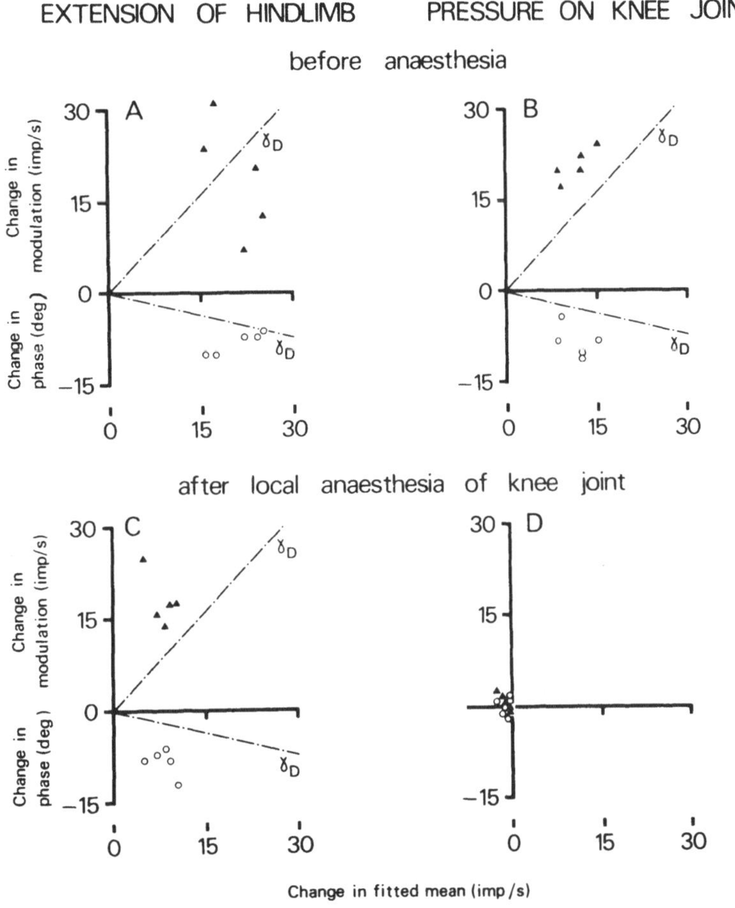

Figure 3. Selective abolition of fusimotor reflex, elicited by pressure applied to the capsule of the contralateral knee joint by local anaesthesia of the joint (intra-articular injection of 20 mg Xylocard). A and B, before local anaesthesia; C and D, after local anaesthesia. A and C, to demonstrate that local anaesthesia of the knee joint only interfered with the fusimotor reflex elicited by stimulation of joint afferent units, but not with the reflex evoked by stimulation of the contralateral limb as a whole.

In all four diagrams, responses of the same primary spindle afferent from triceps during sinusoidal stretching at 1 Hz, 1 mm. For further details, see text.

elicited from confined areas of the contralateral hindlimb, and it seemed possible that such effects were due to the selective excitation of a single type of receptor afferent. In order to confirm this the most likely form of selective natural stimulation was chosen (pressure on joint capsules, stretch or vibration of individual muscles), to demonstrate first that this particular form of stimulation indeed elicited the effect (figure 3B). Then attempts were made to establish whether selective interference with the receptor afferents at issue could abolish the effect. To this end either a particular muscle was denervated, or, as in figure 3D, a given joint was locally anaesthetised (by intra-articular injection of Xylocard) and/or denervated. With the present example it was indeed found that local anaesthesia of the knee completely abolished the reflex provoked by applying pressure to the joint capsule (figure 3D). In order to demonstrate that this was not due to a non-specific loss of reflex responsiveness (caused by removal of some tonic facilitation, or by leakage of the anaesthetic into the circulation), sensitivity tests were performed after the selective interference with the afferents. Thus, for the example of figure 3, the reflex elicited by full extension of the limb was of the same size as before joint anaesthesia (figure 3C compared with figure 3A). However, the qualitative characteristics of the reflex were often altered by the interference procedure (as in figure 3).

Fusimotor reflexes from ipsilateral PBSt muscles

Direct recording from gamma motoneurons to triceps during graded electrical stimulation of peripheral nerves had shown that they received strong excitatory inputs from muscle group II afferents. Moreover, dynamic gamma cells were more frequently excited than static gamma cells (Appelberg *et al.*, 1977; Appelberg *et al.*, in preparation; Noth and Thilmann, 1980; Noth, personal communication). Thus it was investigated as to whether natural stimulation of muscle receptors from PBSt (including spindle group II afferents) could provoke reflex activation of triceps fusimotor neurons. To this end tonic stretch of PBSt within the physiological range of muscle lengths was employed. In contrast to the contralateral reflex studies, the ipsilateral limb was widely denervated, sparing only the nerves to triceps and PBSt. The results may be summarised as follows.

1. Fusimotor reflexes to triceps could indeed be elicited by tonic stretch of PBSt. In two thirds of the cases predominantly dynamic (cf. figure 4A) or mixed (static and dynamic) fusimotor effects were evoked. In the remaining cases a predominantly static fusimotor reflex was provoked.
2. The size of the reflexes from PBSt was considerably smaller than with contralateral stimulation.
3. For a clear manifestation of the reflex, PBSt frequently had to be stretched close to the maximum physiological length.

4. The stretch-induced effects from PBSt were readily abolished by increasing the depth of anaesthesia or by denervation of PBSt.
5. When, contrary to the rule, weak spontaneous e.m.g. activity was present in triceps, it was regularly inhibited by tonic stretch of PBSt, whilst at the same time, dynamic fusimotor neurons were manifestly activated.

Interaction of ipsilateral and contralateral fusimotor reflexes

Since the dynamic fusimotor activity to triceps which was revealed in individual primary spindle afferents could be influenced from both the ipsilateral PBSt muscles and the contralateral hindlimb, an obvious question to ask was, in what way did the two effects interact? In short, two different patterns of reflex interaction were found. First, the ipsilateral and contralateral reflexes were both excitatory, and they summated or facilitated each other. Second, excitation by contralateral stimulation was reduced by stretch of ipsilateral PBSt.

The first type of interaction is illustrated in figure 4. Tonic stretch of PBSt elicited a weak increase in dynamic fusimotor activity in triceps, as revealed in the increase in sinusoidal sensitivity of a primary spindle afferent from triceps (cf. the agreement of the data points of figure 4A with the reference lines of dynamic fusimotor action). Stimulation of the contralateral hindlimb (figure 4B) caused an increase in fusimotor activity, which again appeared to involve mainly dynamic fusimotor neurons, and which was of the same size as with stretch of PBSt. Finally, when the two forms of excitation were combined (figure 4C) the reflex effect was clearly larger than the sum of the individual reflex actions, when the mean size of the effects for the whole series of tests was considered (cf. legend to figure 4). In other primary afferents from triceps, summation of the individual reflexes was found, since their algebraic sum equalled the size of the reflex during combined stimulation.

In other preparations a second type of interaction was seen, when the activation of triceps fusimotor neurons from the contralateral limb was inhibited by tonic stretch of PBSt. Thus a hitherto unknown inhibitory effect of PBSt muscle afferents on triceps dynamic fusimotor neurons was unmasked. So far these two patterns of reflex interaction (excitatory and inhibitory) have not been found operative in one and the same preparation. However, the present sample of interaction cases is still too limited to settle the question of whether simultaneous operation of these two mechanisms is possible.

DISCUSSION

Until recently the investigation of fusimotor reflexes using natural stimulation of peripheral receptors has been largely confined to autogenetic reflex action.

Thus vibration of triceps has been shown to elicit both excitation and inhibition of autogenetic gamma motoneurons (Fromm and Noth, 1976; Trott, 1976; Ellaway and Trott, 1978), and excitation of tendon organ afferents by muscle twitch has been found to cause autogenetic inhibition of gamma efferents (Ellaway *et al.*, 1979; Ellaway and Murphy, 1979). Natural stimulation of cutaneous afferents from various skin regions of the hindlimbs has also

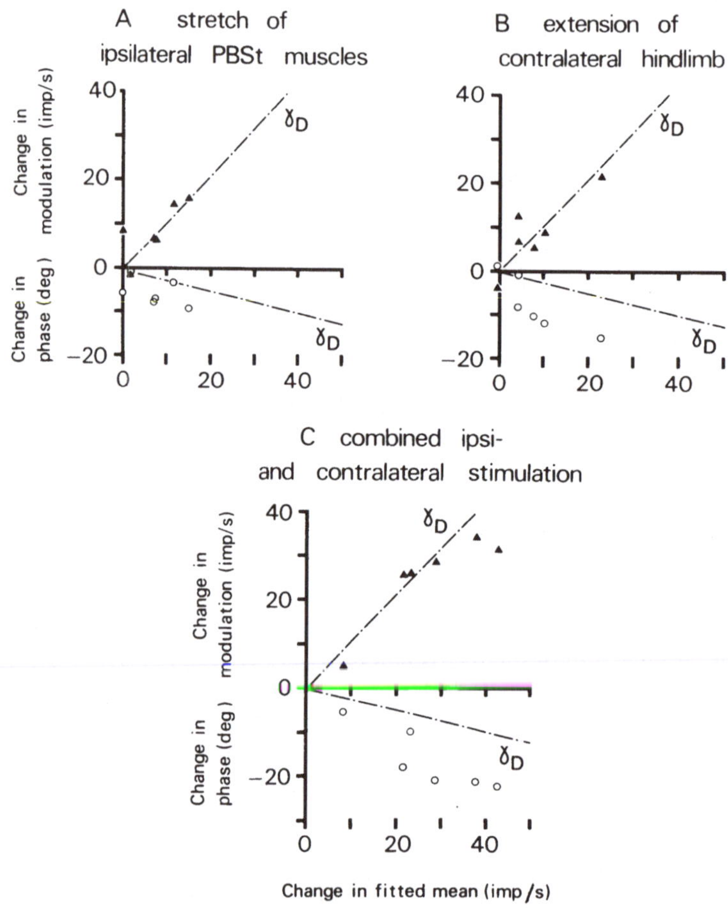

Figure 4. Scatter diagrams to show the facilitation (C) of fusimotor reflexes elicited by stretch of the ipsilateral PBSt muscles (A) and by full extension of the contralateral hindlimb (B). The mean changes in sinusoidal response (at 1 Hz, 1 mm) were, for fitted mean: 7.1 imp/sec (A), 8 imp/sec (B) and 26.9 imp/sec (C); for modulation: 8.5 imp/sec (A), 8.6 imp/sec (B) and 25.1 imp/sec (C); and for phase: −5.5° (A), −7.4° (B), and −16.2° (C). Note that for all three parameters the values of C are considerably larger than the sum of A and B. In all three diagrams, responses of the same primary spindle afferent from triceps. Non-spinalised preparation.

been found to cause either excitation or inhibition of gamma motoneurons to triceps (Eldred and Hagbarth, 1954; Grillner, 1969). All these studies were, however, limited by the fact that the gamma fibres could not reliably be classified as static or dynamic fusimotor efferents.

In the present experiments an indirect method of monitoring fusimotor reflexes has been employed. It had, however, the advantage of permitting a classification of effects as being predominantly dynamic, predominantly static, or mixed static and dynamic. This was achieved by a quantitative comparison between the reflex action on spindle dynamic sensitivity and reference data obtained with controlled electrical stimulation of identified fusimotor fibres (Hulliger *et al.*, 1977a,b). Using this method, a wide peripheral territory from which fusimotor reflexes could be elicited with natural stimulation, has been explored. The results hitherto obtained indicate that for dynamic fusimotor neurons to triceps, this territory is of considerable size and complexity of organisation. From stretch receptors of ipsilateral flexor muscles both excitatory and inhibitory fusimotor reflexes could be elicited, the former confirming the findings of Appelberg *et al.* (1977) with recordings from gamma motoneurons during electrical stimulation of peripheral nerves. Further excitatory reflexes could be evoked by excitation of joint receptors of the contralateral knee and ankle joints, and excitatory inputs were also provided by stretch receptors from contralateral extensor muscles.

At present very little is known about the organisation of the pathways mediating this wide range of fusimotor reflexes to triceps fusimotor neurons. However, the occurrences of both excitatory and inhibitory reflex action with identical stimulation (such as stretch of triceps or PBSt) suggest that parallel reflex pathways must exist. Such an organisation could provide the CNS with a considerable range of options for selective descending control of the excitatory and inhibitory influences converging from the peripheral receptive area on to different pools of fusimotor neurons.

The findings with electrical stimulation of ipsilateral joint, muscle and skin nerves (Appelberg *et al.*, 1977; Appelberg *et al.*, in preparation) and the present results with natural stimulation, suggest that the reflex control of dynamic fusimotor neurons may be quite independent of static fusimotor neurons and skeletomotor neurons. It is particularly noteworthy that in the present experiments powerful and apparently selective activation of triceps dynamic fusimotor neurons could be elicited (especially from the contralateral hindlimb) while static fusimotor neurons appeared not to be influenced, and while triceps was electromyographically completely silent. Indeed, on a few occasions, weak spontaneous e.m.g. activity was even seen to be inhibited by stretch of PBSt at the same time as there was a clear activation of dynamic fusimotor neurons. Thus, in this situation the dominant reflex actions on to the pools of alpha and fusimotor neurons, which were monitored, went in opposite directions. In other words, there was no sign of any rigid linkage between alpha and dynamic fusimotor activity. Thus such linkage or coactivation of skeletomotor and fusimotor activity, if it is at all a general rule, might be restricted to skeletomotor and static fusimotor neurons. This then would leave the CNS considerably more freedom for independent control of the dynamic sensitivity in proprioceptive feedback.

ACKNOWLEDGEMENTS

This work was supported by the Swedish Medical Research Council, Project No. 03873, by Gunvor and Josef Anér Stiftelse, and by the Swiss National Foundation (M.H. Grant No. 831.445.76). We wish to thank Mrs Gerdy Kriström and Mr Göran Westling for valuable technical assistance.

Received on June 9th, 1980.

REFERENCES

Appelberg, B. (1980). Selective central control of dynamic gamma motoneurons utilised for the functional classification of gamma cells, this publication

Appelberg, B., Hulliger, M., Johansson, H. and Sojka, P. (1979). Excitation of dynamic fusimotor neurones of the cat triceps surae by contralateral joint afferents, *Brain Res.*, **160**, 529–532

Appelberg, B., Johansson, H. and Kalistratov, G. (1977). The influence of group II muscle afferents and low threshold skin afferents on dynamic fusimotor neurones to the triceps surae of the cat, *Brain Res.*, **132**, 153–158

Crowe, A. and Matthews, P. B. C. (1964). Further studies of static and dynamic fusimotor fibres, *J. Physiol., Lond.*, **174**, 132–151

Eldred, E. and Hagbarth, K. -E. (1954). Facilitation and inhibition of gamma efferents by stimulation of certain skin areas, *J. Neurophysiol.*, **17**, 59–65

Ellaway, P. H. and Murphy, P. R. (1979). Inhibition of gamma motoneurones by contraction of flexor digitorum longus and triceps surae in decerebrated, spinally transected cats, *J. Physiol., Lond.*, **292**, 23–24P

Ellaway, P. H., Murphy, P. R. and Trott, J. R. (1979). Inhibition of gamma motoneurone discharge by contraction of the homonymous muscle in the decerebrated cat, *J. Physiol., Lond.*, **291**, 425–441

Ellaway, P. H. and Trott, J. R. (1978). Autogenetic reflex action on to gamma motoneurones by stretch of triceps surae in the decerebrated cat, *J. Physiol., Lond.*, **276**, 49–66

Fromm, C. and Noth, J. (1976). Reflex responses of gamma motoneurones to vibration of the muscle they innervate, *J. Physiol., Lond.*, **256**, 117–136

Grillner, S. (1969). The influence of DOPA on the static and the dynamic fusimotor activity to the triceps surae of the spinal cat, *Acta physiol. scand.*, **77**, 490–509

Hulliger, M., Matthews, P. B. C. and Noth J. (1977a). Static and dynamic fusimotor action on the response of Ia fibres to low frequency sinusoidal stretching of widely ranging amplitudes, *J. Physiol., Lond.*, **267**, 811–838

Hulliger, M., Matthews, P. B. C. and Noth, J. (1977b). Effects of combining static and dynamic fusimotor stimulation on the response of muscle spindle primary endings to sinusoidal stretching, *J. Physiol., Lond.*, **267**, 839–856

Noth, J. and Thilman, A. (1980). Autogenetic excitation of extensor motoneurones by group II muscle afferents in the cat, *Neurosci. Lett.*, **17**, 23–26

Trott, J. R. (1976). The effect of low amplitude vibration on the discharge of fusimotor neurones in the decerebrate cat, *J. Physiol., Lond.*, **255**, 635–649

Functional roles of fusimotor and skeletofusimotor neurons studied in the decerebrate cat

W. Z. RYMER*, E. M. POST† AND F. R. EDWARDS‡

SUMMARY

Although the existence of skeletofusimotor innervation of muscle spindles in mammalian muscles is now amply confirmed, the physiological effects of this innervation have not been established. In an attempt to determine such effects, we recorded the discharge of triceps surae primary and secondary spindle receptor afferents, isolated from small dorsal root fascicles in decerebrate cat preparations. Afferent discharge was recorded in quiescent states, and during reflex excitation of the receptor-bearing muscle.

Mechanical or electrical stimulation of contralateral skin or deep tissues consistently evoked increased discharge in soleus and medial gastrocnemius (MG) afferents. During stimulation, afferent discharge rates were reliably observed to exceed those recorded in quiescent states, either when the receptor-bearing muscle was held at constant length, or during muscle stretch and release. These rate increases arose at stimulus intensities insufficient to provoke an EMG response, and the patterns of response were consistent with the activation of both static and dynamic fusimotor neurons. We argue that the absence of EMG activity or active force implies that neither skeletomotor nor skeletofusimotor neuronal activity could have been responsible for the initial rate increases. It follows that fusimotor neurons were responsible, and that they were activated at stimulus intensities insufficient to drive skeletomotor (or skeletofusimotor) neurons.

Approximately 50% of primary and secondary afferents showed some further rate increase with increasing force and in 3/49 endings this discharge consistently increased in a very abrupt manner, as if a new efferent fibre were being recruited. Recordings from MG fusimotor (gamma) fibres confirmed

* Department of Physiology, Northwestern University Medical School, Chicago
† Department of Neurology, Northwestern University Medical School, Chicago
‡ Department of Neurosurgery, State University of New York

that an early increase in discharge rate occurred at stimulus intensities insufficient to activate skeletomotor fibres, and showed that the discharge rate appeared to saturate during stimulation sufficient to activate the muscle. We obtained no evidence supporting continuing recruitment of fusimotor neurons at high force levels.

We propose that increases in spindle receptor discharge arising above the extrafusal threshold are mediated predominantly by skeletofusimotor neurons. More direct support for this hypothesis is now being sought with the aid of spike-triggered averaging techniques, which may permit us to identify the existence of skeletofusimotor innervation under more physiological conditions.

INTRODUCTION

Until quite recently, it was thought that the regulatory actions of the central nervous system on mammalian muscle spindle receptors were mediated almost exclusively by fusimotor neurons. With the recent rapid growth of information supporting the existence of skeletofusimotor (or beta) innervation of mammalian muscle spindles (Bessou *et al.*, 1963; Bessou *et al.*, 1965; Emonet-Dénand and Laporte, 1975; Emonet-Dénand *et al.*, 1975; McWilliam, 1975; Laporte and Emonet-Dénand, 1976; Harker *et al.*, 1977), this position is now clearly untenable. Instead, we are confronted with a seemingly complex control problem in which gamma dynamic, gamma static, beta dynamic and beta static fibres are each potentially capable of influencing muscle spindles. The purpose of this paper is to report briefly on particular features of spindle receptor responses that may be attributed to skeletofusimotor activity, and to advance an hypothesis which may help simplify our thinking about the control of efferent innervation of the mammalian muscle spindle.

METHODS

These methods have been described in detail in recent publications (Rymer *et al.*, 1979; Rymer and Hasan, 1980; Post *et al.*, 1980) and will be summarised only briefly here. Experiments were performed on 40 cats which were anaesthetised with nitrous oxide, oxygen and fluothane, then decerebrated by transection through, or immediately above, the superior colliculus. A small laminectomy, confined to the L6 vertebra, was performed. The L7 spinous process was bared and used to attach a mobile platinum electrode, via a small metal block. A dissection of the left hind limb was performed, and all of the nerves sectioned, except for those innervating the triceps surae muscles. Soleus and medial gastrocnemius (MG) muscle tendons were then separated from the others in the triceps group, detached from the calcaneum, and then clamped to a muscle stretcher which was capable of imposing stretch amplitudes up to 8 mm, over a wide range of velocities. This stretcher also incorporated force and length transducers. The electromyogram (EMG) was measured using intramuscular platinum–iridium or Nichrome wires of 0.002–0.005 in in diameter. EMG signals were amplified, rectified and low-pass filtered.

One hundred and five muscle spindle receptor afferents were isolated from small natural fascicles of the L7 or S1 dorsal roots. The remaining dorsal roots were left substantially intact, preserving reflex activity in the hind limb muscles. In eight experiments, 61 fusimotor fibres were teased from small natural fascicles of the medial gastrocnemius nerve. These fibres were identified by their conduction velocity (see Post *et al.*, 1980) and by the presence of spontaneous discharge.

RESULTS

Range of fusimotor action

It is now evident from a wide variety of human (e.g. Hagbarth and Vallbo, 1968; Vallbo, 1974) and animal studies (e.g. Cody *et al.*, 1975; Prochazka *et al.*, 1976; Loeb *et al.*, 1977; Loeb and Duysens, 1979; Goodwin *et al.*, 1978) that coactivation of skeletomotor and fusimotor neurons is common; this association is usually referred to as 'alpha–gamma coactivation' or 'alpha–gamma linkage' (Granit, 1970). Less well defined are the relative thresholds of these types of neurons to different forms of excitation and the relative dynamic range over which each class of neurons may be modulated. Two apparently contradictory positions exist. On the one hand, earlier studies in reduced animal preparations (Eldred *et al.*, 1953; Wuerker and Henneman, 1963) have shown that fusimotor neurons are activated reflexively by stimuli insufficient to activate skeletomotor (alpha) motoneurons. On the other hand, a number of recent studies reporting on discharge of muscle spindle afferents in intact human subjects (Burke *et al.*, 1978; Hagbarth *et al.*, 1975) have shown that during voluntary activation, spindle afferent discharge increases frequently only after skeletomotor activity has been initiated. While it could be argued that the results reported in intact human subjects are more trustworthy than those obtained in reduced preparations (where non-physiological circumstances might prevail) we will attempt to show that the results in the decerebrate are also relevant, and that the two types of observations may well be compatible.

In our own studies (Post *et al.*, 1980) we have reassessed the effects of increasing cutaneous, muscular and nociceptive stimulation of the contralateral limb on the reflex responses of isometric soleus and MG motor units and muscle spindle afferents. For more than 65% (67/105) of spindle afferents examined during contralateral limb stimulation, the increase in afferent discharge which took place arose at lower stimulus intensities than were required to activate alpha motor neurons. In the remaining afferents (38/105), the contralateral limb stimulation produced either no change in afferent discharge, or simultaneous activation of both EMG and spindle receptor afferents; no case of prior EMG activation was recorded. An important additional finding was that the increase in spindle receptor discharge did not continue much beyond extrafusal threshold. Instead, most receptors tended to show essentially constant discharge once electromyographic activity was discernible, although some receptors showed a modest and apparently separate rate acceleration at higher forces (*see* later results). Since the initial increase in spindle receptor afferent discharge took place in the absence of demonstrable EMG activity, it was probably caused by fusimotor discharge (*see* later results for justification).

The finding of lower thresholds for reflex excitation of fusimotor neurons has also been supported by a study in which the responses to stretch were examined at different levels of fusimotor activation (Houk *et al.*, to be published). Figure 1, which is taken from this study, illustrates the range of variation in the dynamic response that was observed for one primary ending in the course of reflexively induced variations in central excitatory state. The

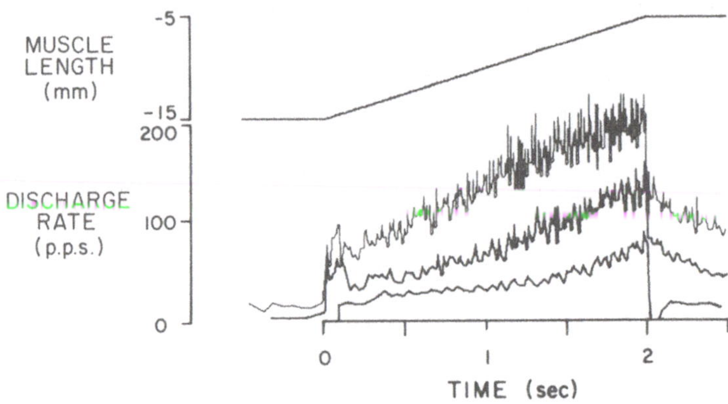

Figure 1. Variations in the size of the dynamic response of a soleus spindle primary ending during reflexively induced changes in spinal excitatory state. The figure depicts three single records of discharge rate, superimposed to facilitate comparison. In each case, the soleus was stretched 10 mm, at a velocity of 5 mm/sec. The lowermost response was recorded in the absence of contralateral stimulation, while the upper two were recorded during increasing stimulation of contralateral skin and deep tissues. No EMG responses accompanied any of the three rate responses shown. This figure is derived from experiments performed in collaboration with Drs J. C. Houk and P. E. Crago.

figure shows that these variations consisted of a systematic increase in both the slope of the response and the size of the initial step increase in discharge. Furthermore, although not documented in the figure, the full range of increase in the size of the dynamic response was traversed before any substantial increase in either the EMG or active force was recorded, implying again that fusimotor neurons were probably responsible. Overall, this figure typifies our general finding which is that most of the apparent range of fusimotor modulation took place in response to levels of sensory stimulation insufficient to activate extrafusal muscle.

The illustrated changes in dynamic response closely resemble the effects induced by electrical stimulation of single gamma dynamic fibres (Crowe and Matthews, 1964). Since it has been clearly established that muscle spindles do not receive dedicated innervation from large diameter efferent fibres (Ellaway *et al.*, 1972), and since the activation of skeletofusimotor neurons would have resulted in the excitation of extrafusal muscle, the absence of EMG or active force provides strong indirect support for a purely fusimotor effect. Based on the features of the dynamic response, it would appear then that gamma dynamic neurons were recruited at low levels of spinal excitation, and induced to traverse their full range of modulation before extrafusal threshold was reached. This limit to the range of rate modulation of the spindle afferent could reflect rate saturation or force-related inhibition of gamma fibres; however, other explanations are certainly feasible (*see* Discussion).

The changes in dynamic response depicted in figure 1 were also accompanied by increases in both the initial discharge rate and in the level of post-ramp discharge. However, in contrast with the features of the dynamic response, these changes are not especially helpful in diagnosing the nature of the fusimotor effects, since they could have been induced by either static or dynamic fusimotor activity. The response of spindle receptors to muscle shortening proved more useful, in that a modest or absent decrease in rate during shortening is a good index of static fusimotor effect (Matthews, 1972). We found that prominent increases in afferent discharge rate recorded under isometric conditions were associated with modest or absent rate reductions during release, implying that static fusimotor innervation was responsible for much of the initial rate acceleration described in the previous section.

It is also of some interest to assess the effects of fusimotor activity in relation to the description of spindle receptor dynamic response presented in an earlier paper in this book (Houk *et al.*, 1981). A useful equation, similar to equation 2 in the cited paper, is

$$r - r_o = \mathrm{K}(x)v^n$$

where r is the discharge rate, r_o is the initial discharge, $\mathrm{K}(x)$ is a function of length, v is stretch velocity and n is the exponent of response. The changes depicted in figure 1, which amount (approximately) to a scaled increase in the magnitude of the dynamic response, could be induced by a change in the value of the scaling function K(x), by an increase in the value of n, or by some combination of these changes. In fact, measurement of K(x) and n values during varying fusimotor states, as well as before and after ventral root section, showed that changes were confined almost completely to the scaling function K(x).

One final implication of these observations warrants emphasis. If the full operating range of fusimotor activation has in fact been traversed by the time that skeletomotor threshold is reached, then any additional rate increases recorded during contralateral stimulation sufficient to induce signficant additional force output is attributable to skeletofusimotor (or beta) activity.

Influence of series compliance

While our observations of lower thresholds for activation of fusimotor neurons would appear to be consistent with previous studies in decerebrate cat preparations (Eldred *et al.*, 1953; Wuerker and Henneman, 1963), these data remain to be reconciled with the recordings obtained in human subjects, in which skeletomotor activity is reported to commence before any change in afferent discharge rate occurs (Burke *et al.*, 1978). Although these differences could well arise from differences in fusimotor responsiveness of the human versus the decerebrate cat preparation, in most studies performed in intact human subjects the presence of an intervening soft-tissue interface probably prevents muscle contraction from being genuinely isometric. The existence of a compliant connection could have influenced the pattern of EMG and rate increase.

Accordingly, the influence of series compliance was tested in the decerebrate cat model by simulating a small compliance in series with the muscle. The compliance was simulated electronically by configuring the muscle stretcher as a slightly compliant spring. Our findings were that, in the presence of this springy connection, the order of stimulus-induced activation of muscle spindle receptors and of the muscle itself was usually reversed (12/14 primary endings), with the onset of the EMG activity now preceding the increase in receptor discharge. However, in circumstances where contralateral stimulation could be graded so as to produce very gradual increases in spinal excitability, spindle receptor acceleration often arose before EMG onset. It should be pointed out that an explanation based upon the effects of compliance would hold only if the onset of activity in skeletomotor and fusimotor neurons was sufficiently close in time so as to permit the more rapid development of extrafusal activity to mask changes in the onset of intrafusal contraction. This temporal proximity has been shown by Vallbo (1971). In sum, our results using compliant connections demonstrate that under some circumstances the effects of intrafusal activation can be masked by internal muscle shortening.

The other major consideration in relation to the observations in human subjects is that increases in spindle receptor discharge arising well above skeletomotor threshold could have been induced by skeletofusimotor activity. Possible skeletofusimotor contributions to spindle receptor output will be addressed in the next section.

Spindle receptor responses during increasing skeletomotor activation

To this point, we have ascribed changes in the response of spindle receptors to increases in fusimotor discharge rate on the presumption that activation of skeletofusimotor neurons would follow rules similar to those described for skeletomotor (i.e. alpha) motoneurons. In other words, our presumption has been that skeletofusimotor neurons differ from ordinary motoneurons solely in that the skeletomotor axon provides an additional branch to the muscle spindle. The evidence for this proposition is not yet very broadly based, and is partly indirect in nature. For example, we do know that pure large fibre innervation of intrafusal fibres does not exist (Ellaway *et al.*, 1972), and that the anatomical features of the collaterally innervated motor units are quite comparable to other motor units in the muscle. It would seem likely that the patterns of afferent projections to skeletofusimotor neurons are probably also comparable to those of skeletomotor neurons. However, the only direct evidence regarding central synaptic connections of identified skeletofusimotor neurons is that of Burke and Tsairis (1977) who report one example of a soleus skeletofusimotor neuron that received typical synaptic projections from homonymous Ia afferents. In summary, although it is likely that skeletofusimotor activation parallels skeletomotor activation quite closely, the patterns of synaptic activation of skeletofusimotor neurons remain to be established.

In an attempt to find components of response attributable to skeletofusimotor activity, we examined the responses of spindle receptor afferents as isometric muscle force was increased by graded stimulation of cutaneous or deep tissues in the contralateral limb. Approximately 50% (56/105) of the spindle receptors examined showed either reductions in discharge rate with increasing force or saturation of discharge, but the remainder displayed increases in discharge above skeletomotor threshold. In most of the latter endings, the increase in discharge was not simply a continuation of the initial rate increase; rather, there was often a region of intervening constant discharge rate even above skeletomotor threshold. In a small number of receptors (8/49), increasing discharge was recorded at quite high force levels, approximating to 30–50% of maximum soleus force output, and in three instances this increase in discharge was very abrupt in nature. An example of such an abrupt increase is depicted in figure 2. The abruptness of rate change suggests that a new spindle efferent fibre was recruited, although it could have been of either fusimotor or skeletofusimotor type. While the occurrence of this sharp rate increase was consistently related to the level of isometric force, this consistency is of itself insufficient to separate a fusimotor or skeletofusimotor mechanism. This is because the recruitment of either fusimotor neurons or skeletofusimotor neurons could take place in some consistent rank order in relation to the level of spinal excitation (cf. Burke *et al.*, 1978).

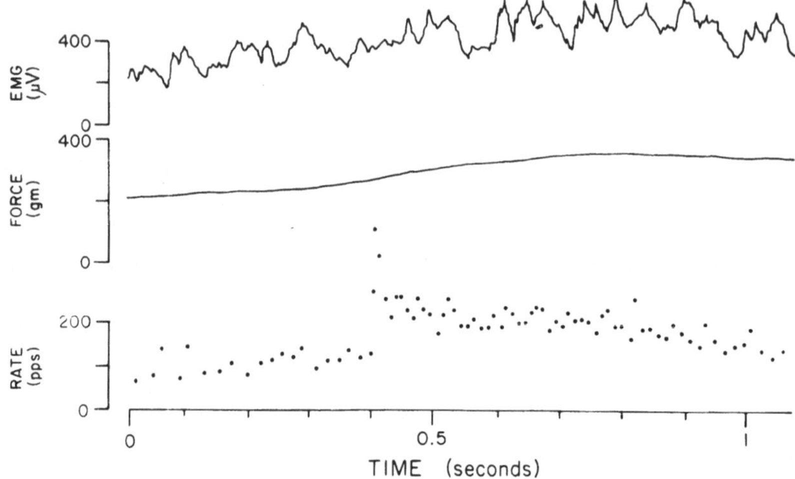

Figure 2. Abrupt increase in discharge of a soleus primary ending during reflexively induced increase in isometric force. Force increase was induced by stimulation of skin and foot-pad of contralateral limb. For this receptor, the rate discontinuity occurred regularly at the same force level.

In order to dissociate fusimotor from skeletofusimotor effects we examined the responses of fusimotor efferent fibres (i.e. gamma fibres) isolated from the peripheral MG nerve (*see* Methods). Sixty-one fibres were isolated from eight decerebrate cats. The response of fusimotor fibres to contralateral limb stimulation was in keeping with the predictions made on the basis of spindle afferent recordings, in that these fibres were activated at lower levels of spinal excitation than were required to activate skeletomotor neurons. Furthermore, these fusimotor fibres did not increase their discharge substantially once the threshold for muscle activation was reached. Most significantly, no fresh fusimotor fibres were seen to be activated beyond the skeletomotor threshold, although such recruitment could conceivably have been obscured by the electrical effect of coincident skeletomotor activation. Because of this technical limitation, the evidence favouring an *exclusive* skeletofusimotor mechanism for the force-related increases in spindle receptor discharge rate is not irrefutable.

Identification of skeletofusimotor activity in physiologically activated muscle

Given the aforementioned limitations of indirect approaches towards the recognition of skeletofusimotor action, a more direct line of experimentation is obviously to be preferred.

One possibility is to make use of the fact that skeletofusimotor neurons jointly innervate intrafusal and extrafusal muscle fibres. As a result of this anatomical arrangement, a descending impulse in a skeletofusimotor axon would be expected to give rise to joint excitation of skeletal muscle and of the muscle spindle. The resulting increase in tension of the intrafusal muscle fibre should then heighten the probability of discharge of the spindle receptor. If we now use the discharge of a receptor afferent derived from such a spindle to provide a synchronisation point for an ensemble average (i.e. a spike-triggered average) of EMG and force changes occurring in the time period spanning the occurrence of each spike, then the existence of skeletofusimotor activity would be expected to induce a distinct correlation between spindle receptor afferent spikes and events in the muscle.

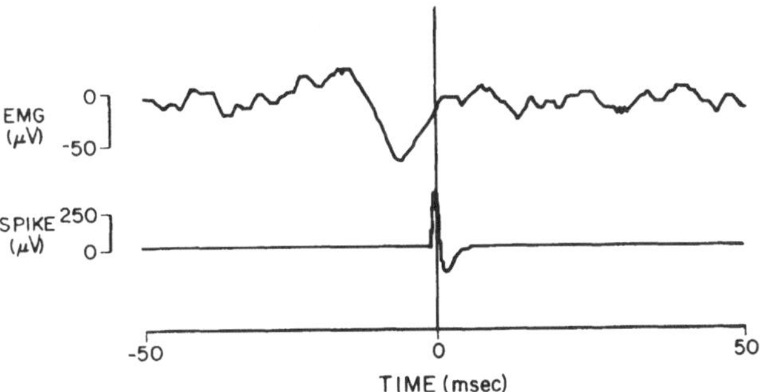

Figure 3. Spike-triggered average (STA) of soleus EMG during reflex excitation of the isometric soleus muscle. Primary spindle afferent discharge was used to trigger the average of ongoing EMG activity, arising from 50 msec before, to 50 msec after, each spike. The average was cumulated on a PDP LSI 11 computer over 800 sweeps. EMG activity was collected with 0.005 in platinum wires, amplified (× 1000), and band pass filtered (8–3000 Hz). Muscle length was 3 mm short of physiological maximum. Upper trace shows correlated EMG wave, commencing about 10 msec before the spike.

In five experiments, we recorded the discharge of 20 primary endings and four secondary endings during reflexively induced variations in isometric muscle force. EMG was recorded with extensively bared intramuscular wire electrodes, and the resulting signals amplified and band-pass filtered. In eight of 20 primary endings we recorded a correlated increase in EMG arising some 5 to 10 msec before the occurrence of the afferent spike; an example of such a correlation is shown in figure 3. This 5–10 msec latency is roughly what would be predicted as an outcome of skeletofusimotor action, because the time taken to induce an EMG response in extrafusal muscle fibres should be less than that required to change intrafusal tension, augment spindle receptor afferent discharge and propagate the impulse to the dorsal root. However, it is also evident from figure 3 that the EMG increase is rather smeared, and its duration is considerably longer than that that would have been expected from an intramuscular differential recording of unitary EMG activity. The reason for this smearing of the EMG is probably that the change in tension of intrafusal fibre induced by skeletofusimotor activity may have been rather gradual, so that the increase in firing probability of the spindle afferent would not be confined to a very narrow time window in relation to extrafusal events. In the circumstances of our experiment, in which the spindle afferent spike is used to drive the average, the temporal relation between EMG and the related afferent discharge might be somewhat variable; hence the slurred EMG potential.

It has been reported recently that many skeletofusimotor neurons are associated with slow twitch muscle fibres, and exert a gamma dynamic type of effect (Bessou *et al.*, 1965: Emonet-Dénand and Laporte, 1975). Since these motoneurons are probably recruited at low levels of excitation, it might be anticipated that the first demonstrable effect of increasing skeletofusimotor activation would be to induce a dynamic type of primary spindle afferent response pattern, developing above the extrafusal threshold. Figure 4 illustrates the changes in the magnitude of the dynamic response that arose during variations in spinal excitation for the same ending as in figure 3. Here are superimposed two response records of force, EMG and discharge rate, one collected without any contralateral stimulation, the other during contralateral stimulation sufficient to induce a substantial increase in initial EMG level. The afferent response to stretch is displayed in the second trace in the form of instantaneous discharge rate, and it shows that the response recorded at the higher force level increased abruptly above the other rate trace, after about 1 mm of stretch. This change in discharge pattern arising during the dynamic phase pattern of activation is unlikely to have been induced by an increase in fusimotor activity, since fusimotor neurons are not strongly excited by muscle stretch (Ellaway and Trott, 1978). It is, on the other hand, exactly what would be anticipated were a dynamic skeletofusimotor neuron to be recruited as a result of stretch-induced muscle afferent excitation.

Although our results with the spike-triggered average technique are still preliminary, it seems possible that the existence of skeletofusimotor activity might well be recognised by the finding of correlated force and EMG response patterns in the spike-triggered average, together with characteristic modifications in spindle afferent response to stretch and release during skeletomotor activation.

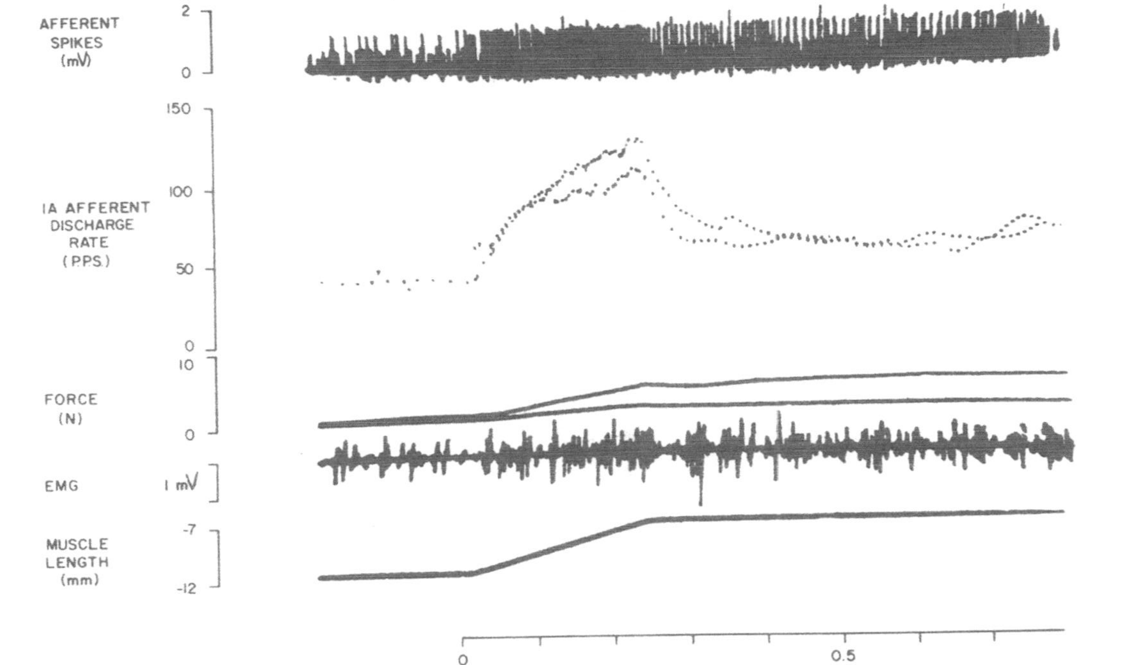

Figure 4. Variations in the dynamic response of soleus spindle primary ending in the course of increasing skeletomotor discharge. Five millimetre stretches were imposed on the muscle at two different levels of excitation, induced by cutaneous stimulation of the contralateral limb. The dynamic response recorded at the higher level of excitation shows a marked upward rate deviation, developing early in the course of the stretch.

DISCUSSION

In the results described, we have concentrated upon the peripheral effects of skeletofusimotor and fusimotor neurons on muscle spindles rather than on the central consequences of skeletofusimotor effects on muscle spindles. The functional consequences of skeletofusimotor connections have been explored theoretically (Houk, 1972), but have been subjected to little direct study. Because of the described limitations in the range of rate modulation of fusimotor neurons, it seems possible that the effects of fusimotor discharge would not be able to compensate adequately for the internal shortening of muscle that accompanies increasing skeletomotor activity. Such shortening would reduce potentially the muscle spindle receptor responsiveness, especially under isometric conditions. In this context, one possible outcome of skeletofusimotor activity might be to maintain the tension on the muscle spindle in circumstances where fusimotor activation is no longer capable of performing this function.

In regard to the apparent saturation of the fusimotor discharge rate, our finding that reflexively induced increases in fusimotor discharge did not continue above skeletomotor threshold could have arisen because the maximum possible discharge rate of the neurons was reached, or because the rate may have been limited by tendon organ mediated inhibition of fusimotor neurons (Ellaway *et al.*, 1979; Ellaway and Murphy, 1980). We are presently unable to distinguish between these possibilities.

Some possible contributions of soft-tissue compliance to the results reported in human subjects have been explored in an earlier section. However, a number of additional factors may have contributed to the findings that EMG activation precedes increases in spindle receptor discharge in apparently isometric human muscles. For example, if a synergist is activated before the receptor-bearing muscle, then spindle unloading resulting from soft-tissue compression may be accentuated. Moreover, if force outputs of several synergists are not precisely matched, then unloading may still arise in one or more muscles, even though all the relevant muscles are contracting. Because of the effects of these compliances, a concurrent change in fusimotor or skeletofusimotor activity would not necessarily induce a change in equatorial tension of the muscle spindle, and spindle receptor discharge might not then increase during increasing efferent excitation. A second factor relates to the times required to activate extrafusal versus intrafusal muscle fibres. For example, following simultaneous spinal activation of skeletomotor and fusimotor neurons, extrafusal muscle fibre contraction proceeds more rapidly than does intrafusal contraction, because of the briefer conduction time required in the larger skeletomotor efferent fibres (cf. Vallbo, 1971). Moreover, an alteration in spindle receptor discharge can only arise after a change in muscle spindle tension has been transduced. For these reasons, it is conceivable that the lower thresholds for activation of fusimotor neurons are sometimes obscured by the combination of compliance effects and time delays described above.

These results suggest that a simplified approach towards the roles of fusimotor and skeletofusimotor innervation is now possible. First, it appears that reflex activation of fusimotor neurons (both static and dynamic) often occurs at lower stimulus intensities than are required to activate skeletomotor neurons. Moreover, our recordings suggest that both types of fusimotor neurons have traversed most of the available dynamic range of rate modulation by the time significant skeletomotor activity is initiated. These threshold relations appear to hold regardless of the form of stimulation used, and they are also evident during spontaneous changes in excitatory state.

Although our results are derived exclusively from the decerebrate preparation, we would suggest that the patterns of control of fusimotor and skeletomotor neurons may well be similar to those seen in intact preparations. While our general scheme must remain quite tentative at this stage, it has the advantage that the need for complex hypotheses requiring independent control of static and dynamic fusimotor neurons is minimised. Furthermore, since it is likely that the rules for recruitment and rate modulation of skeletofusimotor neurons are identical with those regulating skeletomotor neurons (where the 'size principle' is broadly applicable), the options for independent control of skeletofusimotor innervation are also eliminated. In sum, the activation of fusimotor and skeletofusimotor neurons may be inexorably tied to the activation of the motoneuron pool as a whole.

Received on June 9th, 1980.

REFERENCES

Bessou, P., Emonet-Dénand, F. and Laporte, Y. (1963). Occurrence of intrafusal muscle fibre innervation by branches of slow motor fibres in the cat, *Nature, Lond.*, **198**, 594–595

Bessou, P., Emonet-Dénand, F. and Laporte, Y. (1965). Motor fibres innervating extrafusal and intrafusal muscle fibres in the cat, *J. Physiol., Lond.*, **180**, 649–672

Burke, D., Hagbarth, K. E. and Skuse, N. F. (1978). Recruitment order of human spindle endings in isometric voluntary contractions, *J. Physiol., Lond.*, **285**, 101–112

Burke, R. E. and Tsairis, P. (1977). Histochemical and physiological profile of a skeletofusimotor (β) unit in cat soleus muscle, *Brain Res.*, **129**, 341–345

Cody, F. W. J., Harrison, L. M. and Taylor, A. (1975). Analysis of activity of muscle spindles of the jaw closing muscles during normal movements in the cat, *J. Physiol., Lond.*, **253**, 565–582

Crowe, A. and Matthews, P. B. C. (1964). The effects of stimulation of static and dynamic fusimotor fibres on the response to stretching of the primary endings of muscle spindles, *J. Physiol., Lond.*, **174**, 109–131

Eldred, E., Granit, R. and Merton, P. A. (1953). Supraspinal control of the muscle spindles and its significance, *J. Physiol.*, **122**, 498–523

Ellaway, P. H., Emonet-Dénand, F., Joffroy, M. and Laporte, Y. (1972). Lack of exclusively fusimotor alpha-axons in flexor and extensor leg muscles of the cat, *J. Neurophysiol.*, **35**, 149–153

Ellaway, P. H. and Murphy, P. R. (1980). Autogenetic effect of muscle contraction on extensor gamma motoneurons in the cat, *Expl Brain Res.*, **38**, 305–312

Ellaway, P. H., Murphy, P. R. and Trott, J. R. (1979). Inhibition of motoneurones discharged by contraction of homonymous muscle in the decerebrated cat, *J. Physiol., Lond.*, **291**, 425–441

Ellaway, P. H. and Trott, J. (1978). Autogenetic reflex action on to gamma motoneurones by stretch of triceps-surae in the decerebrated cat, *J. Physiol., Lond.*, **276**, 49–66

Emonet-Dénand, F., Jami, L. and Laporte, Y. (1975). Skeletofusimotor axons in hind-limb muscle of the cat, *J. Physiol., Lond.*, **249**, 153–166

Emonet-Dénand, F. and Laporte, Y. (1975). Proportion of muscle spindles supplied by skeletofusimotor axons (β axons) in the peroneus brevis muscle of the cat, *J. Neurophysiol.*, **38**, 1390–1394

Goodwin, G., Hoffman, D. and Luschei, E. S. (1978). The strength of the reflex response to sinusoidal stretch of monkey jaw-closing muscles during voluntary contraction, *J. Physiol., Lond.*, **279**, 81–111

Granit, R. (1970). In *The Basis of Motor Control*, New York, Academic Press, 167–171

Hagbarth, K. E. and Vallbo, A. B. (1968). Discharge characteristics of human muscle afferents during muscle stretch and contraction, *Expl Neurol.*, **22**, 674–694

Hagbarth, K. E., Wallin, B. G., Burke, D. and Löfstedt, L. (1975). Effects of the Jendrassik manoeuvre on muscle spindle activity in man, *J. Neurol. Neurosurg. Psychiat.*, **38**, 1143–1153

Harker, D. W., Jami, L., Laporte, Y. and Petit, J. (1977). Fast-conducting skeletofusimotor axons supplying intrafusal chain fibers in the cat peroneus tertius muscle, *J. Neurophysiol.*, **40**, 791–799

Houk, J. C. (1972). The phylogeny of muscular control configurations. In *Biocybernetics IV* (edited by Drischel and Deffman), Jena, Gustav Fischer Verlag, 125–144

Houk, J. C., Rymer, W. Z. and Crago, P. E. (1980). Nature of the dynamic response and its relation to the high sensitivity of muscle spindles to small changes in length, *Expl Brain Res.*, to be published

Laporte, Y. and Emonet-Dénand, F. (1976). The skeletofusimotor innervation of cat muscle spindle. In *Progress in Brain Research,* Vol. 44: *Understanding the Stretch Reflex* (edited by S. Homma), Amsterdam, Elsevier, 99–105

Loeb, G. E., Bak, M. J. and Duysens, J. (1977). Long term recording from somatosensory neurons in the spinal ganglia of the freely walking cat, *Science, N.Y.*, **197**, 1192–1194

Loeb, G. E. and Duysens, J. (1979). Activity patterns in individual hindlimb primary and secondary muscle spindle afferents during normal movements in unrestrained cats, *J. Neurophysiol.*, **42**, 420–440

McWilliam, P. N. (1975). The incidence and properties of axons to muscle spindles in the cat hind limb, *Q. Jl exp. Physiol.*, **60**, 25–36

Matthews, P. B. C. (1972). *Mammalian Muscle Receptors and Their Central Connections,* Baltimore, Williams and Wilkins, 512

Post, E. M., Rymer, W. Z. and Hasan, Z. (1980). Relation between intrafusal and extrafusal activity in triceps surae muscles of the decerebrate cat: Evidence for beta action, *J. Neurophysiol.*, **44**, 383–409

Prochazka, A., Westerman, R. A. and Ziccone, S. P. (1976). Discharge of single hindlimb afferents in the freely moving cat, *J. Neurophysiol.*, **39**, 1090–1104

Rymer, W. Z. and Hasan, Z. (1980). Absence of force feedback regulation in the soleus muscle of the decerebrate cat, *Brain Res.*, **184**, 203–209

Rymer, W. Z., Houk, J. C. and Crago, P. E. (1979). Mechanisms of the clasp-knife reflex studied in an animal model, *Expl Brain Res.*, **37**, 93–113

Vallbo, A. B. (1971). Muscle spindle response at the onset of isometric voluntary contractions in man. Time differences between fusimotor and skeletofusimotor effects, *J. Physiol., Lond.*, **318**, 405–431

Vallbo, A. B. (1974). Human muscle spindle discharge during isometric voluntary contractions. Amplitude relations between spindle frequency and torque, *Acta physiol. scand.*, **90**, 319–336

Wuerker, R. B. and Henneman, E. (1963). Reflex regulation of primary (annulospiral) stretch receptors via gamma motoneurons in the cat, *J. Neurophysiol.*, **26**, 539–550

Distinctive modes of static and dynamic fusimotor drive in jaw muscles

A. TAYLOR* AND K. APPENTENG*

SUMMARY

This paper describes the results of recordings from fusimotor and alphamotor axons of the cat masseter nerve and from muscle spindle afferents from the jaw closing muscles. Criteria are described for recognising fusimotor from alpha motor axons and their behaviour is examined during cyclic reflex jaw movements under light anaesthesia. Approximately one third of the fusimotor fibres showed an increased and 'sustained' discharge during a movement sequence. The rest showed strong modulation with maximum frequency during the muscle shortening phase and were referred to as 'modulated'. By recording from primary and secondary spindle afferents during separate but very similar experiments it was deduced that the 'sustained' type of response was probably in dynamic fusimotor fibres and the 'modulated' type in static fusimotor fibres. New observations of jaw muscle spindle afferent discharge in freely moving alert cats indicated that the same sort of modulated static fusimotor drive may accompany masticatory movements but that this does not normally overcome the unloading effect of unobstructed shortening. Sometimes, when extra loading slows shortening, a considerable enhancement of spindle afferent firing is seen which could have a significant load compensating effect.

A general proposal is made that one control strategy which can be used is to set the dynamic fusimotor drive to give appropriate incremental sensitivity to stretch and to vary static fusimotor drive as a 'temporal template' of the intended movement.

INTRODUCTION

Soon after the demonstration of the existence of two distinct sets of fusimotor

* Sherrington School of Physiology, St Thomas's Hospital, London

nerve fibres (Matthews, 1962), it seemed possible that they might be used to control independently the length and velocity sensitivity of muscle spindles during normal movements (*see* Matthews, 1964; Taylor, 1972). It was supposed that this would permit a matching of the dynamic characteristics of the spindle-based feedback control system to those of the load being moved. However, subsequent work has failed to confirm separate control of the incremental sensitivity to length and velocity by the static and dynamic fusimotor fibres as required by this idea. Instead, it transpires that there are two main actions of stimulation of single fusimotor fibres, either static or dynamic. First, such stimulation may have a powerful effect on the magnitude of the response to muscle stretch, but is without major effect on its phase (that is on the relative velocity and length sensitivity). Secondly, an important action of fusimotor stimulation is to increase the firing of afferents in just the same way as would an externally applied stretch (*see* Cussons *et al.*, 1977).

The alternative widely discussed concepts of the significance of fusimotor activity in normal movement control, such as the 'length follow-up servo' hypothesis (Merton, 1953) or 'α–γ co-activation' (*see* Granit, 1955; Vallbo *et al.*, 1979) are now unattractive because they fail to incorporate the well-known possibilities for separate static and dynamic fusimotor activation by the central nervous system (*see* Matthews, 1972, p. 538; Appelberg *et al.*, 1975) and are generally unhelpful in explaining evidence from spindle recordings in normally moving animals (Taylor and Cody, 1974; Cody *et al.*, 1975; Goodwin and Luschei, 1975; Prochazka *et al.*, 1976, 1977; Loeb and Duysens, 1979). There is consequently a pressing need for new proposals for advantageous ways in which the potential which exists for independent action of static and dynamic fusimotor neurons may be exploited in natural movement.

This paper introduces recently reported (Appenteng *et al.*, 1980) and new observations made on fusimotor fibres and muscle spindles of the jaw elevator muscles of the cat. Through interpretations of these data, based on established properties of muscle spindles and their motor supply, one possible strategy of control is proposed as follows. Tonic firing of dynamic fusimotor fibres is set at the start of a sequence of movements to give an appropriate sensitivity to stretch of the primary endings. Throughout the ensuing movement, driven essentially via the 'alpha route', there is a static fusimotor discharge modulated in the form of a 'temporal template' of the intended movement such that departures from this are detected and to some extent corrected.

TECHNICAL FEATURES OF THE EXPERIMENTS

Three types of experiment are reported here: (a) fusimotor fibre recordings from filaments of the masseter nerve in lightly anaesthetised cats; (b) spindle afferent recordings under closely similar conditions; and (c) spindle afferent recordings in fully conscious cats eating and drinking normally.

The methods for (a) and (b) have been described recently (Appenteng *et al.*, 1980), but may be summarised as follows.

Cats were lightly anaesthetised with I.P. pentobarbitone followed by I.V. supplements of short-acting barbiturate. A tracheal cannula was inserted and the cat's head supported by a plate screwed and cemented to the skull so as to avoid the strongly noxious stimulation due to the usual head clamps. This was found to be important if reflex jaw movements were to be evoked by mild natural stimuli.

For (a) the animal was supported on its right side, and part of the zygoma removed to expose the masseter nerve and its branches. Fine intramuscular filaments were cut and their central ends dissected to find single efferent units either active spontaneously or in response to various natural stimuli, such as jaw movement, touching the skin, tongue or teeth, or putting fluid in the mouth. Their conduction velocity (CV) was determined either by stimulating in the motor nucleus of the fifth nerve through a previously implanted tungsten electrode or by spike-triggered back averaging from the main nerve about 1 cm central to the cut end of the filament. Jaw movements were recorded by a light compliant strain gauge (Taylor, 1969).

For (b) the animal was supported prone and a small hole made in the skull for stereotaxic location of first-order spindle afferent cell bodies in the mesencephalic nucleus of the fifth nerve (MeV) as described by Cody *et al.* (1972). They were provisionally characterised as primary or secondary in origin by their responses to ramp stretch after I.V. succinyl choline (*see* Cody *et al.*, 1972).

For procedure (c), the technique was similar to that previously reported (Cody *et al.*, 1975) with certain improvements. The hydraulic micro-electrode chamber was directed downwards and backwards by 10° from the vertical to gain access to the denser, more caudal part of the MeV. EMG recording was via Teflon-coated finely braided stainless-steel wires implanted in temporalis, masseter and anterior digastric. An indwelling jugular venous catheter provided for simple, swift anaesthesia by short-acting barbiturate to permit examination of spindle afferent responses to passive jaw movements.

A. Identification of fusimotor fibres in masseter nerve filaments

The fibre diameter spectra for cranial nerves differ from those for hindlimb nerves, so that conduction velocity criteria for identifying fusimotor fibres in masseter nerve had to be checked. In a sample of 55 efferent fibres it was found (Appenteng *et al.*, 1980) that the CV ranged between 15 m/sec and 100 m/sec, with no evidence of separation into two sub-populations. However, the efferent fibres did fall into two groups according to whether or not their firing was significantly enhanced by muscle stretch. Of those so excited ('stretch sensitive'), 30 out of 32 had CVs above 35 m/sec and of those not excited ('stretch insensitive'), 18 out of 23 had CVs below 35 m/sec. In the spinal cord it is known (Eccles *et al.*, 1960) that gamma motoneurons differ from alpha motoneurons in receiving very little excitation from spindle group Ia afferents (*see also* Trott, 1976). It seems likely therefore that the stretch-insensitive, low CV group were fusimotor axons and this conclusion is further supported by differences in resting discharge rate under light anaesthesia (Proske and Lewis, 1972). All the stretch-insensitive units had resting discharge in excess of 11 impulses/sec (ips), whereas only 3 out of 16 of the others had resting discharge above 11 ips. Thus, in one part of the masseter efferent population there was an association of CVs below 35 m/sec, insensitivity to muscle stretch and resting discharge above 11 ips. These units were often strongly excited by natural stimuli in or around the mouth and we regard them as most probably fusimotor in function.

B. Fusimotor action during reflex jaw movements

Animals prepared as above were able to make short runs of cyclic jaw movements in response to placing fluid (water or dilute alcohol) in the mouth. The behaviour of the efferent axons, presumed to be fusimotor by the above criteria, was recorded in 16 animals. Two qualitatively different sorts of behaviour were observed.

The first type, seen in six units, was called 'sustained'. After placing fluid in the mouth there was a pause for 1 or 2 sec, then the fusimotor unit increased its firing frequency as seen in figure 1A to the region of 40–60 ips. Some minor fluctuations could then occur with a variable relation to jaw movements, but generally the frequency remained elevated throughout the movement sequence. The second type of behaviour seen clearly in 15 cases was referred to as 'modulated' because of the large fluctuations in frequency, closely related to the cyclic movements. An example is shown in figure 1B.

The increase in firing frequency occurred principally during the jaw-closing phase when frequencies of up to 135 ips were achieved. Incidentally, the high firing rates reached by these units further distinguishes them from alpha motoneurons which as seen in unitary EMG recordings from masseter and temporalis never exceeded 40 ips in these experiments. A particular pattern of fusimotor discharge, 'sustained' or 'modulated', was consistent for each unit. There was no observed change from one type to another, though the sustained type showed a variable degree of fluctuation, but with no constant relation to the movements.

C. Identification of static and dynamic fusimotor discharge

It is usual to identify fusimotor fibres as static or dynamic by the effects of their stimulation on spindle responses to stretch. However, this was not possible in the present experiments because of the need to record from cut

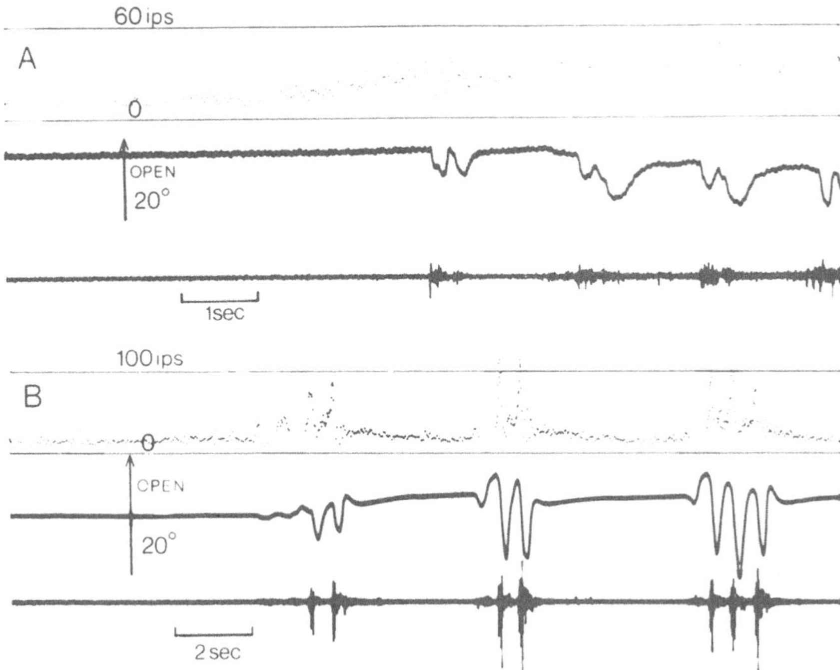

Figure 1. Examples of two types of fusimotor response to the stimulus of placing fluid in the cat's mouth (at the beginning of the record). (A) A 'sustained' response unit. (B) A 'modulated' response unit. Traces from above down: instantaneous firing frequency, jaw movement and masseter EMG.

filaments. An alternative, though indirect, approach was to study spindle afferent behaviour in other experiments set up in closely similar conditions. This was done by recording from the first-order cell bodies of jaw-closing muscle spindles in the MeV in similar lightly anaesthetised cats. Primary and secondary afferents were distinguished by the strong enhancing effect of succinyl choline (200μg—I.V.) on the dynamic index to ramp stretch of primary afferents.

The results are best summarised by figure 2. In A we see a typical response of a spindle primary afferent during an active movement sequence induced by fluid placed in the mouth. Immediately after the onset of the stimulus the spindle firing frequency increased and became more irregular without any movement—clear signs of fusimotor drive. This build-up of firing during the course of about 2 sec resembled in time course the 'sustained' type of fusimotor action, described above. It was only seen in primary afferents and not in secondaries and therefore seems likely to have been due to dynamic fusimotor firing. Some confirmation that dynamic not static fusimotor drive predominated at this stage (point 1 in figure 2A) was shown by the sudden

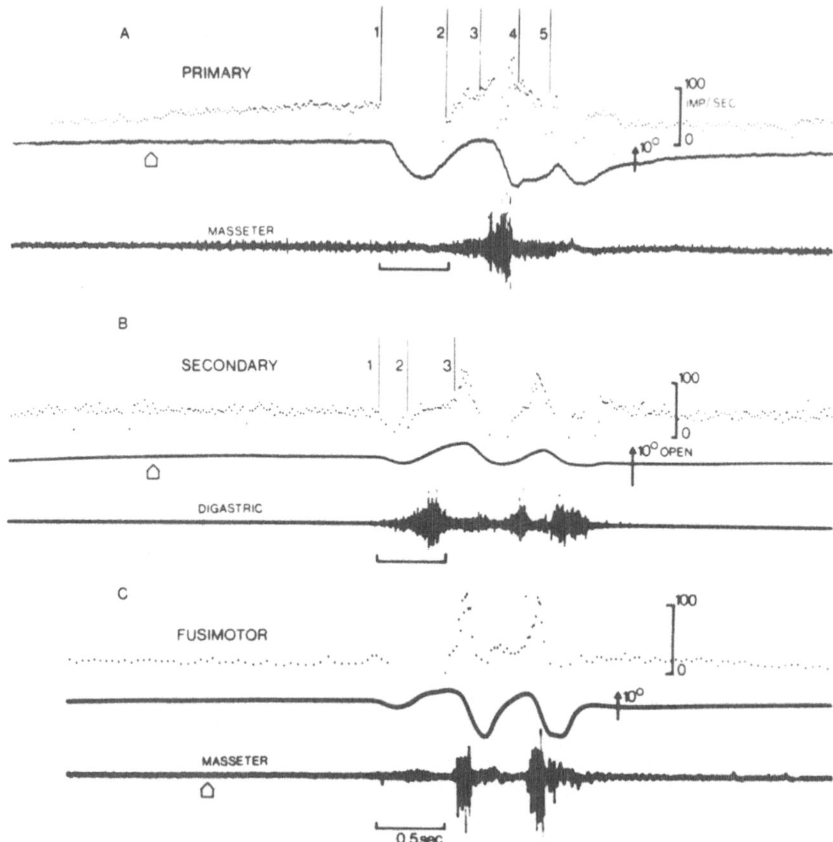

Figure 2. Responses in lightly anaesthetised cat of (A) spindle primary, (B) spindle secondary and (C) 'modulated' fusimotor fibre to masseter. In each case movements were excited by fluid in the mouth (open arrow). The records are from different but closely similar experiments. Traces from above down: instantaneous firing frequency, jaw movements and EMG.

arrest of the spindle primary afferents at the instant when active jaw closing commenced. Following a period of silence during muscle shortening, the spindle then resumed firing (2) when the length had returned about half way to its starting value. Firing then increased roughly in parallel with jaw opening, but then after maximum muscle length had been reached (3) showed a further marked increase in firing starting 0.1 sec before the next onset of shortening. Firing was briefly interrupted at the onset of shortening but resumed to reach a maximum of 150 ips despite the continuation of quite fast muscle shortening. These events between the vertical marks 3 and 4 can be explained by a brief period of increased *static* fusimotor firing since only static fusimotor discharge can cause increased primary afferent firing during shortening (Lennerstrand and Thoden, 1968).

Recordings for muscle spindle secondary afferents (figure 2B) provided confirmation that a static fusimotor burst occurred during the active muscle shortening in these experiments. Notice that there was no rise in frequency after stimulation before the movements began, thus supporting the conclusion that the early increase in figure 2A was due to dynamic fusimotor action. A smooth rise in secondary afferent firing frequency then paralleled muscle stretch in the period 2–3, but there was a marked burst of about 50 ips at the peak of lengthening when the muscle length was essentially constant. Note that as in figure 2A this increase occurred some 0.1 sec before the onset of shortening.

D. Jaw muscle spindle behaviour in alert cats

In the original reports of recordings of jaw elevator muscle spindles in the conscious cat, the most conspicuous feature was the predominance of firing during the muscle lengthening phase and the progressive or total silencing during the shortening phase (Taylor and Cody, 1974; Cody *et al.*, 1975). The question therefore arises as to whether the patterns of static and dynamic fusimotor firing described above in reflex movements are actually present in normal movements or whether they are a consequence of disordered functions due to anaesthesia. It was in fact concluded previously (Cody *et al.*, 1975) from spindle afferent recordings in alert cats that, 'static fusimotor drive can be low or absent during the muscle lengthening phase and yet be enhanced during shortening', and there was also evidence of enhanced dynamic fusimotor firing during natural movements. Also it seems more likely that a normal pattern of fusimotor firing might be slightly altered in timing or in amplitude rather than be totally changed.

To try to settle this question we have performed fresh recordings from spindle afferents in cats eating and drinking normally (*see* Taylor *et al.*, 1980).

Generally the results have been the same as reported previously, namely a fairly simple relationship between jaw movements and spindle afferent firing frequency with increase during lengthening and decrease or silence during shortening. However, on occasions we have seen evidence of the presence of a strong enhancement of static fusimotor drive during muscle shortening. Thus in figure 3A we see during large-scale movements with the cat eating soft food that a spindle primary followed the length (with some velocity component) quite faithfully. Then following the continuation of the record in figure 3B, we see that a remarkable enhancement of the spindle discharge took place in the muscle shortening phases of several cycles. On these occasions it is clear that the rate of shortening was less than usual and that the slowing was due to increased load (presumably tougher food) rather than diminished effort, because there is an increase of EMG activity in masseter muscle.

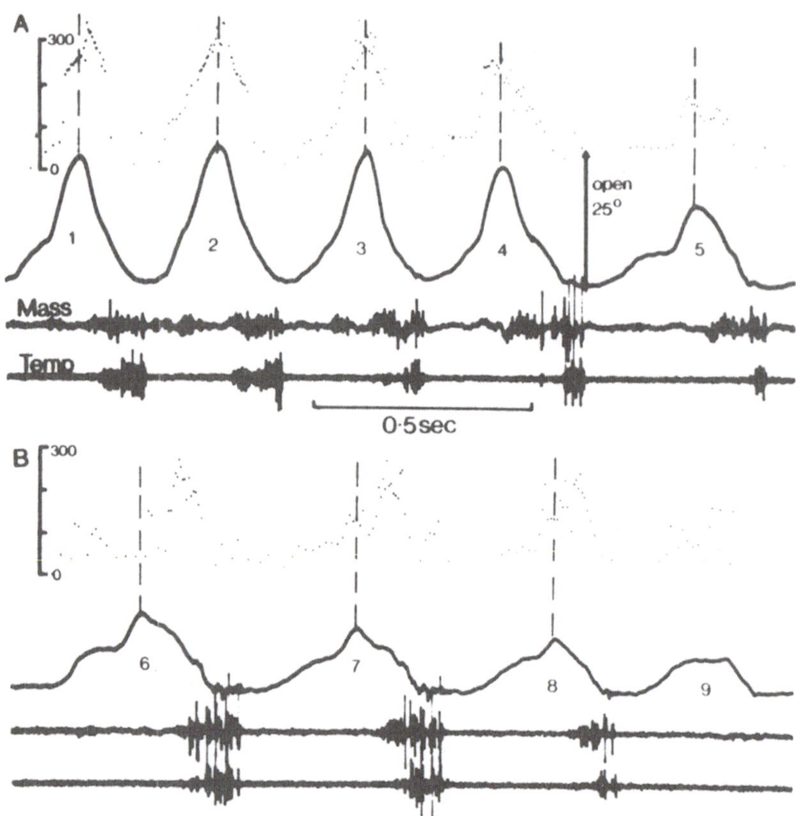

Figure 3. Recordings of a temporalis muscle spindle afferent (probably primary) in a conscious cat eating mixed food. (A) Movements large and fast, spindle behaving like a passive length receptor. (B) Movements slower but with more EMG activity—strong enhancement of spindle firing on shortening phases. Traces from above down: spindle instantaneous firing frequency, jaw movement, masseter and temporalis EMG. Records A and B continuous in time.

This result could be explained only by there being an increase in static fusimotor drive during these periods of shortening. It is proposed that the same sort of modulation of static fusimotor drive is probably occurring in figure 3A, but at the higher speeds of relatively unobstructed shortening here it does not succeed in preventing unloading of the spindle and permits the firing to fall quite smoothly during shortening.

The idea that modulation of fusimotor firing in relation to movement may be a significant occurrence is supported by the record of figure 4. In this the discharge of a spindle primary afferent was first recorded while the cat lapped milk. The animal then ceased drinking for a few seconds and before starting again remained still while looking at the milk. During this period the rhythmic

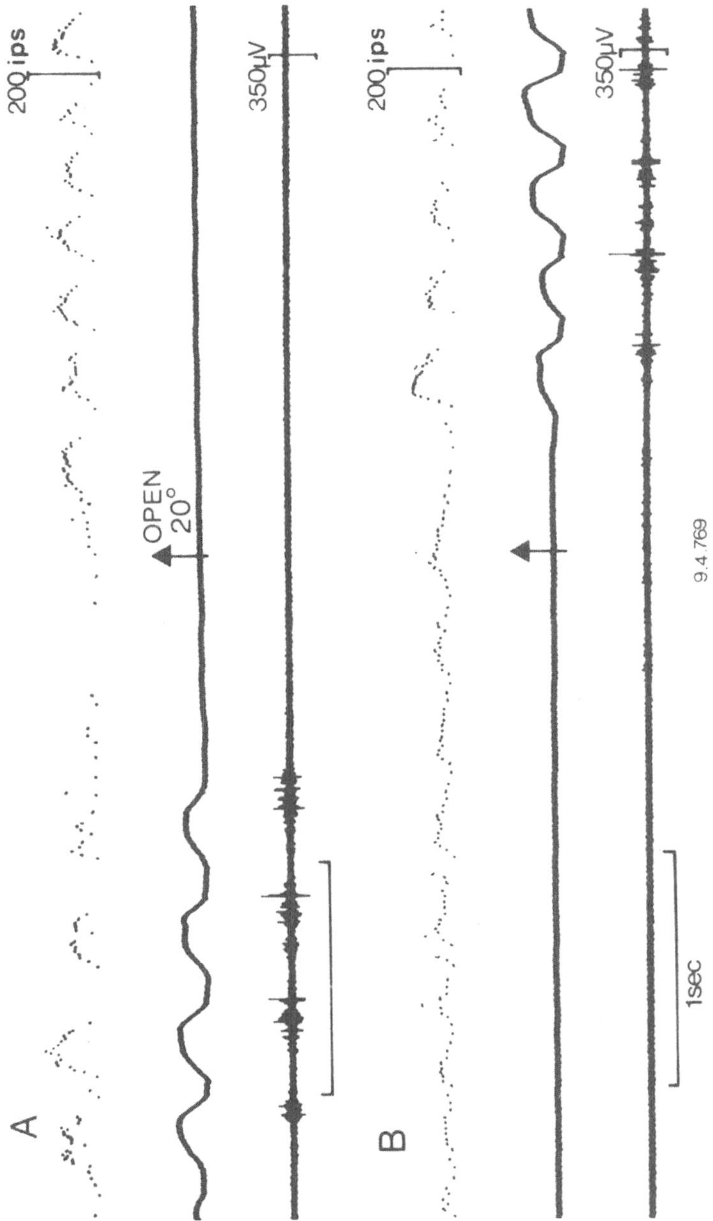

Figure 4. Masseter muscle spindle firing in alert cat lapping milk. Note rhythmic modulation of spindle afferent in latter half of (A) with no movement or EMG. (B) follows (A) with a gap of 5 sec. Traces from above down: instantaneous frequency, jaw movement and masseter EMG.

modulation of the spindle afferent started up again at approximately the same timing as during lapping, though there was no movement and no EMG activity. It seems certain that this must have been due to rhythmic discharge in fusimotor fibres—probably of static type. One supposes that the central rhythmic pattern generator for jaw movement (*see* Dellow and Lund, 1971) must have been driving the fusimotor neurons at a time when the alpha motoneurons were not sufficiently excited to be driven directly or via the spindle afferent activity.

E. Changes caused by anaesthesia

It is evident from the above results that spindle afferent records obtained from fully conscious cats show much less activity during the muscle shortening phase than do records obtained from lightly anaesthetised animals. The reason for this could be either that the static fusimotor discharge is greater than normal during light anaesthesia or that the muscle contraction is smaller and slower than normal. Figure 5 shows that the latter is probably the true explanation. In this case the spindle afferent unit is the same as in figure 4 and it was possible after recording a period of normal lapping (figure 5A) to induce very light anaesthesia with intravenous thiopentone while retaining a stable recording. Lapping movements were then induced by introducing milk into the mouth (figure 5B) while the animal's head was supported by means of the implanted skull plate. Notice that the movements in B are of similar form to those in A, but are rather slower and approximately half the previous amplitude. The EMG of masseter is correspondingly reduced. There is now a conspicuous change in the spindle afferent firing pattern. Instead of a train of low-frequency impulses in each stretching phase as in A we now see a substantial discharge during the shortening phase in B. The obvious explanation is that the anaesthetic has reduced the alpha motor activity more than the fusimotor. Consequently, the burst of static fusimotor activity, which we have deduced above to occur during muscle shortening, now provides a net excitation to the spindle afferent. Any other procedure which could weaken or obstruct the muscle shortening without greatly affecting the modulated fusimotor discharge would be expected to do the same.

GENERAL SIGNIFICANCE OF THESE FINDINGS

The proposal which emerges from the work is that in at least some rhythmic movements, static and dynamic fusimotor neurons have different and distinctive patterns of action, which cannot be profitably described as alpha–gamma co-activation, as generally understood. We have presented some evidence that the dynamic system is activated tonically before and during the movement sequence and that the static system is modulated rhythmically in

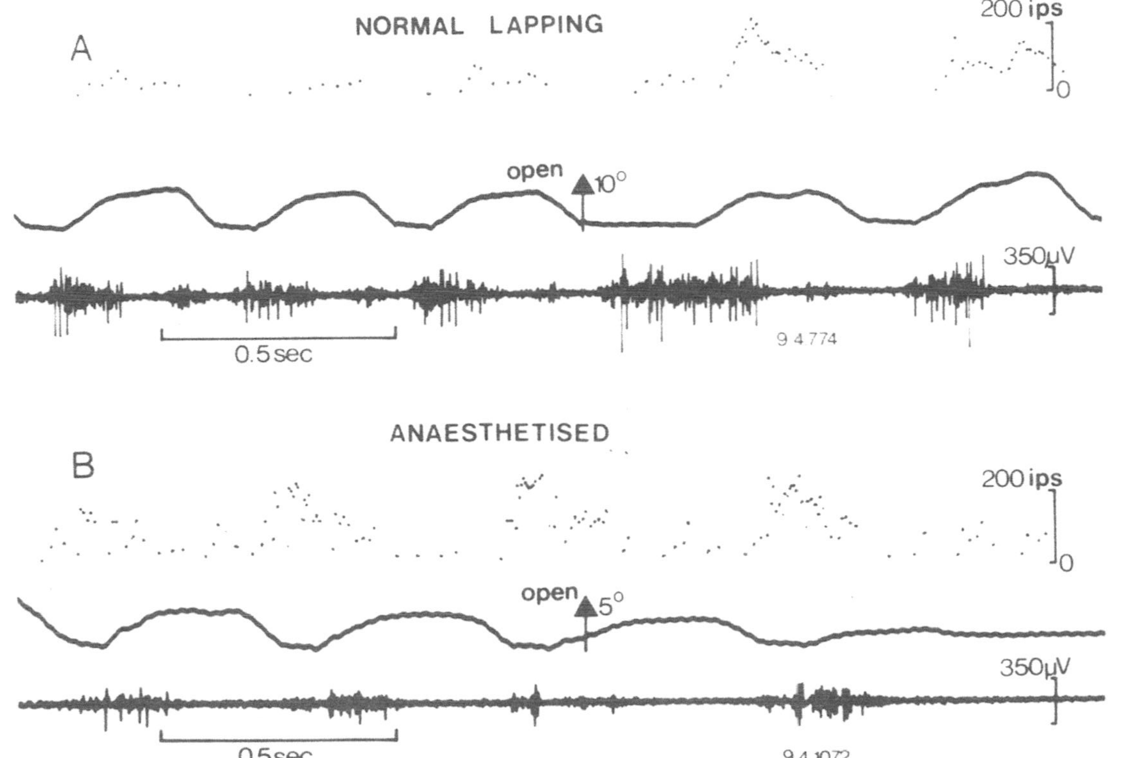

Figure 5. (A) Record of masseter spindle afferent firing in alert cat lapping milk (same unit as in figure 4). (B) Note reversal of spindle behaviour relative to jaw movement after light anaesthesia. Traces as in figure 4. (Note different jaw movement calibration.)

time with the muscle shortening. The static activation is not powerful enough to prevent unloading of the spindle sensory endings during shortening at normal speeds. However, if shortening is slowed by loading or by anaesthesia, then spindle discharge may actually accelerate markedly during muscle shortening. The effect of this would be a significant reflex enhancement of alpha motoneuron drive, since the extra spindle excitation reaches the alpha motoneurons when they are already active. It is suggested that the static fusimotor drive may be regarded as a kind of 'temporal template' of the intended movement. The actual movement is driven principally by direct input to the alpha motoneurons with little effect of negative feedback from the spindles unless the movement is loaded and slowed.

Observations on the intercostal muscles by Sears (1964) and by von Euler and his colleagues (Eklund, von Euler and Rutkowski, 1964; von Euler and Peretti, 1966) are probably most relevant to the present case and are well reviewed by Sears (1973). During spontaneous breathing in anaesthetised cats it was shown that fusimotor neurons appeared to fire in 'tight α–γ linkage' and that the fusimotor drive was sufficient to cause net excitation of spindles in inspiratory muscles during inspiration (Critchlow and von Euler, 1963). Such results were naturally interpreted in terms of the 'length follow-up servo' hypothesis of Merton (1953) although the demonstration of 'central respiratory drive potentials' in the intercostal motoneurons due to descending input made it clear that at least a significant part of the drive to the motoneurons was received direct rather than via the 'gamma loop'. Our present data led us to question whether a net excitation of intercostal spindles actually occurs during natural breathing. It seems very possible that if the spindle afferent recordings could be repeated in the absence of anaesthesia, the muscle shortening might be large enough and fast enough to cause net reduction in afferent activity despite the modulation of gamma firing in time with alpha motoneuron firing, as appears to be the case with the jaw muscles.

In the work of Eklund *et al.* (1964) it was noted that while many of the fusimotor fibres fired phasically with respiration, others fired continuously in tonic fashion. The attempt to identify amongst these which ones were static and which dynamic (von Euler and Peretti, 1966) seems in retrospect to have been unsuccessful and we feel that further examination would be well worth while to see whether the tonic fusimotor fibres in the intercostals might be dynamic and the phasic ones static as is our deduction for the jaw muscles.

In recordings from the cat hindlimb and tail muscle spindles, Prochazka and his colleagues (Prochazka *et al.*, 1976, 1977, 1979) have repeatedly emphasised that spindle behaviour in normal movements most resembles that of stretch receptors in which there is little sign of rapidly fluctuating fusimotor activity. There was evidence of changing dynamic sensitivity from time to time but experiments in which natural muscle shortening was mechanically blocked did not show the enhanced spindle discharge which was to be expected if static fusimotor discharge were increasing at that time. It may be that there is a real difference in the strategy adopted for the use of the fusimotor system in hindlimb muscles as distinct from jaw and intercostal muscles. It should, however, be emphasised that, in the absence of data showing how rapidly modulated static fusimotor drive may interact with rapid shortening movements to determine spindle afferent firing, it is very difficult to interpret the data from natural movements with complete certainty.

It seems likely that the great merit of the evolution of separate static and dynamic fusimotor systems is the flexibility which it might give to movement control. We should not therefore expect to be able to devise one general theory to explain the strategy of control in all situations. The present proposal is that in some natural movements dynamic fusimotor activity is set to a fairly constant level for a particular task, while static fusimotor drive fluctuates in the form of a 'temporal template' of the intended movement. This may well be an advantageous form of control in mastication where quite rapid and precise movements have to be made against thc varying load of foods of unpredictable consistency. No doubt other ways of using the fusimotor systems may be appropriate for other tasks.

ACKNOWLEDGEMENTS

It is a pleasure to acknowledge the collaboration of Dr T. Morimoto in part of the work described here and the technical help of Mr David Earle. The work was supported by the St Thomas's Hospital Research Endowments Fund.

Received on July 5th, 1980.

REFERENCES

Appelberg, B., Jeneskog, T. and Johansson, H. (1975). Rubrospinal control of static and dynamic fusimotor neurons, *Acta physiol. scand.*, **95**, 431–440

Appenteng, K., Morimoto, T. and Taylor, A. (1980). Fusimotor activity in masseter nerve of the cat during reflex jaw movements, *J. Physiol., Lond.*, to be published

Cody, F. W. J., Harrison, L. M. and Taylor, A. (1975). Analysis of activity of muscle spindles of the jaw closing muscles during normal movements in the cat, *J. Physiol., Lond.*, **253**, 565–582

Cody, F. W. J., Lee, R. W. H. and Taylor, A. (1972). A functional analysis of the components of the mesencephalic nucleus of the fifth nerve in the cat, *J. Physiol., Lond.*, **226**, 249–261

Critchlow, V. and von Euler, C. (1963). Intercostal muscle spindle activity and its γ motor control, *J. Physiol., Lond.*, **168**, 820–847

Cussons, T. D., Hulliger, M. and Matthews, P. B. C. (1977). Effects of fusimotor stimulation on the response of the secondary ending of the muscle spindle to sinusoidal stretching, *J. Physiol., Lond.*, **270**, 835–850

Dellow, P. G. and Lund, J. P. (1971). Evidence for central timing of rhythmical mastication, *J. Physiol., Lond.*, **215**, 1–13

Eccles, J. C., Eccles, R. M., Iggo, A. and Lundberg, A. (1960). Electrophysiological studies on γ-motoneurones, *Acta physiol. scand.*, **50**, 32–40
Eklund, G., von Euler, C. and Rutkowski, S. (1964). Spontaneous and reflex activity of intercostal gamma motoneurones, *J. Physiol., Lond.*, **171**, 139–163
von Euler, C. and Peretti, G. (1966). Dynamic and static contributions to the rhythmic γ activation of primary and secondary spindle endings in external intercostal muscle, *J. Physiol., Lond.*, **187**, 501–516
Goodwin, G. M. and Luschei, E. S. (1975). Discharge of spindle afferents from jaw-closing muscles during chewing in alert monkeys, *J. Neurophysiol.*, **38**, 560–571
Granit, R. (1955). *Receptors and Sensory Perception,* New Haven, Yale University Press
Lennerstrand, G. and Thoden, U. (1968). Muscle spindle responses to concomitant variations in length and in fusimotor activation, *Acta physiol. scand.*, **74**, 153–165
Loeb, G. E. and Duysens, G. (1979). Activity patterns in individual hind-limb primary and secondary muscle spindle afferents during normal movement in unrestrained cats, *J. Neurophysiol.*, **42**, 420–440
Matthews, P. B. C. (1962). The differentiation of two types of fusimotor fibre by their effects on the dynamic response of muscle spindle primary endings, *Q. Jl exp. Physiol.*, **47**, 324–333
Matthews, P. B. C. (1964). Muscle spindles and their motor control, *Physiol. Rev.*, **44**, 219–288
Matthews, P. B. C. (1972). *Mammalian Muscle Receptors and Their Central Actions,* London, Arnold
Merton, P. A. (1953). Speculations on the servo-control of movement. In *The Spinal Cord* (ed. G. E. W. Wolstenholme), London, Churchill, 247–255
Prochazka, A., Stephens, J. A. and Wand, P. (1979). Muscle spindle discharge in normal and obstructed movements, *J. Physiol., Lond.*, **287**, 57–66
Prochazka, A., Westerman, R. A. and Ziccone, S. P. (1976). Discharges of single hind limb afferents in the freely moving cat, *J. Neurophysiol.*, **39**, 1090–1104
Prochazka, A., Westerman, R. A. and Ziccone, S. P. (1977). Ia afferent activity during a variety of voluntary movements in the cat, *J. Physiol., Lond.*, **268**, 423–448
Proske, U. and Lewis, D. M. (1972). The effects of muscle stretch and vibration on fusimotor activity in the lightly anaesthetised cat, *Brain Res.*, **46**, 55–69
Sears, T. A. (1964). Efferent discharges in alpha and fusimotor fibres of intercostal nerves of the cat, *J. Physiol., Lond.*, **174**, 295–315
Sears, T. A. (1973). Servo control of the intercostal muscles. In *New Developments in Electromyography and Clinical Neurophysiology* (ed. J. Desmedt), Vol. 3, Karger, Basel, 404–417
Taylor, A. (1969). A technique for recording normal jaw movements in conscious cats, *Med. Biol. Eng.*, **7**, 89–90
Taylor, A. (1972). Muscle receptors in the control of voluntary movement, *Paraplegia,* **9**, 167–172
Taylor, A., Appenteng, K. and Morimoto, T. (1980). Proprioceptive input from the jaw muscles and its influence on lapping, chewing and posture, *Can. J. Physiol., Pharmacol.,* to be published.
Taylor, A. and Cody, F. W. J. (1974). Jaw muscle spindle activity in the cat during normal movements of eating and drinking, *Brain Res.*, **71**, 523–530
Trott, J. R. (1976). The effect of low amplitude muscle vibration on the discharge of fusimotor neurones in the decerebrate cat, *J. Physiol., Lond.*, **255**, 635–649
Vallbo, Å. B., Hagbarth, K.-E., Torebjörk, H. E. and Wallin, B. G. (1979). Somatosensory, proprioceptive and sympathetic activity in human peripheral nerves, *Physiol. Rev.*, **59**, 919–957

A critique of the papers by Ellaway, Murphy and Trott; Appelberg, Hulliger and Sojka; Rymer, Post and Edwards; and Taylor and Appenteng

S. GRILLNER*

This symposium deals with the role of muscle receptors in movement control. Afferent signals, in general, may have distinctly different functions such as:

1. To initiate a movement as in a withdrawal reflex.
2. To provide the central nervous system (CNS) with information about the current status of the motor apparatus, i.e. joint angles, degree of muscle contraction, and so forth. Such information is needed, when the CNS plans, for instance, a fast accurate voluntary movement.
3. To regulate a movement during its progress, i.e. feedback control of some sort. This type of control seems to be the main concern of this meeting.

Movements are of different types, from fast voluntary ballistic to slow tracking, from complex innate movements such as chewing, and somewhat simpler such as swallowing, to very simple, such as a blink reflex or a withdrawal reflex. It is almost certain that information from muscle receptors will not be used in the same way in these different control situations. This session deals with reflex movements. This can presumably be taken to mean simple behaviourally meaningful motor acts such as a withdrawal reflex or swallowing. The word 'reflex' is, however, not very precise, as it also includes an entirely different class of motor effects (3 above), i.e. afferent control or modulation of ongoing simple or complex movements. For instance:

1. The control of the duration of inspiration by stretch receptors in the lungs.
2. The control of the amplitude of limb flexion from hair receptors on the dorsum of the paw during locomotion.

* Department of Physiology III, Karolinska Institutet, Stockholm

3. The traditional autogenetic excitatory and reciprocal inhibitory effects from muscle spindles and the effects from tendon organs.

It is in my opinion unfortunate to use the word 'reflex' both for reflex acts and what could be called 'regulatory reflexes'; perhaps a qualifier should always be used. The situation is somewhat exceptional with regard to reflexes acting on γ-motoneurons in different movements, as the γ-motoneurons are ultimately always part of a regulatory reflex as they control the spindles. The distinction we are trying to make is between a direct activation of γ's as part of a reflex act, as opposed to a regulatory reflex changing the activity of γ-motoneurons during an ongoing movement.

This session on 'reflex movements' contains in fact a blend of these two forms of reflexes. Ellaway *et al.* (1981) deal with autogenetic reflexes, Appelberg *et al.* (1981) with reflexes that may be of this category and Rymer *et al.* (1981) with part of a motor act, the crossed extensor reflex. Taylor and Appenteng (1981) deal finally with the control of chewing, a complex innate movement.

Ellaway *et al.*'s report is, for the larger part, documented in previous papers. Golgi tendon organs in all likelihood contribute part of the autogenetic inhibition to triceps surae γ-motoneurons. I can see no reason, however, to exclude the possibility of contributions from other stretch-sensitive afferents such as non-spindle group II, III and IV afferents, which probably control the clasp knife reaction as suggested by Rymer *et al.* (1979). New data on group I effects of Jankowska, Fetz and McCrea (personal communication) should also be taken into account.

The fact that we thus have autogenetic reflexes to γ-motoneurons is important. Nevertheless, we still know nothing about when and how these reflexes are used. The main objective for further studies must be to find out during which motor behaviours they are used and whether the transmission in the reflex pathways is open all the time, or if a phasic gating occurs. Suitable models could be the flexion reflex, the scratch reflex, or the 'fictive locomotor' models.

α-Motoneurons receive reciprocal Ia inhibition from their antagonists (Laporte and Lloyd, 1952; Eccles *et al.*, 1957; Eccles and Lundberg, 1958). Ellaway argues that this does not apply to γ's as posterior biceps-semitendinosus γ's are not inhibited from the ankle extensor triceps. This reasoning is inappropriate as the relevant antagonist pairs are tibialis anterior and triceps, or vastus and posterior biceps-semitendinosus.

The weak autogenetic excitatory effect from spindle primaries is well documented by Ellaway *et al.* (1981). I would nevertheless like to pose a general question as to how far it is meaningful to go, in describing weak reflex effects. The experimental situation in a passive decerebrate or a spinal preparation may have very little in common with the prevailing conditions in an animal performing one type of movement or another. The fact that a certain weak reflex effect can be demonstrated unequivocally does not necessarily mean that it is significant. It might even result from a coincidence. Research in the last few years has demonstrated that individual neurons with complex input patterns may send branches to widely different parts of the brain. A given neuron could, for instance, be used to transmit polysynaptic descending control signals in one behavioural context and have a completely different role in another, when another input could use this neuron to influence other target cells. During different active behaviours certain 'lines' may be open and others completely closed. In the experimental state of a passive animal, many potential polysynaptic transmission 'lines' may instead be 'half open' or 'half closed'.

In 1969 I noted that triceps surae γ-motoneurons in spinal cats . . . 'could be influenced by manipulation of the muscles from the level of the spinalisation and further caudally; also from the periosteum of the vertebrae, the iliac crests and sacrum. Even pressure applied to the last few millimetres of the tail influenced the discharge rate as well as manipulation of the rectal thermometer and an increased pressure in the bladder. It is difficult to find any particularly meaningful pattern in this modification of discharge rate obtained from such a large receptive field' (Grillner, 1969a) and further, 'The functional significance of these large receptive fields might be to provide a background for the descending control, which will "rule" through an inhibition of facilitation of (neurons in) the reflex path' (Grillner, 1969b). It might thus be important always to ask what role a so-called 'reflex effect' may have.

This brings me on to the paper of Appelberg *et al.* These authors seek to find out if individual γ's receive very specific reflex effects. Their main finding is that an extension of the contralateral limb excites dynamic γ's to the triceps surae, with little effect on static γ's. This may be a specific effect of limb position, that somehow could be used in a movement control. However, a less interesting possibility is that the full extension of the limb will activate many stretch-sensitive elements in joints, muscles and skin, as opposed to their degree of activity in the middle pendant position. If so, a crossed extensor reflex sub-threshold for the α's could come about. Such weak crossed reflex effects are already known to act preferentially on dynamic γ's (Alnaes *et al.*, 1965). The fact that pressure applied to the knee joint capsule also gives such effects may also be taken to indicate that they could be part of the same general crossed reflex pattern. The ipsilateral excitatory effects on the dynamic γ-drive in two thirds of the observations cannot be judged with respect to specific mechanisms. It would seem important to evaluate these effects in a situation close to a normal behaviour.

Rymer *et al.* investigated crossed effects, without defining the exact type of stimulation used. This is important for a comparison with previous studies, which have shown that different stimulation parameters can cause selective effects on either dynamic (γ_D) or static (γ_S) γ-motoneurons. They obtained marked effects on γ_S and γ_D, when the α's were still unaffected as judged by the absence of EMG. This situation is uncertain as the α's may or may not have received EPSPs at the same time as the γ's. The only difference might thus be that α's remain sub-threshold for a longer period of time. This criticism applies to all similar studies from Eldred *et al.* (1953) and onwards, and has been expressed several times earlier. To deal with the problem it is obviously necessary to have an indicator of the α-motoneuron excitability, such as (1) an intracellular α-motoneuron recording, (2) monosynaptic reflexes (somehow) or (3) background EMG activity, which Rymer has often used in other studies.

One interesting finding was that the degree of γ-activity was already at its highest level of discharge at a stage when the α's started to be recruited. This may suggest a marked difference between the reflex pathways to γ's and α's. The excitatory current produced by the reflex input may, however, be the same and have the same time course in both γ's and α's. A different threshold for the α's and the γ's could then explain the findings (input resistance, etc.). If, for instance, 90% of the excitatory current must be reached before the first α is recruited, only the range from 90% to 100% is available to recruit the remaining α's. In that case, the discharge rate of the γ's may appear virtually unchanged.

It was useful to see that a change in the series compliance of the tendon at its attachment could change an apparent γ-lead in the reflex contraction to an α-lead, when a primary ending is used to monitor the γ-activity. Nevertheless I feel a little bit uncomfortable when I read that the data on the γ-lead from crossed extensor reflexes in decerebrate cats should be reconciled with data from human slow voluntary contractions. At best I would hope that the results should agree with the corresponding studies of crossed extensor reflexes in humans, but I find no particular reason to expect that they should agree exactly with results from, for instance, a human voluntary finger or leg movement.

Taylor and Appenteng (1981) suggest that dynamic and static γ's are used in a separate way in the jaw-closing muscles during chewing. The dynamic γ's should be switched on tonically, whereas the static ones should be phasically active, together with the α's. The degree of activity in the static γ's should, on the other hand, be so low that it could not prevent the unloading during normal active muscle shortening. To this interpretation a few comments may be relevant.

1. The putative dynamic γ's recorded in their figure 1A show in addition to a tonic increase a clear bout of phasic activity with each movement.
2. The coactivation* of α's and static γ's became quite striking with chewing against a presumed increase in resistance (Taylor and Appenteng,

* The word 'coactivation' is used here as a purely descriptive term. It has no relation to different theoretical frameworks in which the word may have been used.

1981, figure 3). The authors explain this as an effect of the slowing of the movement, but as the EMG also increases, there may in addition be an 'automatic' increase in the phasic, static γ-drive (cf. Vallbo, 1981). Furthermore it cannot theoretically be excluded that we are dealing with a separate static γ-command that in a certain context (e.g. chewing peanuts rather than porridge) could switch on a phasic increase in static γ-drive and in other situations diminish the degree of γ-coactivation.

It is interesting to note that in alert animals the degree of α–γ-coactivation is much less apparent in locomotion and chewing than could be expected from previous experiments on decerebrate or decorticate walking (Severin *et al.*, 1967; Perret and Cabelguen, 1980) or chewing animals. One obvious explanation for this is the one that Taylor and Appenteng (1981) seem to lean towards, i.e. that it is technically difficult (or impossible) to reveal a moderate static coactivation in a rapidly shortening muscle or a dynamic increase in a muscle that only shows shortening or a very moderate lengthening. The view suggested is thus that there is at least in chewing normally a simultaneous activation of static γ's and α's. There is another perhaps more challenging explanation (in addition to the above) for the finding that α–γ-coactivation usually appears more marked in decerebrate or anaesthetised preparations. It might reveal a genuine quantitative difference. The basic innate motor programmes may incorporate an automatic strong α–γ-coactivation by analogy with the even older pattern of an obligatory extra and intrafusal coactivation, as seen, for example, in amphibians. In the alert mammal, on the contrary, the CNS could possibly utilise the new sophistication in phylogeny, a separate fusimotor system, to bias the spindle activity to be appropriate for the actual behavioural context.

ACKNOWLEDGEMENT

Support has been obtained from the Swedish Medical Research Council (Project No. 3026).

Received on September 1st, 1980.

REFERENCES

Alnaes, E., Jansen, J. K. S. and Rudfjord, T. (1965). Fusimotor activity in the spinal cat, *Acta physiol. scand.*, **63**, 197–212

Appelberg, B., Hulliger, M., Johansson, H. and Sojka, P. (1981). Reflex activation of dynamic fusimotor neurons by natural stimulation of muscle and joint receptor afferent units. This publication

Eccles, J. C., Eccles, R. M. and Lundberg, A. (1957). The convergence of monosynaptic excitatory afferents on to many different species of α-motoneurones, *J. Physiol., Lond.*, **137**, 22–50

Eccles, R. M. and Lundberg, A. (1958). Integrative pattern of Ia synaptic actions on motoneurones of hip and knee muscles, *J. Physiol., Lond.*, **144**, 271–298

Eldred, E., Granit, R. and Merton, P. A. (1953). Supraspinal control of the muscle spindles and its significance, *J. Physiol., Lond.*, **122**, 498–523

Ellaway, P. H., Murphy, P. R. and Trott, J. R. (1981). Autogenetic effects from spindle primary endings and tendon organs on the discharge of gamma motoneurons in the cat. This publication

Grillner, S. (1969a). The influence of DOPA on the static and the dynamic fusimotor activity to the triceps surae of the spinal cat, *Acta physiol. scand.*, **77**, 490–509

Grillner, S. (1969b). Supraspinal and segmental control of static and dynamic γ-motoneurones in the cat, *Acta physiol. scand.*, Suppl. No. 327, 1–34

Laporte, Y. and Lloyd, D. P. C. (1952). Nature and significance of the reflex connections established by large afferent fibres of muscular origin, *Am. J. Physiol.* **169**, 609–621

Perret, C. and Cabelguen, J. M. (1981). Main characteristics of the hindlimb locomotor cycle in the decorticate cat, with special reference to bifunctional muscles, *Brain Res.*, **187**, 333–352

Rymer, W. Z., Houk, J. C. and Crago, P. E. (1979). Mechanisms of the clasp-knife reflex studied in an animal model, *Brain Res.*, **37**, 93–113

Rymer, W. Z., Post, E. M. and Edwards, F. H. (1981). Functional roles of fusimotor and skeletofusimotor neurons studied in the decerebrate cat. This publication

Severin, F. V., Orlovsky, G. N. and Shik, M. L. (1967). Work of the muscle receptors during controlled locomotion, *Biofizika* (Engl. transl.), **12**, 575–586

Taylor, A. and Appenteng, K. (1981). Distinctive modes of static and dynamic fusimotor drive in jaw muscles. This publication

Vallbo, Å. B. (1981). Basic patterns of muscle spindle discharge in man. This publication

Cutaneous and proprioceptive reflex effects on intact muscle efferents and afferents

P. BESSOU*, M. JOFFROY* and B. PAGÈS*

INTRODUCTION

Some groups of afferent and efferent fibres of a muscle nerve appear to function as part of a servoregulatory loop controlling muscle contractions (Matthews, 1972). The knowledge of the internal functioning of such loops comes from the single unit analysis of the activity, generally recorded at the level of spinal roots or of peripheral nerve, of neurons forming a part of these loops. Action potentials are recorded from nerve axons with microelectrodes or conventional hook electrodes. Using the latter type of recording, it is necessary to split spinal rootlets or to tease a nerve in order to prepare filaments fine enough so that the discharge of a single element functionally identified is prominent in the record and can be distinguished from all other activity on the basis of amplitude and shape (Hunt, 1951). Such a method leads entirely or partially to an impairment of the functioning of the peripheral control loop. In addition, experiments are often performed on curarised animals or after extensive denervation of limbs.

The recorded activity affords direct information concerning the function of the investigated neuron and may indicate the behaviour of related neurons; for instance, a recording of muscle spindle discharge may indirectly permit an evaluation of motor activity (γ and/or β) acting upon intrafusal muscle fibres (see, for instance, Eldred *et al.*, 1953; Hagbarth and Vallbo, 1967, 1968).

Inferences made about the internal functioning of servoregulatory loops through a collection of unitary samples may be open to criticism. Inferences may be biased because at any time the activity of only one and not of all the neural elements linked in succession in the loop is observed; in addition, in many instances the experimental techniques used result in the loop being opened and indirect evidence may be interpreted in different ways.

* Laboratoire de Physiologie, Faculté de Médecine, Toulouse

The work in this paper has been performed in order to investigate the internal functioning of intact neural loops between a muscle and its spinal centre in decerebrated preparations that underwent minimal denervation and did not receive any curarising drug. This aim was achieved by directly and simultaneously assessing the segregated activity of α and γ efferent axons and of group I and group II afferent fibres which together form a representative sample of the totality of the nerve fibres of similar function in muscle nerves.

METHODS

Two electrodes placed 4 mm apart on a thin intact branch of the lateral gastrocnemius (LG) nerve provided two monopolar recordings. An electronic device allowed the simultaneous detection of both sensory and motor nerve activity according to the direction of impulse propagation. The afferent and efferent action potentials could then be segregated electronically according to their conduction velocities (Joffroy, 1975, 1980). The separate activity of group I and group II afferent fibres and of α and γ axons was displayed as a mean frequency of firing and was considered representative of the activity of the homonymous pool constituting the whole LG nerve. The fidelity of the equipment is high (at least 95% of potentials detected) as long as the total firing frequency remains under 2500 impulses/second (imp/sec).

To examine the reflex patterns induced by stimulating various structures, muscle action potentials have been recorded by electrodes placed on the surface of the triceps surae muscle. In this manner it was possible to observe not only the activity of the gastrocnemius muscles (composed mainly of type A muscle fibres) but also the activity of the soleus muscle (exclusively composed of type B fibres) (Henneman and Olson, 1965).

RESULTS

The number of active fibres of each group recorded from the branch of the LG nerve varied from one preparation to another, particularly because the number of fibres constituting the branch varied, as did the condition of the decerebrate animals. When the triceps surae were at physiological length, and under resting conditions, the mean firing frequency in each group of axons in the intact branch of LG muscle nerve varied considerably in the course of an experiment, along with variations in rigidity. The extent of the variations in 35 experiments has been found to be 20–250 imp/sec for group I fibres, 40–400 imp/sec for group II fibres, 50–250 imp/sec for γ axons and 0–50 imp/sec for α axons. Similarly the amplitude and the latency of changes in the firing frequency of each group induced either by stimulating the sural nerve or by stretching and releasing the triceps surae have been found to vary widely. In results described here only the chronological order in which the changes in firing frequency occurred will be taken into account.

During sural nerve stimulation the ankle was firmly maintained at a fixed angle so that the length of the triceps surae was kept constant. Generally a single stimulus to the contralateral or ipsilateral sural nerve would respectively induce a crossed extensor or a flexor reflex, provided that both group Aδ and C fibres were stimulated. However, sometimes repetitive stimulation (3–10 stimuli, 10–50/sec) of Aα and Aβ fibres alone might be effective. Since reflex responses were generally weak and short-lasting when using single stimuli and consequently difficult to discriminate from spontaneous variations of activity, repetitive stimulation (2–5 stimuli, 5–20/sec) of the sural nerve was employed.

Crossed extensor reflex

During crossed extensor reflexes, the onsets of which were indicated about 100 msec after the first stimulus by an increase in the electromyographic activity of triceps surae, development of activity was recorded from α axons of the branch of the LG nerve. The latency of this α activity varied between 400 msec and 600 msec and its duration usually lasted 3–5 sec. The explanation for the discrepancy in timing between the electromyographic response of triceps surae and α responses of LG probably lies in the nature of the motor units brought into activity: the units activated by the smallest α motoneurons of soleus becoming active first and the units activated by the largest nerve fibres of LG being recruited successively (Henneman *et al.*, 1965).

Sural nerve stimulation sufficient to elicit a crossed extensor reflex always reflexively increased γ discharge to LG. γ responses were recorded just in advance of, or at the same time as, the beginning of the electromyographic activity of triceps surae and regularly preceded LG α responses by 200 msec at least. The γ firing rate, after this increase, returned very slowly (10 sec) to the initial level.

An increase in the firing frequency of group I afferent fibres started approximately 100 msec after the beginning of the γ response, this change lasting several seconds. Since at that time the α firing of the branch of the LG nerve had not started, the increase in group I afferent fibre activity has to be considered as resulting exclusively from the activation of muscle spindle primary endings caused by the γ response. However, owing to the very high sensitivity of the primary endings to any mechanical event, an action caused by the activity generated in the underlying soleus muscle (as indicated by the triceps surae electromyogram (EMG)), cannot be excluded. The rise in the firing frequency of group II afferent fibres starts shortly (30–50 msec) after the onset of the increase in group I firing; it can be definitely attributed to the activation of muscle spindle secondary endings. At the time of the emergence of the α LG response the increase in group II activity, as well as that of group

I, does not undergo a weakening—on the contrary, a reinforcement is observed. It may be supposed that in addition to activation by contraction of intrafusal muscle fibres, secondary endings of spindles may have been stimulated by the asynchronous contraction of extrafusal motor units of lateral gastrocnemius muscle.

Reciprocal inhibition during flexor reflexes

On suitably stimulating the sural nerve a flexor reflex was observed ipsilaterally. During this reflex the electromyographic activity of triceps surae successively underwent inhibition and activation. EMG suppression started 150–200 msec after the first stimulus and lasted 150–200 msec. Afterwards a long-lasting (5–10 sec) activation was observed that often began before the cessation of the stimulation.

Likewise the α-axon discharge in the branch of the LG nerve was successively inhibited (150–200 msec after the first stimulus) and then activated. However, inhibition and activation lasted longer (500–600 msec) and shorter (5 sec), respectively, than similar features observed in the electromyographic activity of triceps surae. The duration disparity of both inhibition and activation of the LG α firing frequency and of the triceps surae EMG was in agreement with the fact that the large α motoneurons innervating LG muscles have a higher threshold than the smaller ones innervating soleus.

The mean firing frequency of γ axons was never depressed; on the contrary, it underwent an increase, the onset of which appeared during the inhibition of α motoneurons and occurred approximately 150 msec before the post-inhibitory facilitation of the LG α neurons.

An increase in group I afferent activity always appeared before the beginning of the rebound in the LG α axon activity. When this group I activity took place before the beginning of the firing increase of γ axons, it could be attributed to the isometric relaxation of LG. The activity of group II afferent fibres underwent an increase that began about 30 msec after the onset of increase of firing frequency of group I afferent fibres.

Some important features may be established from observing these changes of activity in the branch of lateral gastrocnemius nerve with special reference to the γ activity:

1. Afferent volleys in the sural nerve always produced an increase in the mean firing frequency of γ axons whatever the reflex pattern produced in the motor units of the LG muscle.
2. γ axon activation was brought about both by the increase in activity of tonically firing axons and by the recruitment of previously silent ones. At this time there is no evidence of simultaneous inhibition of any γ motoneurons. Generally, under resting conditions, among γ axons displaying tonic firing the fast-conducting γ-axon population is less

evident than the slow-conducting γ-axon population, numerous fast-conducting γ axons being silent. The reflex activation of the fast-conducting γ-axon population began later than that of the slow-conducting γ-axon population. All these observations are consistent with the idea that the discharge of small motoneurons is induced more easily than that of large ones.

3. During reflex responses α–γ coactivation was observed in the nerve branch of LG muscle, but LG γ activation was always earlier than LG α activation. With suitable stimulation even isolated γ responses could be elicited.

Myotatic reflex

The action of stretching and releasing the triceps surae on the activity of the branch of the lateral gastrocnemius nerve has been studied. Changes of muscle length were achieved by flexing the foot at the ankle; both of these anatomical structures were previously de-afferented by tightly encircling the distal part of the leg by a wire, but excluding the Achilles tendon. When the ankle was flexed from 120° to 90° in approximately 1 sec (phasic stretch) the electromyographic recording from triceps surae clearly indicated, following a delay of about 30 msec, the onset of the stretch reflex (phasic stretch reflex); small motor units began to discharge, followed by large ones. While the ankle was maintained at 90° during static stretch (5–15 sec), less-intense electromyographic activity was observed (tonic stretch reflex). When the ankle was extended from 90° to 120° in approximately 0.5 sec the electromyographic activity immediately stopped.

The α axons of the branch of the lateral gastrocnemius nerve were activated about 500 msec after the beginning of the phasic stretch; the α discharge in the LG nerve branch reached a peak when the phasic stretch reflex of the triceps surae was at its maximum and usually ended shortly after the phasic stretch and always before the muscle was released.

The γ axons responded to the stretch of triceps surae either with excitation or with inhibition, or with both excitation and inhibition. The excitation was characterised by a steep increase in the firing frequency during the dynamic stretch followed by a plateau from the onset of the static stretch. With muscle release the initial level of mean firing frequency returned with a slow decay (about 6 sec). When inhibition was elicited, there was a slow decrease in the γ firing frequency from the beginning of the static phase of the stretch. At the release of the stretch there was a fast return within 1 sec to the initial level of mean firing frequency. In the mixed pattern, the γ-axon discharge increased steeply during the dynamic stretch, and then, at the beginning of the static phase of the stretch, slowly decreased to reach a steady state below the initial level of mean firing frequency. At the time of the muscle release there was a sudden increase above the resting discharge followed by a slow decay.

The pattern of responses to muscle stretch and release observed during experiments depended upon the resting level of γ activity. Muscle stretch induced a pattern of excitation in preparations with a low degree of γ activity and a pattern of inhibition in preparations with a high level of γ activity. The mixed pattern was found most frequently. The γ motoneurons of the branch of the LG nerve underwent excitatory and inhibitory influences according to the size of their axons—the fast-conducting γ axons mostly underwent inhibition, whereas the slow-conducting γ axons mostly received excitation. When excitation was prevalent the γ responses always began earlier than the responses of the α neurons of the branch of the lateral gastrocnemius nerve and at approximately the same time as the electromyographic responses of the triceps surae.

The increase in group I afferent fibre activity in response to a mechanical stretch started immediately. During static stretch the group I activity adapted. Usually as soon as the muscle was released there was no pause in the group I discharge, but often, on the contrary, a rebound in activity was seen, which corresponded with the rebound in activity of γ axons when the mixed pattern of response was in evidence.

The increase in activity of group II afferent fibres began hardly any later than the increase in group I activity. The slope of the increase in activity was less important in group II axons than in group I axons. There was either no adaptation or only a very small one of group II discharges during the static stretch. Upon muscle release the decrease in the activity of group II was moderated by the γ discharge at that time and even exhibited a short firing increase when there was a γ rebound of activity.

ACKNOWLEDGEMENTS

This research was supported by INSERM ATP No. 297661 and by DGRST DN.P63 No. 79.7.1064.

Received on June 2nd, 1980.

REFERENCES

Eldred E., Granit, R. and Merton, P. A. (1953). Supraspinal control of the muscle spindles and its significance. *J. Physiol., Lond.*, **122,** 498–523

Hagbarth K. E. and Vallbo Å. B. (1967). Mechanoreceptor activity recorded percutaneously with semi-microelectrodes in human peripheral nerves, *Acta physiol. scand.*, **69**, 121–122

Hagbarth K. E. and Vallbo Å. B. (1968). Discharge characteristics of human muscle afferents during muscle stretch and contraction, *Expl Neurol.*, **22,** 674–694

Henneman E. and Olson C. B. (1965). Relations between structure and function in the design of skeletal muscles, *J. Neurophysiol.*, **28,** 581–598

Henneman E., Somjen G. and Carpenter D. O. (1965). Functional significance of cell size in spinal motoneurons, *J. Neurophysiol.*, **28,** 560–580

Hunt, C. C. (1951). The reflex activity of mammalian small nerve fibres, *J. Physiol. Lond.*, **115,** 456–469

Joffroy M. (1975). Méthode de discrimination des potentiels unitaires constituant l'activité complexe d'un filet nerveux non sectionné, *J. Physiol., Paris*, **70,** 239–252

Joffroy M. (1980). Une nouvelle méthode d'analyse in situ de l'activité des fibres nerveuses afférentes et efférentes; application à l'étude du système fusimoteur. Thèse doctorat d'Etat-Sciences Université Paul Sabatier, Toulouse

Matthews P. B. C. (1972). *Mammalian Muscle Receptors and Their Central Actions*, London, Edward Arnold, 546–611

Autogenetic and antagonistic group II effects on extensor gamma motoneurons of the decerebrate cat

J. NOTH*

SUMMARY

The influence of group II muscle afferents on ipsilateral gamma motoneurons supplying the triceps surae were investigated in decerebrate cats. A new technique of isolating functionally single gamma efferents in intact ventral root filaments was applied. Of 30 gamma efferents exhibiting a resting discharge, two thirds were markedly facilitated by electrical stimulation of group II fibres running in the nerve to the triceps surae (GS). The predominant effect of group II fibres of the deep peroneal (DP) nerve was inhibition. Stimulation of the nerve to the posterior biceps-semi-tendinosus (PBST) muscle was ineffective within the group I range but resulted in a strong fusimotor facilitation when the stimulus strength exceeded the group I level (12 out of 15 cells). Some functional aspects of the described fusimotor reflexes and some implications of the involvement of secondary muscle spindle afferents are discussed.

INTRODUCTION

Recent work with recordings from functionally single muscle spindle afferents in unrestrained cats and in man has promoted our knowledge about the role of muscle receptors involved in various types of movements. Yet at present we can only speculate about the changes in fusimotor activity which influence the activity of muscle spindle afferents during such motor tasks. The interpretation of these results is indeed so difficult that it seems impossible to draw unambiguous conclusions from these studies. Primary spindle afferents, for instance, can be influenced by overall length changes of their own muscle, by contractions of motor units within the muscle and by intrafusal contractions

*Department of Neurophysiology, University of Freiburg

induced by static and dynamic fusimotor fibres. For a full account of the functional role of the muscle spindle system in motor control, we therefore need to know more about the fusimotor outflow to the muscle spindles during various motor tasks. This paper is concerned with the particular aspect of the control of fusimotor activity by reflexes evoked from group II muscle afferents (for a general review, see Murthy, 1978).

For technical reasons, autogenetic group II effects on gamma motoneurons are difficult to study. Thus it has remained an open question as to whether secondary spindle endings can facilitate their own gamma motoneurons, as they do their own alpha motoneurons in extensors (Matthews, 1969; Kirkwood and Sears, 1975; Kanda and Rymer, 1977). Intracellular recordings from a limited number of gamma motoneurons suggest that triceps surae gamma motoneurons receive oligosynaptic group II facilitation by single-shock electrical stimulation of the homonymous muscle nerve (Appelberg *et al.*, 1977) and it was thus thought desirable to investigate the influence of repetitive electrical group II stimulation on the discharge rate of gamma motoneurons, i.e. the net group II effect. This has now become possible with a new method of isolating functionally single gamma efferents in intact ventral root bundles (Noth and Thilmann, 1980). With regard to the autogenetic fusimotor effects, this technique allows for the first time a comparison between the electrically induced reflexes and those evoked by stretch and vibration of the homonymous muscle. Furthermore, the responsiveness of a gamma motoneuron to a range of other stimuli can also be tested, because the recording condition normally remains stable for hours. In this paper the results obtained with electrical stimulation within the group II range of the nerve to the triceps surae (GS), the deep peroneal (DP) nerve, and the nerve to the posterior biceps-semitendinosus (PBST) are presented.

METHODS

Thirty gamma efferents supplying the triceps surae were investigated in decerebrate cats. All units exhibited a resting discharge. The cats were paralysed and their left hindlimb was denervated sparing only triceps and the hip muscles. The triceps muscle was freed from surrounding tissue and the Achilles tendon was cut and tied to a stretching device. The gamma efferents were isolated in small natural ventral rootlets, which were left in continuity until the peripheral destination of the efferent was determined. This was done using the antidromic collision technique. Thus by the occurrence of a collision block it was established that both the spontaneous orthodromic discharge and the antidromically elicited potential were recorded from the same fibre (see inset B of figure 2). After the identification, the small isolated filament and the rest of the ventral roots L6–S2 were cut in order to eliminate unwanted antidromic fusimotor effects during electrical stimulation of muscle nerves.

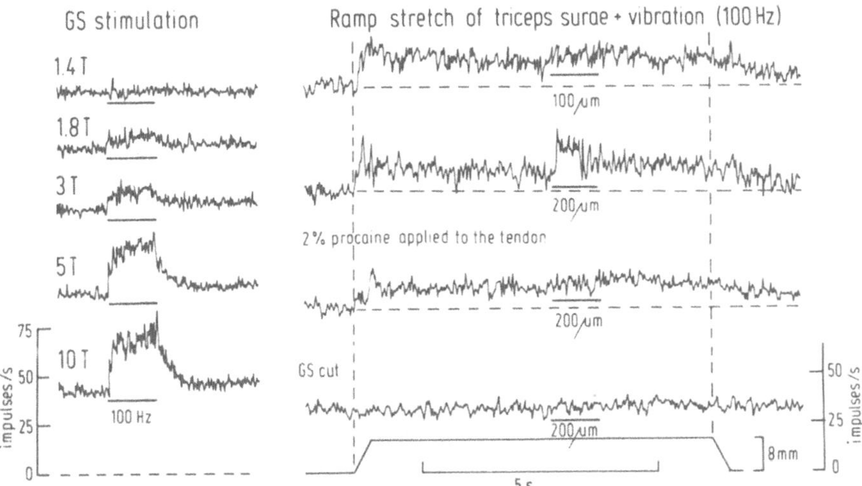

Figure 1. Response of GS gamma efferent (38 m/sec) to repetitive electrical stimulation of the ipsilateral GS nerve (left records), and to stretch and vibration of the triceps surae (right records). Traces are averages (n = 10) of the cell's instantaneous frequency. Electrical and vibratory stimuli (100 Hz for 1 sec) are marked by a horizontal bar below each record. Stimulus strength given relative to threshold group I. 8 mm ramp stretches of the paralysed triceps (−10 mm to 2 mm maximum physiological length) are indicated by schematic drawing (bottom trace).

RESULTS

Of the 30 triceps gamma motoneurons, 20 were markedly facilitated by electrical stimulation of the GS nerve within the group II range (cf. Noth and Thilmann, 1980). The threshold of this reflex ranged from 1.6 to 2.0T (T being the threshold of the compound group I action potential recorded from the dorsal roots L7 and S1). This was best seen with cells lacking a group I effect. Such a cell is depicted in figure 1. In addition to the electrically evoked facilitation, this cell responded to ramp stretch of the triceps with a tonic facilitation. Control experiments showed that this stretch-induced facilitation was not strongly, or at least not exclusively, mediated by stretch-sensitive receptors located within the tendon. Thus the right-hand records of figure 1 show that procainisation of the tendon abolished the facilitation which could be evoked by the squeezing of the tendon or by muscle vibration at 200 μm. However, it did not markedly reduce the stretch-induced effect. With a few cells, a mixed facilitatory and inhibitory response was observed, which manifested itself as a steep increase in activity at the onset of the repetitive nerve stimulation (100 shocks/sec for 1 sec). This was then followed by a rapid decline during the ongoing stimulation. The decline was not exclusively attributable to a fast decay of facilitation, because in some cases the discharge rate clearly fell below the resting level. This finding agrees well with the results of Appelberg *et al.* (1977) with intracellular recording from GS gamma motoneurons under otherwise comparable conditions. They observed autogenetic early group II EPSPs followed by IPSPs which were evoked at a stimulus strength of 5T.

The main group II effect of the antagonistic DP nerve was inhibition, which is in agreement with an earlier study (Fromm *et al.*, 1976). The antagonistic group II inhibition appeared in a few cells without a low-threshold group I fusimotor effect. But, as a rule, triceps gamma motoneurons receive strong antagonistic group I inhibition (Fromm *et al.*, 1976), so that the additional group II inhibition expressed itself by a further reduction of activity when the stimulus strength was raised above 2T. In contrast to the earlier work, some cells were facilitated by stimulation of antagonistic group II fibres.

Recent work (Appelberg *et al.*, 1977, 1979, and in preparation) suggests a specific fusimotor reflex from ipsilateral group II afferents on to dynamic gamma motoneurons to the triceps surae. Such fusimotor responses were readily obtained by electrical stimulation of the PBST nerve and by stretching the ipsilateral PBST muscles. But since these results were either obtained with micro-electrode recording of PSPs or inferred from Ia afferent responses

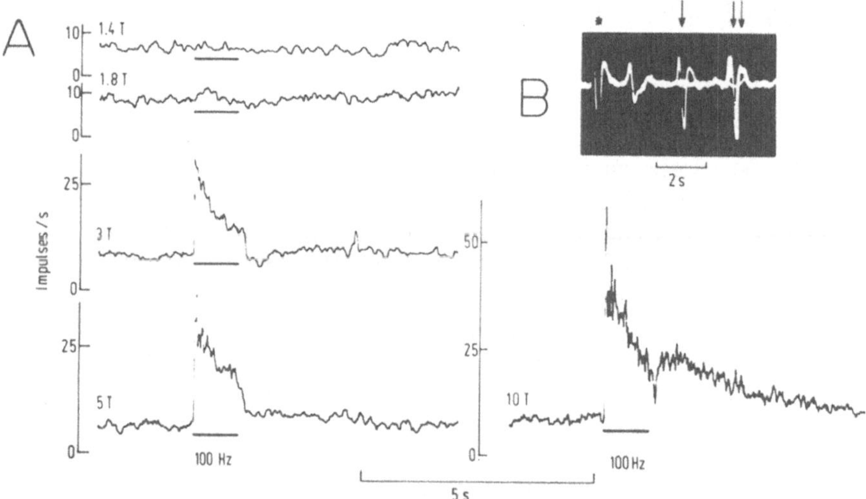

Figure 2. (A) Response of GS gamma efferent (26 m/sec) to repetitive electrical stimulation of the PBST nerve (A). Stimulation and recording corresponds to that of figure 1. Inset (B) exemplifies the identification of gamma efferents (same cell as depicted in (A). Five responses to single-shock stimulation (2.2 V) of the GS nerve are superimposed, recorded in a fine ventral rootlet left in continuity. Asterisk marks the stimulus artefact. Note that the spontaneous orthodromic potential blocks the antidromically elicited one (double arrow), which proves that they run in the same filament. For further details see text.

to sinusoidal muscle stretches, it remained to be seen as to whether or not similar effects could be revealed when the impulse activity of gamma efferents was monitored.

Of 16 cells tested in the present study with electrical stimulation of the ipsilateral PBST nerve, 13 indeed were strongly facilitated when the stimulus strength exceeded 2T. Since notable group I effects were never encountered, the threshold of the group II fusimotor reflex could be assessed with higher accuracy than with the fusimotor reflexes described above. The records of figure 2 are representative in this respect and show that the threshold lies around 1.8 for this particular cell. The post-excitatory rebound which was first seen at 10T suggests that higher threshold muscle afferents exerted a qualitatively different effect from that seen at 5T and below. The results on PBST nerve stimulation are summarised in figure 3. Again it is seen that the threshold is centred around 1.8T where group II fibres are known to be excited, and that the largest increment in reflex action appears between 1.8T and 3.0T. Since only weak and inconsistent responses were obtained from the nerve to the anterior biceps semi-membranosus muscles, the excitatory effects from group II afferents on triceps gamma motoneurons seem to be muscle specific. The present result, on the other hand, does not favour the view (Appelberg *et al.*, 1977) that only dynamic gamma motoneurons were

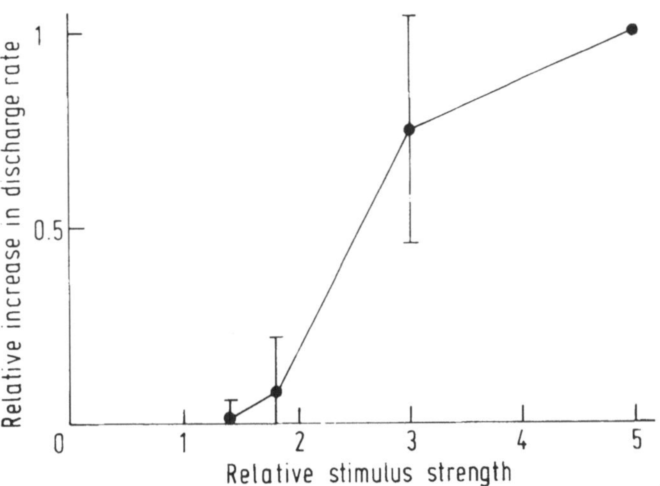

Figure 3. Summarised effect of ipsilateral PBST stimulation on 13 triceps gamma efferents. Ordinate gives the mean increase in discharge rate ± S.D. relative to that measured at 5T. In case of a phasic–tonic response (compare figure 2), the mean change in firing rate during the repetitive stimulation (100 shocks/sec for 1 sec) was read off by eye. On the abscissa, the stimulus strength is plotted relative to the group I threshold. Mean absolute increase in firing rate at 5T was 23.3 ± 16.6 imp/sec (± S.D., n = 13). The three cells without PBST effect are omitted from this graph.

involved as target cells in this reflex, because a bias towards preferential isolation of dynamic gamma efferents was unlikely in the present decerebrate preparations.

DISCUSSION

In order to assess the functional significance of the described fusimotor group II reflexes, it would be desirable to know more about the receptor type involved. Although group II receptors other than secondary spindle endings seem to be rare in the triceps surae (Boyd and Davey, 1968), a few non-spindle group II fibres could nevertheless exert a strong fusimotor effect. Even the fact that the electrically evoked facilitation was paralleled by stretch-induced facilitation in the same gamma efferent does not strictly prove a contribution or sole responsibility of secondary spindle afferents, since non-spindle group II receptors may also be sensitive to muscle stretch (Paintal, 1960). In the case of the group II facilitation evoked by PBST nerve stimulation, the situation is similar. Nevertheless, the low threshold of the fusimotor group II reflex, which was around 1.8T, and the steep rise of the reflex action between 1.8 and 3.0T (figure 3), makes it most likely that the main afferent input stemmed from the secondary spindle afferent fibres.

In any case, stretching of the homonymous muscle and of other closely related limb muscles can induce activation or inhibition of extensor gamma motoneurons which, in turn, will influence the gain around the spindle loop for small and large movements (Hulliger *et al.*, 1977). As far as the autogenetic group II reflex is concerned, it closely parallels the recently postulated autogenetic group II facilitation of triceps surae alpha motoneurons. Yet any involvement of static gamma motoneurons, as is indeed suggested by the present results, would establish a positive feedback loop, resulting in a high fusimotor activity. Such a loop may be useful for presetting the fusimotor activity to a high level in order to prepare the muscle for a certain motor task (i.e. during a specific phase of the step cycle). But an effective inhibitory control of this loop has to be postulated (Appelberg, 1981) which should be able to interrupt the loop when necessary. If, in a pathological state, such an inhibitory control were permanently withdrawn, a steady hyperactivity of gamma motoneurons could result, and the possibility should be considered that such a mechanism contributes to the muscle rigidity seen in the extensors of the decerebrate cat.

ACKNOWLEDGEMENTS

This work was supported by grants of the Deutsche Forschungsgemeinschaft, SFB 70. The author is grateful to Dr M. Hulliger for advice and help with the English version of the manuscript.

Received on June 30th, 1980.

REFERENCES

Appelberg, B. (1981). Selective central control of dynamic gamma motoneurons utilised for the functional classification of gamma cells, this publication

Appelberg, B., Hulliger, M., Johansson, H. and Sojka, P. (1979). Excitation of dynamic fusimotor neurones of the cat triceps surae by contralateral joint afferents, *Brain Res.*, **160,** 249–307

Appelberg, B., Johansson, H. and Kalistratov, G. (1977). The influence of Group II muscle afferents and low threshold skin afferents on dynamic fusimotor neurones to the triceps surae of the cat, *Brain Res.*, **132,** 153–158

Boyd, J. A. and Davey, M. R. (1968). *Composition of Peripheral Nerves*, Edinburgh, Livingstone

Fromm, C., Noth, J. and Thilmann, A. (1976). Inhibition of extensor γ-motoneurons by antagonistic primary and secondary spindle afferents, *Pflügers Arch.*, **363,** 81–86

Hulliger, M., Matthews, P. B. C. and Noth, J. (1977). Static and dynamic fusimotor action on the response of Ia fibres to low frequency sinusoidal stretching of widely ranging amplitudes, *J. Physiol., Lond.*, **267,** 811–838

Kanda, K. and Rymer, W. Z. (1977). An estimate of the secondary spindle receptor afferent contribution to the stretch reflex in extensor muscles of the decerebrate cat, *J. Physiol., Lond.*, **264,** 63–87

Kirkwood, P. A. and Sears, T. A. (1975). Monosynaptic excitation of motoneurones from muscle spindle secondary endings of intercostal and triceps surae muscles in the cat, *J. Physiol., Lond.*, **245,** 64–66P

Matthews, P. B. C. (1969). Evidence that the secondary as well as the primary endings of the muscle spindles may be responsible for the tonic stretch reflex of the decerebrate cat, *J. Physiol., Lond.*, **204,** 365–393

Noth, J. and Thilmann, A. (1980). Autogenetic excitation of extensor γ-motoneurones by group II muscle afferents in the cat, *Neurosci. Lett.*, **17,** 23–26

Paintal, A. S. (1960). Functional analysis of group III afferent fibres of mammalian muscles, *J. Physiol., Lond.*, **152,** 250–270

SECTION 3
MUSCLE AFFERENT DISCHARGE IN VOLUNTARY MOVEMENT

Overview

Attempts to find out how the fusimotor system is used in controlling normal movements have mainly centred on making single unit recordings of spindle afferents and deducing from them and movement records what static and dynamic activities must be present. Three main experimental approaches have given useful data of this kind. The introduction of recording unit activity with metal microelectrodes in human nerves by Hagbarth and Vallbo (microneurography) has yielded much interesting data over the past thirteen years and this is reviewed separately by those authors. Equivalent data in animals were first obtained in 1973 from jaw muscle spindle afferents recorded in mid-brain in cats and monkeys, and by Prochazka and his colleagues who in 1975 were able to achieve the same result by implanting microwires in dorsal roots in the lumbo-sacral region. Loeb's group independently developed a similar technique by implanting wires in the dorsal root ganglia. The resulting data from human and animal experiments have given somewhat different views of the relationship of alpha activity to fusimotor activity. Human experimental data have been interpreted as generally supporting the idea of alpha–gamma coactivation as a principle of motor organisation. On the other hand, the animal data seem to favour more independence of action of the two systems, but according to Prochazka and Wand, with the spindles usually responding more obviously to length changes than to skeletomotor-coupled fusimotor changes. Two presentations from each camp restate these somewhat contrasting views and provide new data. Video films were shown by Hagbarth; Loeb and Hoffer; and Prochazka and Wand. The critiques which followed each pair of presentations gave a good opportunity for considered discussion to be presented. Some of the differences may have been explained but others remain to stimulate the various groups to return to their laboratories. The feeling was certainly generated amongst the audience that the two different types of experiment were studying rather different movements and that the strategy of fusimotor involvement need not necessarily be the same in both cases. It also emerged that deduction of fusimotor activity patterns from spindle activity and movement is not a simple matter.

The papers in this section were read at the symposium, with the exception of the second paper of Prochazka and Wand, the video recordings of which formed part of a demonstration.

Muscle spindle function during normal and perturbed locomotion in cats

G. E. LOEB* AND J. A. HOFFER*

SUMMARY

The available evidence about the behaviour of muscle spindle afferents during normal movements obtained in this and other laboratories is reviewed.

The data taken as a whole do not appear compatible with any single pattern of fusimotor control.

We suggest that spindles may be grouped into functional subsets based on the tasks being performed by their muscles of origin. At least two such sets have been identified during treadmill locomotion: (1) flexor-task spindles in which afferent firing persists during rapidly shortening, unopposed contractions (presumably mediated by gamma-static coactivation), and (2) extensor-task spindles in which afferent activity is dominated by stretch sensitivity (probably velocity enhanced by gamma dynamic activity) during isometric and lengthening contractions.

New data are presented demonstrating that cutaneous afferent stimulation causes rapid and large modulations of spindle sensitivity, presumably through complex gamma motoneuron reflexes.

REVIEW

Of all our senses, proprioception is, perhaps, the poorest understood. Producing 'sensibility' rather than 'sensation', its signals are rarely, if at all, directly consciously perceived (*see* Roland, 1978, for review). Worse, we have little quantitative data regarding normal receptor function in intact, behaving organisms. The classical physiological techniques successfully employed in the study of end-organ activity in other senses are not suitable for an

* Laboratory of Neural Control, National Institute of Neurological and Communicative Disorders and Stroke, NIH Bethesda

environment of movement and mechanical stress. Certain proprioceptive end-organs such as the Golgi tendon organ, skin stretch receptors, and, less certainly, the joint receptors, have mechanical structures simple enough that we feel somewhat confident inferring their normal function from recordings obtained in reduced, anaesthetised preparations. This is clearly not the case for the muscle spindles, each of which is a highly complex assembly of specialised afferent endings responsive to several kinds of intrinsic, as well as extrinsic, mechanical events. Most problematic is the fact that the nature of the spindle afferents' responses to extrinsic factors such as length and velocity of stretch can be potentially completely dominated by the activity of the several kinds of intrafusal muscle fibres, whose motoneurons receive inputs from afferent, segmental and descending pathways (*see* Murthy, 1978, for review).

Nevertheless, theories of how spindles might be used in intact animals have arisen, based largely on concepts of feedback and servo-control which were developed by, and borrowed from, electrical engineering in the 1950s. Such theories, derived from limited observations under restricted conditions, have tended to be simple and global on the assumption that spindles in general must subserve some universal function such as the regulation of muscle length (Merton, 1953) or follow some simple rule such as 'alpha–gamma coactivation' (Granit, 1970). As preparations have become available which revealed the activity of other populations of spindles during other limited behaviour, the difficulty of fitting all the data to any single theory has become apparent.

We have no alternative simple or universal rule or role to offer. However, our experience to date suggests that it might be useful first to consider what the muscles are trying to accomplish before considering how the spindles help them to do it. During locomotion, functionally pure extensors (anti-gravity muscles) are mostly active while lengthening or nearly isometric (Grillner, 1975). This spring-like activity is an energy efficient way to generate the large forces (but low work output) required to maintain a non-accelerating forward motion in which the body weight must still be transferred completely from side to side (Walmsley *et al.*, 1978). The loads and their rates of change are highly dependent on gait speed. In contrast, most functionally pure flexors shorten actively and rapidly against minimal loads (essentially inertial mass of the limb). The rates of shortening and loads are much less dependent on gait speed since swing time tends to stay constant (Grillner, 1975). However, trajectory variations have great behavioural significance. It would seem necessary for the level of extensor pre-activation to be influenced by anticipation of the foot placement position, which could be obtained by monitoring the progress of the flexor muscles. It seems inappropriate to design a sense organ with multiple, independent modes of control and then use it in the same way to regulate both of the above tasks.

Activity recorded from extensor muscle spindles has suggested little modulated fusimotor influence during locomotion. Spindle endings responded when and to the degree to which the muscle was stretched and they fell silent when the muscle rapidly shortened (Prochazka *et al.*, 1976, 1977; Loeb and Duysens, 1979). The possibility of some gamma dynamic enhancement of velocity sensitivity has been difficult to assess without knowing the velocity sensitivity of the endings when de-efferented; the data contain little to suggest any strong alpha–gamma static coactivation. In contrast, human microneurography studies have shown strong spindle primary activation under isometric and active shortening conditions, which appeared to be generated in fixed proportion to the intensity of voluntary extrafusal recruitment (Vallbo, 1970; Hagbarth and Vallbo, 1968; Hagbarth *et al.*, 1975). We have elsewhere (Loeb and Duysens, 1979), and will here, show examples of spindle activity in flexor muscles which are also consistent with such strong alpha–gamma-static 'coactivation'. Recording and unit identification techniques have been described elsewhere (ibid. and Loeb *et al.*, 1977). The spindle afferents reported here all had similar conduction latencies, but we cannot state whether these were primaries or secondaries. Recently perfected techniques for obtaining accurate axonal conduction velocities were not available at the time these recordings were made (Hoffer *et al.*, 1979).

RECENT RESULTS

One of the difficulties of comparing spindles in different muscles is that length changes are usually not comparable and differences in extrafusal fibre types and distributions from which EMG records must be obtained make level of effort comparisons unreliable. The biarticular and bifunctional anterior thigh muscles, rectus femoris (RF) and sartorius (SA) pars lateralis (knee extensor portion), provide an interesting test case. Both muscles extend the knee and flex the hip and both are active in the stance (extension) and swing (flexion) phases of gait (our records show RF active in swing mostly for faster gaits). Since the hip movements cause larger length changes than knee movements, activation during the stance phase occurs as the muscles lengthen; activation during the swing phase occurs as the muscles rapidly shorten.

Figures 1 and 2 show spindle afferent activity from RF and SA pars lateralis, respectively, occurring only during the stance phase, when the muscles were actively lengthening. Inspection of the records suggests that the extent of stretch and velocity of stretch were both important factors but that the level of extrafusal activation (demonstrated by the EMG) was not well correlated with spindle activation. (This was particularly true for the sartorius spindle, which was completely silent during the large extrafusal activation accompanying the flexion phase shortening contraction.) The responses to low-level, single-sural nerve shocks (1 mA, 0.1 msec, via implanted bipolar nerve cuff) are revealing. The low-rate, irregular spindle discharge during quiet standing was interrupted by the reflex responses to the stimuli, which

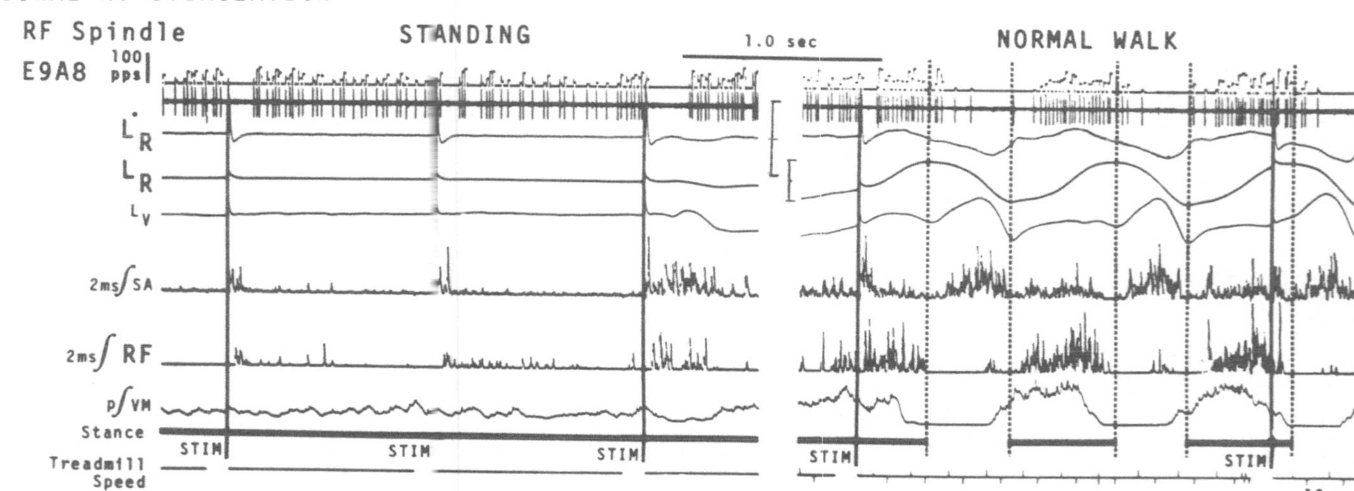

Figure 1. Activity from RF spindle afferent and kinesiological records during quiet standing and normal walking, both interrupted periodically by single electrical stimuli to sural nerve (solid vertical lines). Traces from bottom up: treadmill speed (10 cm per pair); stance phase of ipsilateral hindlimb (from 60 field/sec videotape, dotted vertical lines at foot fall and lift); Paynter filter integration (50 msec time constant) of vastus medialis (VM) bipolar implanted EMG; pulsed integration (2 msec bins, *see* Bak and Loeb, 1979) of RF EMG; pulsed integration of SA EMG; length of vasti muscles (length gauge implanted between greater trochanter and proximal patellar margin, stretch up); length of RF/SA muscle group (gauge implanted from anterior iliac crest to proximal patellar border, calibration shows normal walking excursion of RF from 82 to 92 mm, rest length 87 mm, based on cadaver measurements at replicated limb positions); velocity of RF stretch (electronic time derivative of length, calibrated at ± 1 rest length per second); unprocessed microelectrode signal; instantaneous frequencygram of RF spindle afferent.

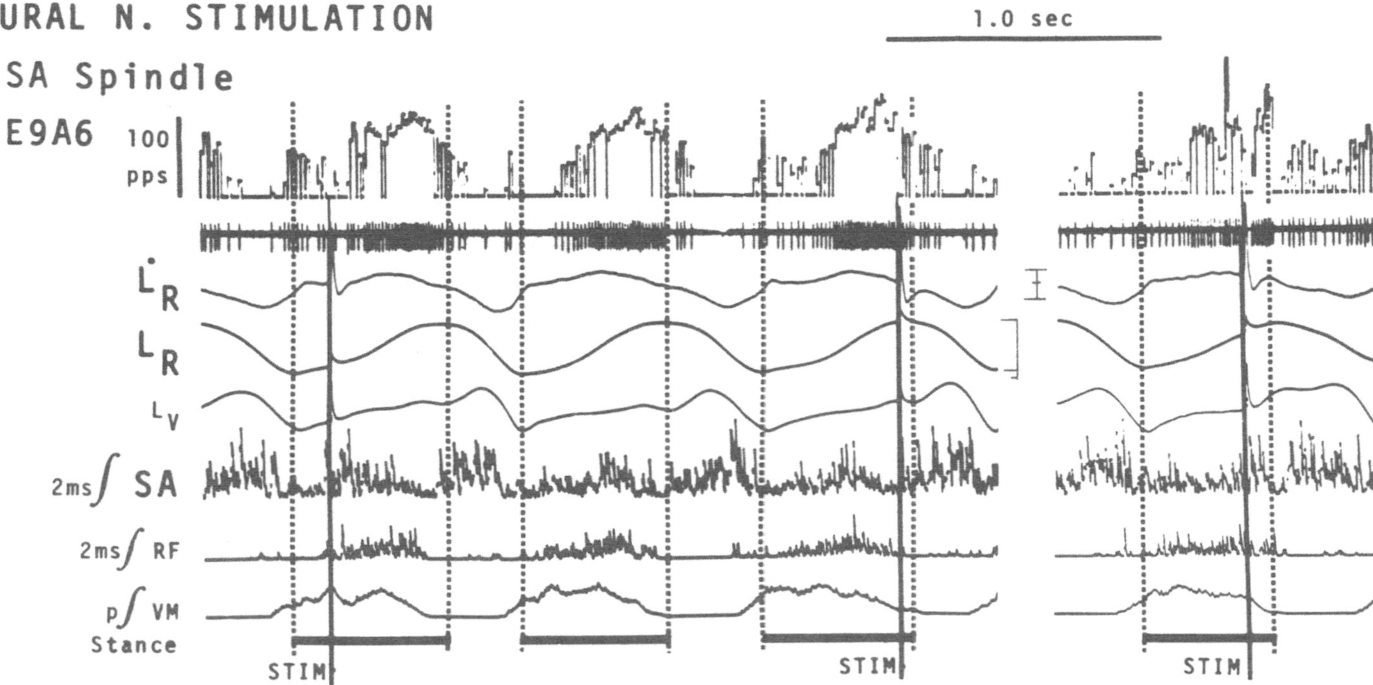

Figure 2. Unit activity from SA spindle during walking with periodic sural nerve stimuli, traces as in figure 1. RF/SA length (L_R) calibration shows normal walking excursion of SA pars lateralis (103–127 mm, 101 mm rest length); velocity calibration is ± 1 SA rest length per second.

consisted of low-amplitude extrafusal activation of the nearly isometric muscle (length gauge deflections are stimulus artefacts as shown by zero latency and confirmed on videotape). During walking, the same stimuli caused similar spindle pausing, but mostly when the spindle was being stretched rapidly. These records suggest that this spindle received little gamma static input and that even this (during standing) tended to be inhibited during reflexly generated extrafusal activity. The stretch velocity sensitivity was probably enhanced by gamma dynamic input which was also inhibited during reflexive muscle activation.

One anomaly remains, however. Following the third stimulus in figure 2, the spindle silencing immediately gave way to a large rebound accompanying the large extrafusal reflex and rapid knee flexion. The subsequent flexion phase limb trajectory was quite abnormal and brisk spindle firing persisted during muscle shortening. While the early flexion firing could be interpreted as a response to gamma static coactivation unmasked by the decreased shortening velocity, the later activity suggests a persistently changed fusimotor programme. A similar late swing burst was seen in another SA (knee extensor part) spindle in an animal with an exaggerated flexion as a result of a limp.

The activity of another SA spindle afferent shown in figure 3 reveals an entirely different pattern. The standing sequence is the same as shown in figure 1 for the simultaneously recorded RF spindle, but the activity of this SA spindle was accelerated by the stimulation rather than inhibited, consistent with alpha–gamma coactivation. The walking sequence also suggests that a powerful gamma static input maintained spindle firing during the rapid active shortening during flexion. However, there was little spindle acceleration following the same stimuli given during walking despite large extrafusal reflex activation (although EMG may not reflect activity of the truly homonymous motor pool; *see* below). Activity from this spindle during a fast trot has been published elsewhere (Loeb, 1980) and shows that even when the swing phase shortening exceeded three rest lengths per second, the spindle activity continued at 100–150 pps without even decelerating. The stance phase active lengthening, also about three rest lengths per second peak velocity, resulted in about the same spindle firing rates. (A similar pattern was recorded recently from an ending identified as a primary from SA, knee flexor part.) The sustained moderate firing rate (even in the face of extrafusal reflexes) would be well suited for the detection of unexpected perturbations of muscle length in either direction.

It might be asked whether these spindles represent extremes of a continuous distribution or a real dichotomy in spindle behaviour (and presumably fusimotor activity) within a single muscle. Records of five other spindles from the RF and SA muscles have not demonstrated any intermediate

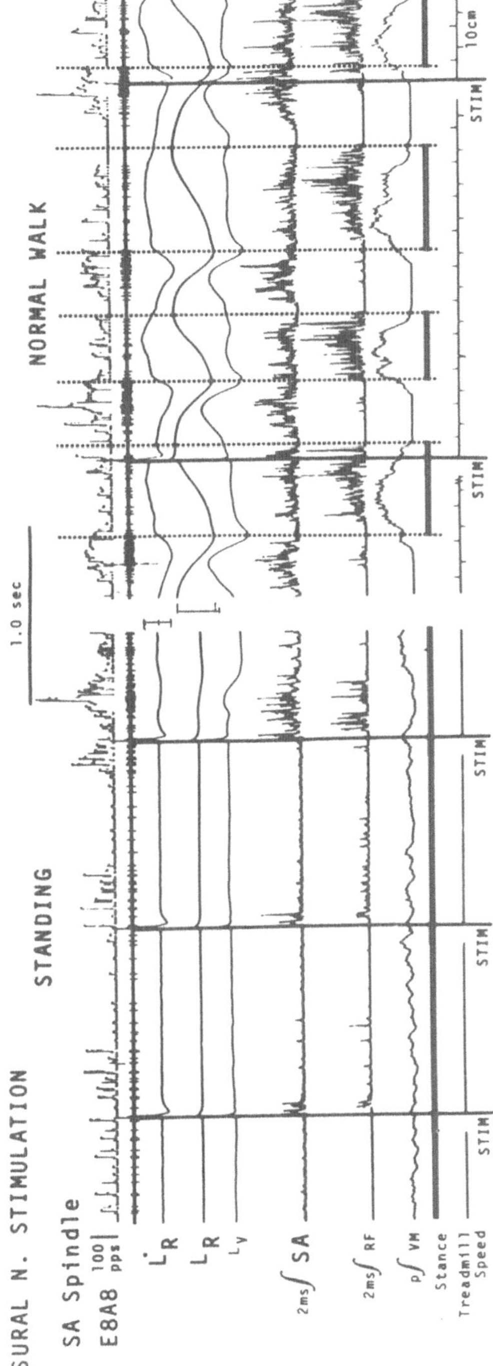

Figure 3. Unit activity from another SA spindle from the knee extensor part, recorded simultaneously with the RF spindle shown in figure 1, traces as in figure 1 with length and velocity calibration for SA excursions as in figure 2.

patterns. More intriguing, though, are recent recordings of alpha motoneurons from the same muscles suggesting that a similar dichotomy obtains extrafusally: motoneurons contributing to the flexor burst in SA are not recruited during extension and those active during extension are silent during flexion (Hoffer *et al.*, 1980). Eccles and Lundberg (1958) suggested a functional division of the SA motoneuron pools projecting to the knee flexor and knee extensor parts of the muscle. Our recordings of EMG from the knee extensor part show bursts of activity in both phases (but the knee flexor part is active only during flexion). Our data showing two kinds of spindle activity in the knee extensor part of the muscle might then reflect two complete and separate motor control networks (alpha motoneurons, gamma motoneurons and spindle afferents) within the same gross anatomical structure.

DISCUSSION

Data from a preparation similar to ours led Prochazka *et al.* (1979) to suggest that spindle afferents, as a rule, are dominated by passive stretch sensitivity when their muscles undergo length changes greater than 0.2 rest lengths per second, and are dominated by fusimotor effects only for slower, near-isometric conditions. While this is an acceptable interpretation of figures 1 and 2 and other extensor spindles we have studied, it directly conflicts with figure 3 and other flexor spindles we have reported (Loeb and Duysens, 1979). From human microneurography, Burke and colleagues (*see* Burke, 1980, for review) have concluded that, as a rule, fusimotor neurons are coactivated in proportion to extrafusal motoneurons so that spindles are not good length transducers during extrafusal activity. This would account for neither of our extensor-like spindles reported here nor the abrupt fusimotor modulations inferred from the reflex responses. However, a modification of this statement might be considered: alpha–gamma coactivation could occur in both kinds of spindles during voluntary movements, but this would involve gamma dynamic effects in extensors and gamma static effects in flexors.

In attempting to resolve these differences, we would point out that Prochazka's group has recorded mostly from extensor or bifunctional (potentially extensor) muscle afferents, whereas Burke's group has recorded mostly from flexors and not during locomotory reflexes which might reveal gamma dynamic effects. It should also be kept in mind that inferences about fusimotor activation rest on untested assumptions about 'passive' spindle properties. We are currently working on techniques to demonstrate the response properties of chronically recorded units to automatically replicated limb trajectories during deep barbiturate anaesthesia.

Be that as it may, neither the normal activity of these spindles nor their responses to cutaneous stimuli seem likely to be explained by any single rule of fusimotor control. In attempting to find some pattern in these and other such data, we have noted similarities among some bifunctional muscle spindles and pure flexor spindles, and among other bifunctional spindles and pure extensor spindles (Loeb, 1980). A similar dichotomy was noted by Feldman *et al.* (1977) during scratching movements in decerebrate cats. The coexistence of different spindle utilisation patterns in bifunctional muscles would support the notion that there are identifiable motor tasks which call for particular modes of fusimotor control. However, these identified activity patterns certainly do not exhaust the possibilities and probably not the reality. The increasingly apparent diversity of the fusimotor control apparatus, including various types of gamma and beta motoneurons, is consistent with our previous suggestion (Loeb and Duysens, 1979) that the muscle spindles are independent sense organs capable of being used for control of a wide variety of different motor tasks.

ACKNOWLEDGEMENT

We thank present and former members of the Laboratory of Neural Control whose work and ideas have contributed to this project, including M. J. Bak, R. E. Burke, J. Duysens, W. B. Marks, M. J. O'Donovan, C. A. Pratt and B. Walmsley.

Received on June 8th, 1980.

REFERENCES

Bak, M. J. and Loeb, G. E. (1979). A pulsed integrator for EMG analysis, *Electroenceph. clin. Neurophysiol.*, **47**, 738–741

Burke, D. (1980). The activity of human muscle spindle endings in normal motor behavior, *International Review of Physiology: Neurophysiology IV* (Edited by R. Porter), Baltimore, University Park Press

Eccles, R. M. and Lundberg, A. (1958). Integrative pattern of Ia synaptic actions on motoneurones of hip and knee muscles, *J. Neurophysiol.*, **144**, 271–298

Feldman, A. G., Orlovsky, G. N. and Perret, C. (1977). Activity of muscle spindle afferents during scratching in the cat, *Brain Res.*, **129**, 192–196

Granit, R. (1970). *The Basis of Motor Control*, New York, Academic Press

Grillner, S. (1975). Locomotion in vertebrates: Control mechanisms and reflex interactions, *Physiol. Rev.*, **55**, 247–304

Hagbarth, K.-E. and Vallbo, A. B. (1968). Discharge characteristics of human muscle afferents during muscle stretch and contraction, *Expl Neurol.*, **22**, 674–694

Hagbarth, K.-E., Wallin, B. G. and Löfstedt, L. (1975). Muscle spindle activity in man during voluntary fast alternating movements, *J. Neurol. Neurosurg. Psychiat.*, **38**, 625–635

Hoffer, J. A., Loeb, G. E., O'Donovan, M. J. and Pratt, C. A. (1980). Unitary activity patterns during walking confirm the existence of two functionally distinct classes of motoneurones in sartorius muscle of cat, *J. Physiol., Lond.*, to be published

Hoffer, J. A., O'Donovan, M. J. and Loeb, G. E. (1979). A method for recording and identifying single motor units in intact cats during walking, *Soc. Neurosci. Abst.*, **5**, Abstract 1248

Loeb, G. E. (1980). Somatosensory unit input to the spinal cord during normal walking, *Canad. J. Physiol. Pharm.*, to be published

Loeb, G. E., Bak, M. J. and Duysens, J. (1977). Long-term unit recording from somatosensory neurons in the spinal ganglia of the freely walking cat, *Science*, **197**, 1192–1194

Loeb, G. E. and Duysens, J. (1979). Activity patterns in individual hindlimb primary and secondary muscle spindle afferents during normal movements in unrestrained cats, *J. Neurophysiol.*, **42**, 420–440

Merton, P. A. (1953). Speculations on the servo-control of movement. In *The Spinal Cord* (edited by G. E. W. Wolstenholme) London, Churchill, 247–255

Murthy, K. S. K. (1978). Vertebrate fusimotor neurones and their influences on motor behaviour, *Prog. Neurobiol.*, **11**, 249–307

Prochazka, A., Stephens, J. A. and Wand, P. (1979). Muscle spindle discharge in normal and obstructed movement, *J. Physiol., Lond.*, **287**, 57–66

Prochazka, A., Westerman, R. A. and Ziccone, S.P. (1976). Discharges of single hindlimb afferents in the freely moving cat, *J. Neurophysiol.*, **39**, 1090–1104

Prochazka, A., Westerman, R. A. and Ziccone, S. P. (1977). Ia afferent activity during a variety of voluntary movements in the cat, *J. Physiol., Lond.*, **268**, 423–448

Roland, P. E. (1978). Sensory feedback to the cerebral cortex during voluntary movement in man, *Behav. Brain Sci.*, **1**, 129–171

Vallbo, A. B. (1970). Discharge patterns in human muscle spindle afferents during isometric voluntary contractions, *Acta physiol. scand.*, **80**, 552–566

Walmsley, B., Hodgson, J. A. and Burke, R. E. (1978). Forces produced by medial gastrocnemius and soleus muscles during locomotion in freely moving cats. *J. Neurophysiol.*, **41**, 1203–1216

Independence of fusimotor and skeletomotor systems during voluntary movement

A. PROCHAZKA* AND P. WAND†

SUMMARY

Variations in the sensitivities of primary and secondary muscle spindle afferents were observed in cats performing voluntary movements. The time course of fusimotor action was deduced from these variations. In unobstructed movements, there was evidence of steady, relatively low levels of both static and dynamic action. In imposed movements involving slight to moderate resistance, eight of the nine primary endings studied showed evidence of increased, steady, dynamic action. This may therefore represent a class of movement which is most often associated with a known change in spindle sensitivity in the normal animal. In imposed movements involving substantial resistance, further fusimotor action of both types was implicated. Although there was some evidence for e.m.g.-linked increases in fusimotor action in two of the primaries, it was not possible to decide whether or not these were examples of α–γ linkage. We conclude that much of the functionally important fusimotor action occurring during normal movements in cats is independent of skeletomotor activity.

INTRODUCTION

Have mammals evolved fusimotor neurons separately from skeletomotor neurons only to activate them strictly in synchrony? This question has now come to a head, as a result of new data from normally behaving animals, which suggest that skeletomotor and fusimotor neurons may often be activated quite independently of each other by the CNS (Taylor and Cody, 1974;

* Sherrington School of Physiology, St Thomas's Hospital Medical School, London
† Max Planck Institut für Experimentelle Medizin, Göttingen

Prochazka *et al.*, 1979; Prochazka, 1980; Prochazka and Wand, 1980b; Loeb and Duysens, 1979; Loeb and Hoffer, 1980).

The concept of a strict linkage of skeletomotor (α) and fusimotor (γ) neurons (Granit, 1955) has received its most influential support from the human neurography studies of Hagbarth (1980), Vallbo (1980) and their co-workers. Thus Burke *et al.* (1979) stated that, 'Human studies indicate that, during a voluntary contraction, there is a very tight, possibly even "hard-wired" linkage between the skeletomotor and fusimotor drives'. Granit (1979) speaks of 'an ubiquity of alpha–gamma linkage that elevates it to a principle in motricity'.

Yet abundant evidence for selective activation of skeletomotor and fusimotor neurons in reduced animal preparations exists (Granit *et al.*, 1955; Alnaes *et al.*, 1965; Bergmans and Grillner, 1969; Post *et al.*, 1978). Furthermore, Appelberg (1980) has demonstrated that electrical stimulation in the rubral region of the mesencephalon activates dynamic fusimotor neurons without any simultaneous effects on static fusimotor or alpha motoneurons. The implication of strict alpha–gamma linkage would be that these various independent pathways would never be utilised in voluntary movement.

Fusimotor action during normal movements has almost exclusively been deduced from patterns of firing of muscle spindles. Matthews (1972) has listed the available criteria for deducing static and dynamic fusimotor action. An elevation of spindle firing at constant muscle length indicates increased fusimotor action of unspecified type. This is the criterion most often used in human neurography. The stability of implanted dorsal root electrodes in cats has allowed the study of spindle discharge during rapid movements of a great variety (Prochazka, 1980; Loeb and Hoffer, 1980). This raises the possibility of deducing specific fusimotor action from changes in the spindle afferent sensitivity to the length changes occurring during normal movements and during imposed movements in the conscious and anaesthetised animal.

ELECTRONIC MODELS OF THE SPINDLE TRANSDUCING MECHANISM

If an electronic model could be found which transformed the signal from a muscle length transducer in just the same way as a passive spindle ending transduced the length signal, then systematic discrepancies between the two transduced signals during voluntary movements would indicate fusimotor action.

The simplest such model is a linear filter with additive length and velocity terms (Matthews and Stein, 1969; Poppele and Bowman, 1970; Hasan and Houk, 1975). Since spindle endings show a considerable degree of nonlinearity, particularly in respect of their responses to widely ranging velocities (Houk *et al.*, 1980) a linear filter is clearly a limited model. However, as will be shown, provided that comparisons are restricted to movements in which the length variations are similar, a linear filter is useful, both in detecting changes in spindle sensitivity, and in identifying velocity responses in spindle firing which are not immediately obvious given the length signal alone.

Figure 1 (left column) shows the responses to ramp-and-hold stretches of a spindle primary in the absence of fusimotor stimulation (top), and during stimulation at 70 Hz of a single static and dynamic fusimotor fibre, respectively (after Brown and Matthews, 1966). The length signal was passed through a linear filter having a zero at 0.6 Hz (and seven poles at 80 Hz to suppress higher frequency components). Figure 1 (right column) shows the same responses, this time plotted against the filtered length signal. The resulting relationships have recognisably different slopes and biases. The slopes are a measure of sensitivity, and thus static stimulation can be seen to

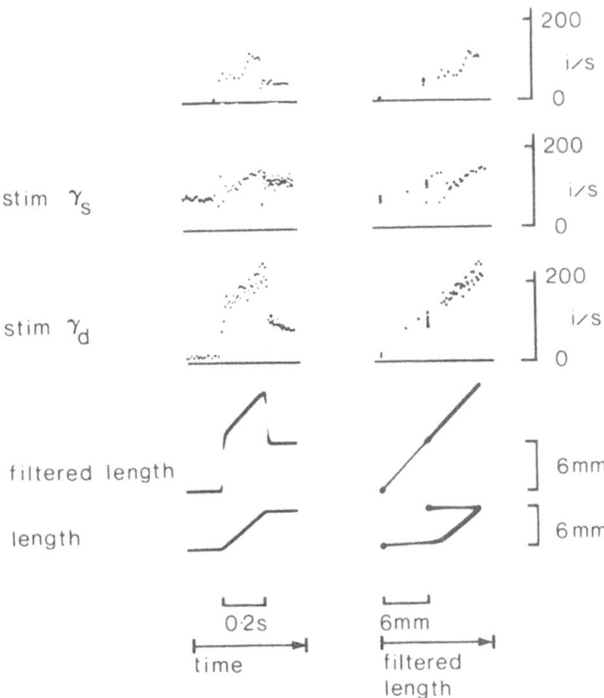

Figure 1. *Left column:* responses of a spindle primary to ramp-and-hold stretches, in the absence of fusimotor stimulation (top), and during static (γ_s) and dynamic (γ_d) action. Data after Brown and Matthews (1966). The 'filtered length' signal was obtained by passing the ramp-and-hold length signal through a linear filter. *Right column:* sensitivity plots; same data, with the filtered length signal as the independent variable plotted along the horizontal axis. The slopes of the relationships between spindle firing and filtered length are measures of sensitivity. Horizontal scale indicates time-invariant component of filter response only.

have led to a smaller sensitivity than that in the absence of fusimotor stimulation, whereas dynamic stimulation led to a larger sensitivity.

GENERAL COMMENTS ON TECHNIQUES

For reliable comparisons of spindle sensitivity during voluntary movements and during anaesthesia, the following prerequisites were required of afferent recording sessions.

(a) Clear single-unit recordings.
(b) Clear afferent identification. In our experience, the use of suxamethonium during anaesthesia is essential for an unambiguous identification of spindle afferents. In many cases, tendon organs respond readily to muscle stretches, taps and vibration, and may even pause in their discharge during electrically evoked muscle twitches, if in-parallel motor units happen to be stimulated. Clearly, a tendon organ mistaken for a spindle would provide very convincing records of e.m.g.-linked firing (Prochazka and Wand, 1980a).
(c) Appropriate length monitoring. The sensitivity plots described above rely on accurate, linear, length monitoring between muscle origin and insertion. Our technique now relies on the attachment of an external length gauge to fine, flexible wires issuing percutaneously from bony fixation points at the ischium, femoral epicondyle, tibial tuberosity and calcaneum. Even the smallest deviation from a true length record causes non-linearity in the sensitivity plots. Thus great care must be taken during the implantation to ensure that the fixation wires emerge in line with the muscle, and that they slip easily in the subcutaneous space.
(d) Appropriate e.m.g. Standard unipolar 0.3 mm diameter concentric needles are inserted percutaneously. In our experience, skin areas can be readily found, through which the electrode may be inserted, virtually without the animal noticing. The electrode may then be placed so as to record from the part of the muscle whose mechanical stimulation leads to a maximal afferent response.

SENSITIVITIES OF SPINDLES IN VOLUNTARY MOVEMENTS: SPINDLE PRIMARIES

Figure 2 shows the responses of an ankle extensor spindle primary during voluntary flexion–extension movements (figure 2A) and during deep Epontol (Bayer) anaesthesia (figure 2B). Figure 2C shows the afferent's responses to three muscle twitches evoked by electrical stimuli applied through the e.m.g.

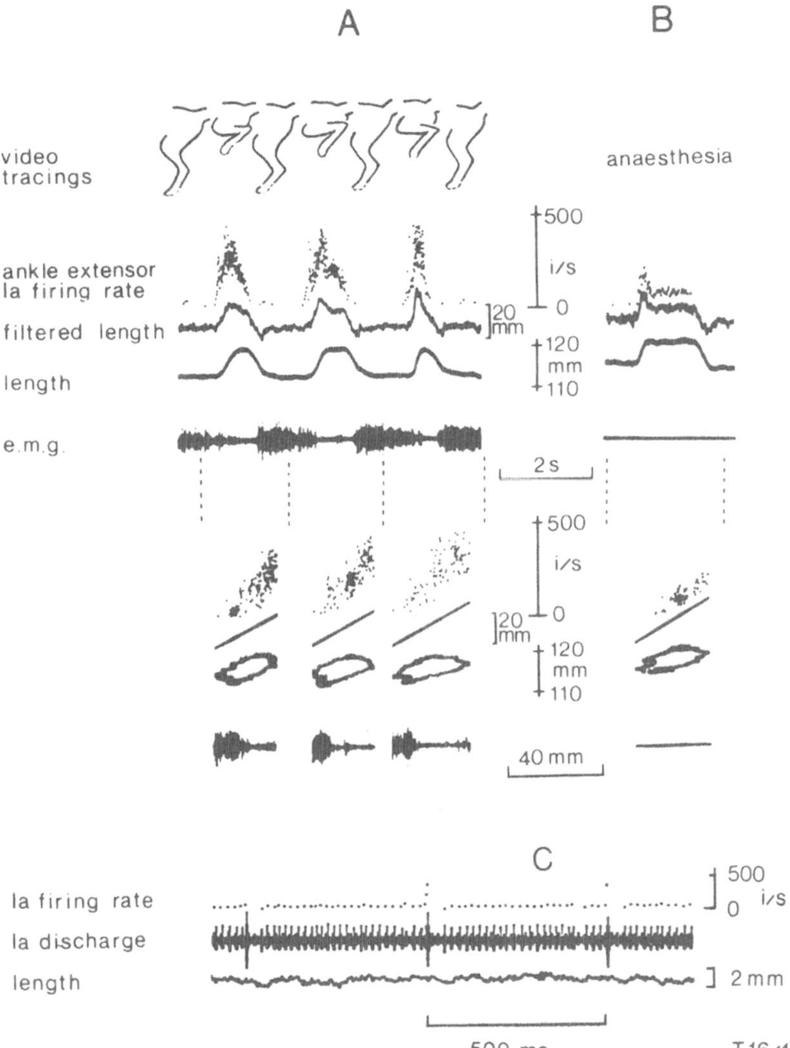

Figure 2. Ankle extensor spindle primary. (A) Recorded during three voluntary flexion–extension movements; sensitivity plots for each movement appear beneath the appropriate segment, as defined by the vertical dashed lines. (B) Responses to length changes during deep anaesthesia. (C) Responses to three electrically evoked muscle twitches.

needle electrode. This electrode had been inserted into lateral gastrocnemius in a region of maximal spindle sensitivity, and small manipulations of the electrode were shown to influence readily the firing of the spindle. Suxamethonium (100 μg/kg i.v.) caused an increase in the dynamic index of this afferent from 120 to 160 i/s.

The modulation of afferent firing rate in the unobstructed voluntary movements of figure 2A were strongly related to variations in the muscle length signal. However, the filtered length signal (linear filter, with one zero at 0.8 Hz, seven poles at 80 Hz) shows an even closer relationship, with phase advances and certain transient features emerging clearly.

To obtain the sensitivity plots beneath each segment of record, the filtered length signal was simply used to modulate the x-input of a standard storage oscilloscope. Comparing any one of the three sensitivity plots of figure 2A with that of figure 2B, the slope in figure 2B is always the smaller. Given that the range of lengths and velocities in A and B was similar, these data, taken together with the results above from figure 1, suggest that dynamic fusimotor action was present during the movements of figure 2A. Furthermore, the heightened sensitivity occurred during periods of e.m.g. silence. This alone does not rule out a purely skeletomotor-linked fusimotor action, since the after-effects of fusimotor activity can be of long duration (Brown *et al.*, 1969). However, three further observations do militate against linkage in this case. Firstly, the lack of tonic firing during e.m.g. activity is hard to reconcile with the occurrence of a powerful burst of dynamic fusimotor discharge. Secondly, the sensitivity of the afferent was similar whether preceded by strong or weak e.m.g. activity (or indeed even by the absence of e.m.g. in cases of consecutive flexions without intervening stance). Thirdly, imposed movements in the absence of detectable resistance or e.m.g. were associated with similar, or even greater, sensitivities.

Figure 3 shows a comparison of the sensitivity of the same afferent during (A) unobstructed voluntary movements, and (B) imposed movements involving moderate resistance (as judged from the e.m.g. and a simultaneously recorded video film). The very high sensitivity evident during the imposed movements indicates a further increase in dynamic fusimotor action over that in the unobstructed voluntary movements. As will be seen, the most consistent finding in this work so far, is that the sensitivity of spindle primaries has nearly always been at its highest in imposed movements involving slight to moderate resistance (i.e. movements such as those in figure 3B). Again, tonic (rather than linked phasic) action was likely, as the sensitivity often remained high in the absence of detectable e.m.g. (not shown in the figure).

Figure 4A shows the firing of a knee flexor spindle primary during a sequence involving three imposed extensions which the cat resisted moderately, followed by a powerful, obstructed flexion movement, followed in turn by an unobstructed extension. The sensitivity of the afferent during the three imposed movements (figure 4A: sensitivity plot) was clearly much higher than that in figure 4C, during deep Epontol anaesthesia.

It should be pointed out that the peak velocities in figure 4C were approximately double those in the analysed segment of figure 4A. Assuming a velocity dependence of $V^{0.3}$ (Houk *et al.*, 1980), a halving of velocities in

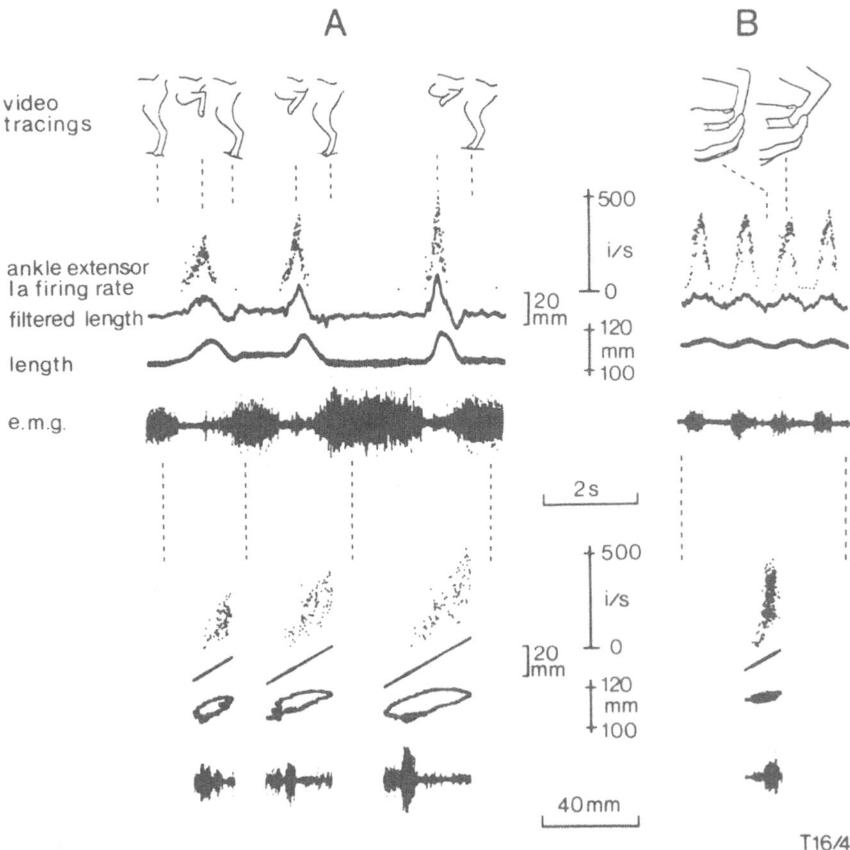

Figure 3. Same afferent as in figure 2. (A) Three voluntary flexion–extension movements with corresponding sensitivity plots. (B) Four imposed flexion–extension movements, with a single sensitivity plot for the whole sequence. Slight resistance is evidenced by the stretch-evoked e.m.g.

figure 4C might have resulted in an increase in slope of (at most) 25%. Nevertheless, considering the very large differences in slope in figures 4A and 4C, the basic interpretation, that dynamic fusimotor action was present during the imposed movements, seems justified.

A comparison between the sensitivity plots of figures 4B and 4C indicates a moderately increased sensitivity during the unobstructed flexion–extension movement over that during anaesthesia. Linkage between fusimotor and skeletomotor activation could not be ruled out for this afferent, as the continued firing during rapid shortening in figure 4A, seen consistently in subsequent trials, could be explained either by tonic static fusimotor action throughout, or by e.m.g.-linked action during the shortening.

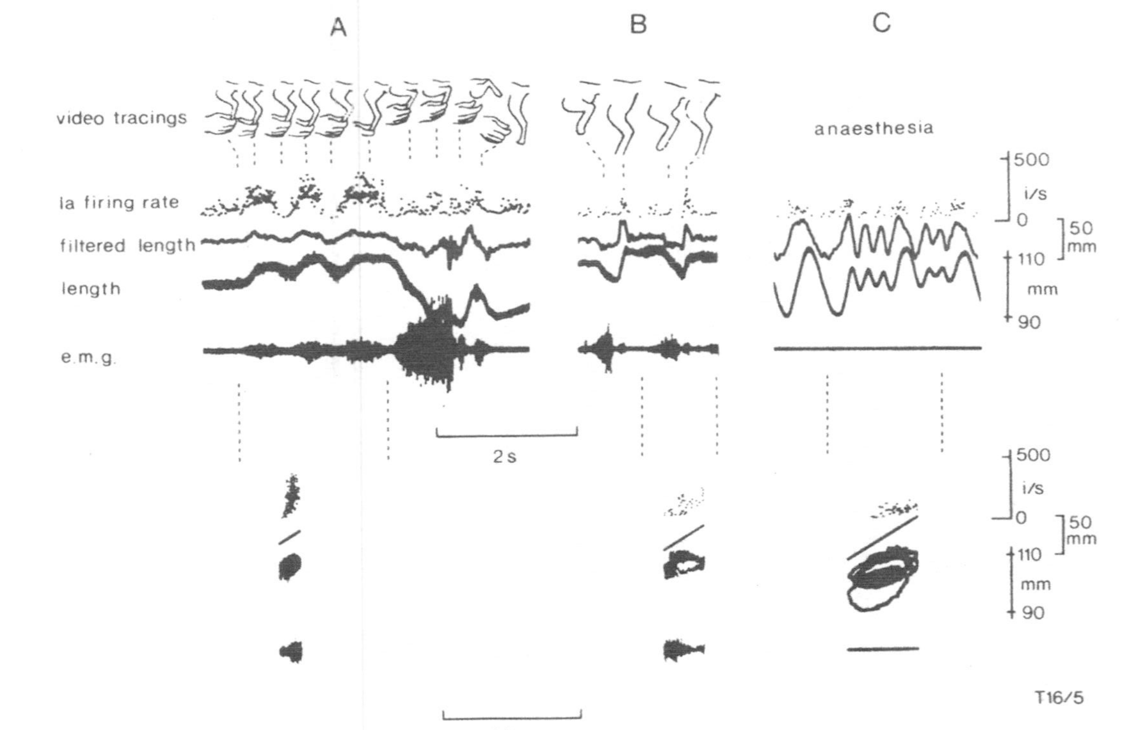

Figure 4. Knee flexor spindle primary. (A) Three imposed extension–flexion movements, followed by powerful voluntary flexion–extension; sensitivity plot applies to the first three movements only (between vertical dashed lines). (B) Two voluntary, unobstructed flexion–extension movements; sensitivity plot applies to second movement only. (C) Imposed length changes during deep anaesthesia.

Figure 5 shows another example of spindle primary responses during imposed movements. The muscle stretching in each case was associated with the development of e.m.g., and with high peak firing rates of the primary afferent. Sensitivity plots (not shown) indicated that the afferent was significantly more sensitive than during subsequent deep Epontol anaesthesia. At the arrows, bursts of e.m.g. occurred during the imposed muscle

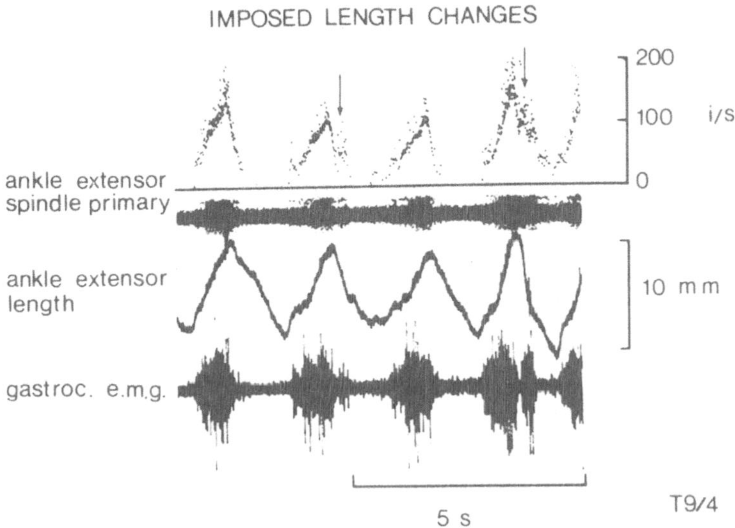

Figure 5. Ankle extensor spindle primary (not the same as in figures 1 and 2). Four imposed flexion–extension movements (muscle lengthening upwards). At the arrows, bursts of e.m.g. occurred, with associated increases in spindle discharge during muscle shortening.

shortening, and these were clearly associated with increased spindle discharge. Although it is tempting to ascribe this pattern of response to phasic co-activation of fusimotor and skeletomotor neurons, it is necessary to draw attention to the persistence of the spindle firing after the termination of the second e.m.g. burst. An equally plausible explanation of the response is therefore that a change in the animal's state of arousal led to a co-activation of skeletomotor and fusimotor neurons, but by virtue of independent central control, their activation was not terminated simultaneously. This example serves to illustrate the difficulty in deciding whether specific cases of e.m.g.-linked spindle firing do or do not represent examples of α–γ linkage.

SECONDARY ENDINGS

Spindle secondary endings are affected only slightly, if at all, by dynamic fusimotor action. Static fusimotor action, on the other hand, can both greatly increase their firing 'bias' at constant muscle length, and increase their sensitivity to length changes (Jami and Petit, 1978). Figure 6 shows the firing of a tail abductor spindle secondary during five powerful skeletomotor contractions, the first four of which occurred while the animal's tail was firmly

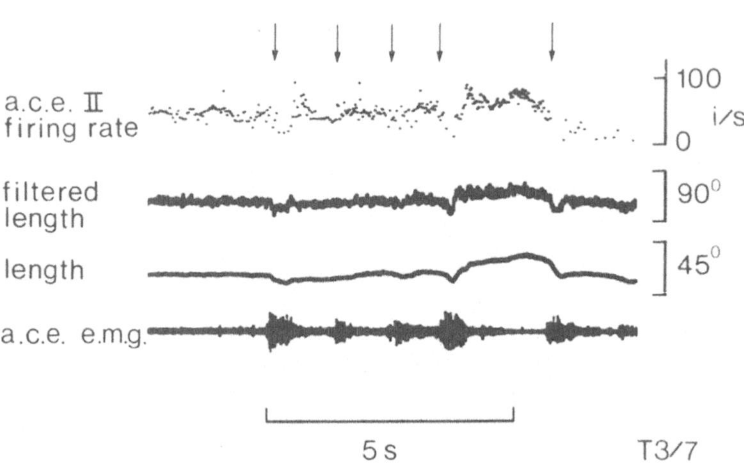

Figure 6. Abductor caudalis externis spindle secondary. Five powerful skeletomotor contractions (arrowed), occurring while the animal's tail was firmly held by experimenter (muscle lengthening upwards).

held in a fixed position by the experimenter. There is little evidence in this record of e.m.g.-linked changes in spindle bias. Again, the e.m.g. was recorded with a concentric needle electrode introduced into the part of the muscle whose mechanical stimulation led to the largest spindle responses. Electrical stimulation through the electrode (figure 7C) gave clear pauses in the spindle discharge. Suxamethonium (200 μg/kg i.v.) had no detectable effect on this afferent's maximum dynamic index.

Thus for this ending there was little evidence of skeletomotor-linked fusimotor action under these conditions. However, a comparison between the sensitivity plots of the afferent during imposed movements involving little resistance (figure 7A), and during imposed movements in deep Epontol anaesthesia (figure 7B), reveals a definitely higher sensitivity in the former case. A similar elevated sensitivity of this afferent was present during unobstructed voluntary movements (*see* Prochazka, 1980, figure 7). All of these results suggest a relatively tonic, static fusimotor action on this ending both during voluntary and during imposed movements.

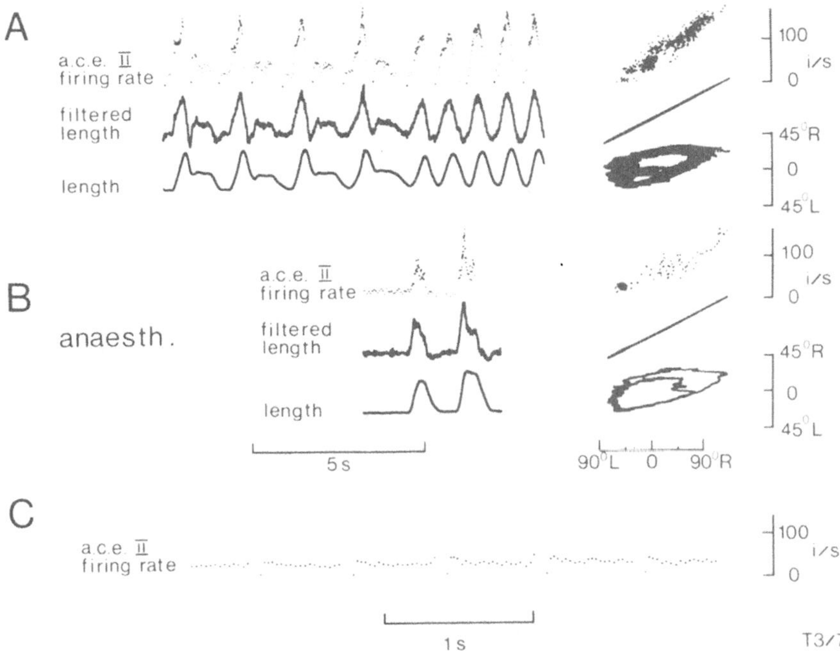

Figure 7. Same afferent as in figure 6. (A) Imposed changes in muscle length (lengthening upwards); sensitivity plot on the right applies to the whole segment. (B) Same during deep anaesthesia. (C) Responses to electrically evoked muscle twitches.

Recordings from an ankle extensor secondary ending (Prochazka *et al.*, 1979, figure 4) led to a similar negative result with regard to evidence for significant skeletomotor–fusimotor linkage.

OVERVIEW OF THE RESULTS

1. Unobstructed voluntary movements. Three primaries showed evidence of tonic dynamic action, three primaries and one secondary of tonic static action, and three primaries and one secondary showed little change in sensitivity or bias when compared with their responses during anaesthesia.
2. Imposed movements involving slight to moderate resistance. Eight primaries showed evidence of strong tonic dynamic action, and one primary and one secondary of weak, tonic static action.
3. Imposed movements involving substantial resistance. Two primaries showed evidence of strong dynamic action. Three primaries and two secondaries showed evidence of static action. In two of these primaries,

there was some evidence for phasic increases in fusimotor action during e.m.g. bursts. Data involving this sort of movement were not obtained for the remaining four primaries in the sample.

DISCUSSION

The clearest finding from the above sensitivity studies is the evidence for tonic, predominantly dynamic, fusimotor action during imposed movements which the animals resisted, at most, moderately. If this result finds confirmation in further studies, it may be the first instance of a category of movement which is reliably associated with a known change in spindle sensitivity.

Unobstructed movements of a fairly predictable nature (such as stepping and tail-wagging) were associated with sensitivities indicating low-level, tonic fusimotor action of mixed type. There was no compelling evidence to indicate that while dynamic fusimotor action was sustained, static action was modulated in parallel with alpha activity (a possibility suggested by Grillner *et al.*, 1969, Matthews, 1972 and recently by Appenteng *et al.*, 1980).

Loeb and Hoffer (1980) have proposed a strong linkage of skeletomotor and static fusimotor neurons in flexors, and a linkage of skeletomotor and dynamic fusimotor neurons in extensors. This proposal stems from chronic recordings in which it was not possible to differentiate between primary and secondary endings. Whereas static fusimotor action enhances the sensitivity of secondary endings (Jami and Petit, 1978), it reduces that of primary endings. Thus, without knowledge of the type of afferent involved, it is not clear how the type of fusimotor action could be deduced. Loeb and Hoffer also argue that in our own recordings, we have only studied extensor, or 'potentially extensor', muscles. According to Sherrington (1910), the bifunctional hamstring muscles posterior biceps femoris and semitendinosus (from which we have recorded afferents) are excited in the flexion-reflex, in contrast with crureus, vasti, gastrocnemius and soleus, which are inhibited. Thus according to Sherrington's scheme of classification at least, we have recorded from both flexor and extensor afferents. In spite of these points, the general idea that the proportion of dynamic to static fusimotor action which a spindle receives is related to whether it is located in an extensor or a flexor, must certainly remain a possibility.

α–γ Linkage versus independent fusimotor control

The evidence presented above for independent control of fusimotor and skeletomotor neurons does not in any way disprove the existence of a degree of linkage between them. However, the modulatory strength of skeletomotor-linked fusimotor action seemed, in most of our recordings, to be low or undetectable, when compared with the modulatory strength of the variations in muscle length. In contrast, sustained variations in spindle sensitivity (and therefore in presumed fusimotor action) which were apparently unrelated to skeletomotor activation, were the most striking features of the recordings.

Powerful, obstructed contractions occurring during imposed limb movements did, in some afferents, result in raised firing levels consistent with e.m.g.-linked fusimotor action. However, as discussed earlier in relation to figure 5, sudden arousal could quite conceivably result in a coactivation of skeletomotor and fusimotor neurons, through completely independent pathways. Testing of a more subtle kind is therefore required to clarify the mode of fusimotor control in movements of this nature.

Comparisons with human neurography data

Our notion of a fusimotor system whose main effects are independent of skeletomotor activation are at odds with the interpretations of afferent recordings in humans. In comparing the data, it must be recognised that both sets of experiments have their disadvantages. Cats do not produce specified types of movement on request. The description of a movement as being 'voluntary' is more ambiguous when applied to cats than to humans. In human neurography, on the other hand, suxamethonium has not been used in identifying afferents (identification has relied upon the observation of responses to mechanical and electrical stimuli. In our own experience these stimuli do not allow a reliable differentiation between spindle and tendon organ afferents). The speed of movements in the human experiments has generally not exceeded 0.2 resting lengths per second, a restriction dictated by the requirements for electrode stability. In this context, possibly the most relevant physical difference between the experiments concerns the relative freedom of movement. Thus it is conceivable that for unrestricted movements involving a normal sequencing of postural and voluntary muscular contractions, a much greater independence of the fusimotor and skeletomotor systems might also become apparent in humans.

ACKNOWLEDGEMENTS

This work was supported by MRC Grant G.80/0040/2N and the Deutsche Forschungsgemeinschaft, SFB 33.

Received on June 28th, 1980.

REFERENCES

Alnaes, E., Jansen, J. K. S. and Rudjord, T. (1965). Fusimotor activity in the spinal cat, *Acta physiol. scand.*, **63**, 197–212

Appelberg, B. Selective midbrain control of dynamic gamma motoneurons, this publication

Appenteng, K., Morimoto, T. and Taylor, A. (1980). *J. Physiol., Lond.*, to be published

Bergmans, J. and Grillner, S. (1969). Reciprocal control of spontaneous activity and reflex effects in static and dynamic γ-motoneurones revealed by an injection of DOPA, *Acta physiol. scand.*, **77**, 106–124

Brown, M. C., Goodwin, G. M. and Matthews, P. B. C. (1969). After-effects of fusimotor stimulation on the response of muscle spindle primary afferent endings, *J. Physiol., Lond.*, **205**, 677–694

Brown, M. C. and Matthews, P. B. C. (1966). On the subdivision of the efferent fibres to muscle spindles into static and dynamic fusimotor fibres. In *Control and Innervation of Skeletal Muscle* (ed. B. L. Andrew), Dundee, Thomson, 18–31

Burke, D., Hagbarth, K.-E. and Skuse, N. F. (1979). Voluntary activation of spindle endings in human muscles temporarily paralysed by nerve pressure, *J. Physiol., Lond.*, **287**, 329–336

Granit, R. (1955). *Receptors and Sensory Perception,* New Haven, Yale University Press

Granit, R. (1979). Interpretation of supraspinal effects on the gamma system. In *Reflex Control of Posture and Movement, Prog. Brain Res.*, **50** (eds. R. Granit and O. Pompeiano), Amsterdam, Elsevier, 147–154

Granit, R., Holmgren, B. and Merton, P. A. (1955). The two routes for excitation of muscle and their subservience to the cerebellum, *J. Physiol., Lond.*, **130**, 213–224

Grillner, S., Hongo, T. and Lund, S. (1969). Descending monosynaptic and reflex control of γ-motoneurones, *Acta physiol. scand.*, **75**, 592–613

Hagbarth, K.-E. (1980). Fusimotor and stretch reflex functions studied in recordings from muscle spindle afferents in man, this publication

Hasan, Z. and Houk, J. C. (1975). Analysis of response properties of de-efferented mammalian spindle receptors based on frequency response, *J. Neurophysiol.*, **38**, 663–672

Houk, J. C., Rymer, W. Z. and Crago, P. E. (1980). Nature of the dynamic response and its relation to the high sensitivity of muscle spindles to small changes in length, this publication

Jami, L. and Petit, J. (1978). Fusimotor actions on sensitivity of spindle secondary endings to slow muscle stretch in cat peroneus tertius, *J. Neurophysiol.*, **41**, 860–869

Loeb, G. E. and Duysens, J. (1979). Activity patterns in individual hindlimb primary and secondary muscle spindle afferents during normal movements in unrestrained cats, *J. Neurophysiol.*, **42**, 420–440

Loeb, G. E. and Hoffer, J. A. (1980). Muscle spindle function during normal and perturbed locomotion in cats, this publication

Matthews, P. B. C. (1972). *Mammalian Muscle Receptors and Their Central Actions,* London, Arnold

Matthews, P. B. C. and Stein, R. B. (1969). The sensitivity of muscle spindle afferents to small sinusoidal changes of length, *J. Physiol., Lond.*, **200**, 723–743

Poppele, R. E. and Bowman, R. J. (1970). Quantitative description of linear behaviour of mammalian muscle spindles, *J. Neurophysiol.*, **33**, 59–72

Post, E., Rymer, W. Z. and Hasan, Z. (1978). Relation between extrafusal and intrafusal activity in the decerebrate cat model: a role for beta fibres, *Soc. Neurosci. Abstr.*, No. 958

Prochazka, A. (1980). Muscle spindle function during normal movement. In *MTP Int. Rev. Physiol.*, Neurophysiology IV (ed. R. Porter), Baltimore, University Park Press, to be published

Prochazka, A., Stephens, J. A. and Wand, P. (1979). Muscle spindle discharge in normal and obstructed movement, *J. Physiol., Lond.*, **287**, 57–66

Prochazka, A. and Wand, P. (1980a). Tendon organ discharge during voluntary movements in cats, *J. Physiol., Lond.*, **303**, 385–390

Prochazka, A. and Wand, P. (1980b). Fusimotor action during normal movements, deduced from variations in muscle spindle sensitivity, *J. Physiol., Lond.*, to be published

Sherrington, C. S. (1910). Flexion-reflex of the limb, crossed extension-reflex, and reflex stepping and standing, *J. Physiol., Lond.*, **40**, 28–121

Taylor, A. and Cody, F. W. J. (1974). Jaw muscle spindle activity in the cat during normal movements of eating and drinking, *Brain Res.*, **71**, 523–530

Vallbo, A. B. (1980). Basic patterns of muscle spindle discharge in man, this publication

A critique of the papers by Loeb and Hoffer and Prochazka and Wand

K.-E. HAGBARTH*

We have now heard about the results obtained independently by two research groups studying patterns of fusimotor control in the leg muscles of awake cats. Even though the results and conclusions are not quite consonant there is one common denominator: the indications that the skeletomotor and fusimotor systems are capable of independent action. This is where the findings in cats seem to differ from the findings in man, which Vallbo and I have described.

We cannot definitely exclude the possibility that the discrepancy depends on a species difference. One such difference seems to be that the spindle afferents can be made to fire with higher frequencies in cats than in man. In some of the records shown by Prochazka and Wand the spindle firing rates went up to 400–500 Hz. Such firing rates are not seen in man, where the instantaneous frequency plots seldom show rates higher than 50 Hz. It is only in response to vibration that the primaries in man can be driven up to rates as high as 200 Hz or more. In this connection we should also recall that the Ia afferents have a larger diameter and higher conduction velocity in cats than in man. With such structural and functional differences in the peripheral part of the system it would not be very surprising if there were some differences also in the patterns of fusimotor control.

Another possible reason for the discrepancies between the findings in cat and man is, of course, that to a large extent, different types of motor acts have been studied. Whereas Vallbo and I have not studied the patterns of fusimotor control in human gait, Loeb, Prochazka and co-workers have not studied the patterns in cats requested either to relax completely, to contract a specific muscle or to perform voluntary movements according to instruction.

But my main task as a critic is not to make assumptions regarding species differences or differences related to various types of motor behaviour, but to examine critically the animal findings and to search for alternative interpretations. We have learnt from our recordings in man to look with caution

*Department of Clinical Neurophysiology, University Hospital, Uppsala

upon reports suggesting lack of α–γ coactivation in various motor acts. Even during isometric conditions one can see individual spindle endings which decelerate rather than accelerate during voluntary contraction of the receptor-bearing muscle, and we think there are two common reasons for this. One is that the static fusimotor neurons supplying that particular spindle need a relatively strong voluntary effort or contraction strength to be activated, so the ending tends to be unloaded by weak contractions. Another common reason is that the contraction of the receptor-bearing muscle is accompanied by even stronger contractions in neighbouring synergistic muscles, and the negative effect of this mechanical unloading can, for a given ending, exceed the positive effect of an intrafusal contraction.

Under isotonic conditions, when the receptor-bearing muscle is allowed to shorten during the contraction, there are even greater chances that the negative effect of the muscle shortening exceeds and conceals the effect of the fusimotor drive. In many of the records shown in the two previous papers, we saw how during alternating gait movements the spindle firing went down to a minimum during the active shortening phases when the e.m.g. activity was optimal. That is certainly not good evidence for a lack of α–γ coactivation; it may instead be regarded as a good example of how efficiently the negative effect of a rapid muscle shortening may counteract and conceal the effect of a concurrent static fusimotor drive.

But a main point in the report by Prochazka and Wand is that during the stretch phases in the gait cycle, when the e.m.g. activity is minimal, the stretch sensitivity of the spindles seems to be higher than during similar stretch movements imposed during anaesthesia. This is taken as evidence for dynamic fusimotor action during the gait movements. I would like to put forward the following alternative interpretation. The records show that during the gait cycle the stretches occur at about the same time as the preceding muscle contractions decline. And we know that the primary endings tend to fire during the declining phase of a muscle contraction. So, what I am suggesting is that the very high firing rates seen during the stretch phases of the gait cycle are not due to dynamic fusimotor action, but to the fact that in this case the spindles are exposed to the combined excitatory effects of a sudden stretch movement and a suddenly declining contraction.

During the imposed stretch movements in awake cats the spindles apparently respond much more vigorously to stretch than they did during anaesthesia. But most of the records show that in contrast to the anaesthetised cats, the awake animals resisted the stretch movement. So in the awake cats the combined excitatory effects of α-linked fusimotor drive and muscle stretch may have occurred. It is stated by Prochazka and Wand that the high stretch sensitivity of the spindles was also seen in the absence of detectable resistance or e.m.g. activity, but we all know how difficult it is to be sure that a muscle is completely relaxed.

Loeb and Hoffer present evidence indicating a strong linkage of skeletomotor and static fusimotor neurons in flexors, and a linkage of skeletomotor and dynamic fusimotor neurons in extensors. And they suggest that the discrepancies between the findings in alert cats and man may be due to the fact that mostly extensor muscles have been studied in cats and mostly flexor muscles in man. It is true that for technical reasons we often choose to record from such muscles as the finger or wrist flexors in the arm or from the anterior tibial muscle in the leg. But we have also recorded from such muscles as the calf muscles in the leg or the wrist extensors in the arm. Also in these latter extensor muscles we find signs of static fusimotor action during voluntary contractions. Still, it remains a possibility that the proportion of static and dynamic fusimotor action is different for spindles in flexor and extensor muscles. But, for the sake of argument, I would like to suggest an alternative interpretation of the findings. In all phases of the cat gait the legs are held in a more or less flexed position, so on the whole the extensor muscles are more stretched than the flexors. Now, at least in man, it is true that if one wants to see spindle endings showing signs of unloading during contraction it is advisable to let the contraction take place in a muscle that is held in a stretched position so that as many spindles as possible exhibit a static resting discharge when contraction starts. Only when there is such a static stretch discharge is it possible to detect those endings which are unloaded by weak contractions and which require stronger motor efforts to be activated. In passively shortened muscles, on the other hand, where most of the spindles are silent, one cannot see the negative unloading effect of the contraction but only the positive accelerative effect of the fusimotor drive. Could this be a reason for Loeb and Hoffer tending to see flexor spindles accelerating and extensor spindles decelerating during contraction?

Received on July 31st, 1980.

A critique of the papers by Loeb and Hoffer; and Prochazka and Wand

ÅKE VALLBO*

DIFFERENCES IN MUSCLE SPINDLE DISCHARGE DURING NATURAL MOVEMENTS IN CAT AND MAN

The technique of recording from single afferents in intact, unanaesthetised and freely moving cats is to my mind a marvellous achievement. The incredible stability of the recording which allows the study of proprioceptive mechanisms while the animals are moving has resulted in many interesting findings and a lot more is to be expected. The papers by Loeb and Hoffer (1981) and by Prochazka and Wand (1981) describe fundamental and important aspects of the system in action and the findings elucidate facets of the system which have not been studied with the previously available method of recording from intact human subjects.

I would like to focus my comments on one point, namely the apparent difference between responses of spindles in man and cat. A considerable emphasis has been put on differences between the two species with regard to discharge patterns of spindle afferents during natural motor activity. In man the impulse rate in spindle afferents runs reasonably parallel with the skeletomotor activity as estimated from the EMG activity, provided there are no large movements (for details, *see* Vallbo, 1981). This finding suggests there is, in a broad sense, a parallelism between fusimotor and skeletomotor activity in man. (The term α–γ-linkage is deliberately avoided in this context because it is both inadequate and indistinct.) In contrast, spindle discharge rate seems to be more influenced by muscle length and velocity of movement than by fusimotor activity in intact cats. Moreover, in these experiments the impulse rate often bears no obvious relation to the amount of skeletomotor activity. These points have been stressed in the two papers presented at this symposium (Loeb and Hoffer, 1981; Prochazka and Wand, 1981) as well as in a number of earlier papers.

*Department of Physiology, University of Umeå.

Before discussing the difference between findings from the two species I would like to draw attention to a few aspects of the work on human subjects. A prominent aim in these studies has been to arrive at a description of the basic pattern of spindle discharge under standardised conditions with the purpose of producing a basis for studies of more complex motor activities. The current interpretation of spindle function in man is based on recordings from several hundred afferents collected over a period of about 15 years. Right from the beginning when this technique was introduced it was a surprise—and it is probably right to say a disappointment—to find no clear sign of the independent fusimotor activation which was to be expected from a wealth of findings in acute animal experiments. It is true that scattered observations have been made which do not conform to the basic pattern—and an example will be given below—but there is a striking agreement between the various groups with regard to the basic finding that a relatively close coactivation of the skeletomotor and fusimotor systems characterises motor activities in man. Identification of the afferents is based on a variety of tests used in combination, as described in each paper, although a fair proportion of units had to be discarded when a proper classification of the ending could not be achieved. To estimate the level of probability of the conclusions based on human experiments, these points might have some additional interest besides the detailed accounts of the experiments which are given in each paper. In particular it seems perfectly sound to reject the possibility that the description of the basic pattern of spindle discharge in man is erroneous due to contamination by non-spindle afferents.

If we consider only qualitative aspects, spindle responses may be characterised as passive or active in the sense that passive responses are dominated by the mechanical input to the sense organ, i.e. muscle length and changes of muscle length, whereas active responses are dominated by fusimotor effects. Cat spindles are often described as passive in this sense whereas human spindles are described as active. However, it should be pointed out that both response patterns are seen in human subjects. Figure 1 shows the most common pattern—an active response—in this case during an isometric contraction. The impulse rate in the spindle afferent ran reasonably parallel with the skeletomotor activity, indicating an increased fusimotor drive during the skeletomotor contraction.

Incidentally, the second test contraction in figure 1 illustrates a finding of diagnostic relevance. The burst of impulses on the falling phase of the contraction indicates that this unit is a spindle afferent rather than a tendon organ because the discharge increased during a phase when the active tension was falling and the internal length was increasing. This phenomenon has been seen in some tests in a fair number of spindle primary afferents, but it is not a rule, not even when external shortening of the muscle is allowed. Further examples may be seen in figures 2 and 3.

Figure 2 illustrates a passive response during an isometric contraction. The

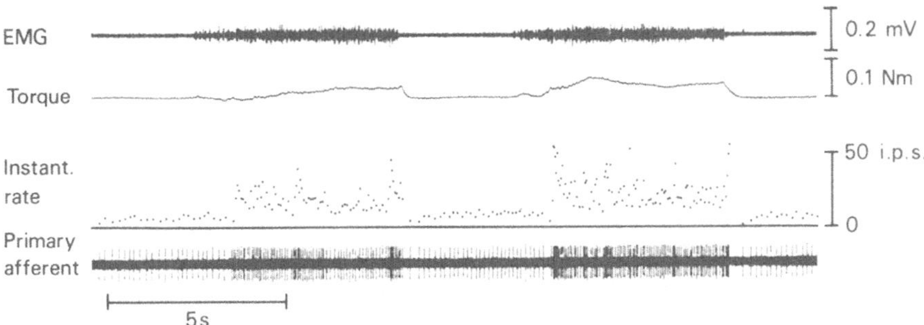

Figure 1. Discharge of a primary afferent from a muscle spindle during isometric voluntary contractions. The ending was located in the ring-finger portion of the flexor digitorum muscles on the forearm, and the torque was produced by flexion of the ring finger alone. EMG was recorded with a needle electrode which was inserted at the level of the ending. Fusimotor effects gave rise to an active response pattern of the afferent.

spindle discharge decreased when the parent muscle was contracting, indicating that the ending was unloaded by the internal length changes. Clearly, there was not enough fusimotor drive to compensate for the mechanical effects in this case, whereas in the tests of figure 1 the fusimotor drive was strong enough to over-compensate for the unloading effect.

Although the contractions of figures 1 and 2 were isometric, the records may serve to illustrate the finding that both passive and active spindle responses, as defined above, are encountered in human subjects. Hence, there is no absolute difference between the two species in this respect. Indeed, there may even be a shift from one pattern to the other in the response of a single afferent. This is illustrated in figure 3 where recordings from a spindle afferent are shown when the subject performed a number of grossly similar isotonic movements with identical loads. All six contractions were done within 2 min. An active pattern is seen in the top records, whereas the records in the lower row illustrate passive patterns. A rather intermediate type of response is present in the middle row. Two kinds of interpretations may be offered to explain such findings. One is that there was a general shift in α–γ-balance. In the sample records of figure 3 there would be a relatively strong fusimotor drive during contraction in the tests of the top row resulting in a reasonable parallelism between spindle discharge rate and amount of EMG activity. In the tests of the bottom row the fusimotor drive would be weak or absent and the relation between spindle discharge rate and EMG activity was poor as in many records from moving cats. It was not possible to define the decisive factor giving rise to the shift in this experiment, but it is tempting to speculate about such factors. Are they related to an altered motor strategy by the subject, to a change in his concentration on the task, or to fatigue? These and related hypotheses have been tested in qualitative experiments but positive support has not been found. However, before going into speculations of this kind it should be asked whether records like those of

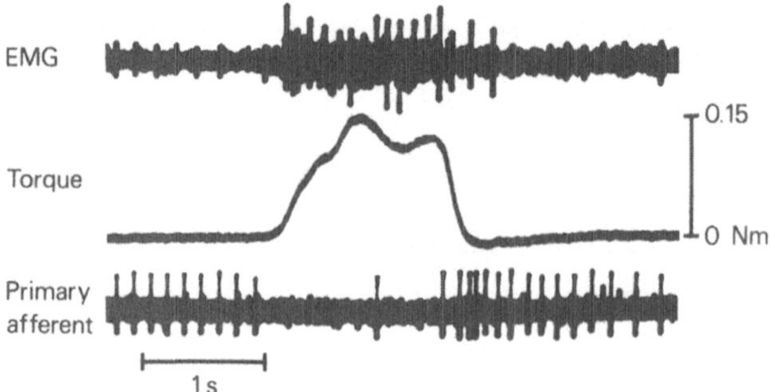

Figure 2. Discharge of an afferent from a muscle spindle during an isometric voluntary contraction. The ending was located in the portion connected to the fifth finger of the extensor digitorum muscles on the forearm, and the torque was produced by extension of the fifth finger alone. EMG was recorded with surface electrodes in the vicinity of the ending. Mechanical unloading gave rise to a passive response pattern of the afferent.

figure 3 really demonstrate unequivocally a general shift in α–γ-balance. There is another and probably less-exciting interpretation, namely that the change in the response pattern seen in this afferent reflects a local variation in fusimotor output. If it is generally true that the fusimotor drive is largely restricted to the active muscle portions (as suggested by studies on human subjects (Vallbo, 1981)), then a change from an active to a passive firing pattern may occur if the muscle tissue surrounding the spindle is less active in some tests than in others. In a complex motor system where several muscles are engaged in the particular joint movement which is studied, it might be particularly difficult to discriminate, in the response of a single afferent, between fusimotor effects of a general nature and local effects related to variations in the design of the skeletomotor output. However, this does not mean that the possibility of a flexible α–γ-balance in man is rejected. On the contrary, we are constantly looking for clear evidence of such a flexibility (cf. Vallbo, 1981) while realising that there are several pitfalls that may give rise to a false support of this idea (cf. Vallbo *et al.*, 1979). In fact, more convincing evidence of a flexible balance between the skeletomotor and the fusimotor systems has recently been reported in man (Vallbo and Hulliger, 1979; Burke *et al.*, 1980), although very dramatic effects have not been seen.

This reasoning leads to a fundamental question. Which factors determining the response pattern of a muscle spindle afferent can be experimentally defined? The investigators who have been working with human subjects have probably paid more attention to systematic studies on this point than the groups working with cats. A simple scheme may be presented as a schematic summary of the main points. Three factors seem to be relevant for the type of response pattern that a spindle afferent will exhibit during a motor task: (1)

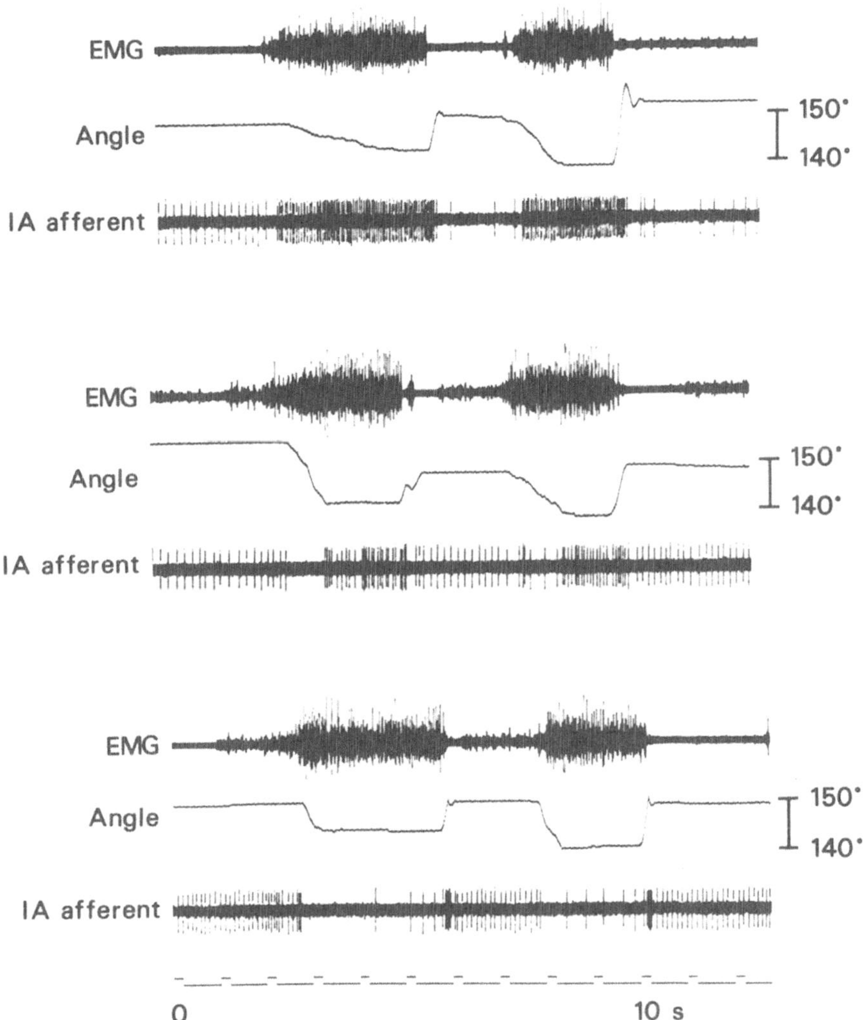

Figure 3. Discharge of a primary afferent from a muscle spindle during voluntary movements. The load was identical in the six contractions. It consisted of a constant torque load of about 3% of the maximal voluntary contraction force and a small inertial load of about ten times the inertia of the finger. The ending was located in the ring-finger portion of the flexor digitorum muscles on the forearm and the movement was performed with the ring finger alone. EMG was recorded with a needle electrode inserted at the level of the ending.

vicinity of the ending to the active muscle portion, (2) intensity of the skeletomotor activity, which is, of course, often intimately related to the load on the muscle and (3) rate of change of muscle length. The balance between these three factors seems crucial for which of the two inputs (either the mechanical or the fusimotor), will dominate to produce passive or active response patterns. It should be added that different spindle endings apparently have different thresholds for an active response in terms of intensity of contraction of the parent muscle (Burke *et al.*, 1978). In addition, there is recent evidence that several factors such as the kind of load on the muscle, the length of the muscle, and the cutaneous and vestibular input, may play some role in influencing spindle discharge (Vallbo and Hulliger, 1979; Burke *et al.*, 1980).

On the basis of this scheme it seems reasonable to offer at least a partial explanation for the finding that the pattern of discharge in spindle afferents is described as more passive in cat and more active in man. A combination of a true species difference, and differences in test procedures and type of motor activity studied, might account for some lack of agreement.

All evidence indicates that the proportion of continuously discharging spindle afferents with relaxed muscles is smaller in man than hindlimb muscles of the cat (Vallbo, 1974). Thus, the mechanical threshold of spindles with relaxed muscles seems to be higher in man than in cats. Afferents with high mechanical threshold which are silent in the first place and which do not exhibit an active response pattern during contraction would more easily escape attention or be rejected because they are difficult to identify properly. The higher threshold and the relatively small amplitude of movements employed in experiments on man might account for a low proportion of passive spindle afferents. In cats, active movements are studied mostly when muscles work against relatively small loads, the movements are fairly fast and of large amplitudes. A small load implies a relatively moderate activation of the muscles which in turn would tend to favour a passive spindle response, and the large-amplitude movements would evoke prominent stretch responses particularly if the mechanical threshold of the unit is low. This interpretation implies that the apparent difference between the two species with regard to spindle tendency to produce active or passive responses might be of a quantitative rather than a qualitative nature. The interpretation also implies the general hypothesis that the afferent signal from spindles under natural conditions are of two kinds, that the two occur simultaneously from different sets of spindles and that they may preferentially occur in different kinds of motor acts (cf. Loeb and Hoffer, 1981). One signal explicitly describes muscle length and velocity of movement, whereas the other carries more complex information which might reflect demand as well as accomplishment. It seems of great interest to elucidate such an hypothesis in future experiments.

There is another difference between the experiments with the two species which might be worth keeping in mind when the findings are compared. In human experiments interest is focused on single movements or contractions while intense efforts are made to have the subject relax his muscles before and after each separate contraction. The purpose of such a strategy is, of course, to compare the functional state of the system when the muscle is relaxed and when it is working. In contrast, studies on cats have been mainly concerned with cyclic movements when the muscle activity is waxing and waning. Such movements might be associated with a different and more continuous fusimotor engagement, which could be related to the presence of a continuous skeletomotor activity of low intensity compensating for reactive forces, or there could be a fusimotor after-effect from one phase of the rapid movement to the next.

It seems that it would be of great interest to see, in the future, studies of motor activities which are as similar as possible in order to be able to make more adequate comparisons between the two species. Moreover, it would be interesting to see more quantitative data which would allow a clear description of the average response and the range of variation within a sample of afferents under standardised conditions so as to be able to estimate the weight that a limited number of recordings may carry.

ACKNOWLEDGEMENTS

This work was supported by the Swedish Medical Research Council, project No. 2075. I would like to thank Dr B. Appelberg for valuable discussions.

Received on September 1st, 1980.

REFERENCES

Burke, D., Hagbarth, K.-E. and Skuse, N. F. (1978). Recruitment order of human spindle endings in isometric voluntary contractions, *J. Physiol., Lond.*, **285,** 101–112

Burke, D., McKeon, B. and Westerman, R. A. (1980). Induced changes in the thresholds for voluntary activation of human spindle endings, *J. Physiol., Lond.*, **302,** 171–181

Loeb, G. E. and Hoffer, J. A. (1981). Muscle spindle function during normal and perturbed locomotion. This publication

Prochazka, A. and Wand, P. (1981). Independence of fusimotor and skeletomotor systems during voluntary movements. This publication

Vallbo, Å. B. (1974). Afferent discharge from human muscle spindles in non-contracting muscles. Steady state impulse frequency as a function of joint angle, *Acta physiol. scand.*, **90,** 303–318

Vallbo, Å. B. (1981). Basic patterns of muscle spindle discharge in man. This publication

Vallbo, Å. B., Hagbarth, K.-E., Torebjörk, E. and Wallin, G. (1979). Somatosensory, proprioceptive and sympathetic activity in human peripheral nerves, *Physiol. Rev.*, **59,** 919–957

Vallbo, Å. B. and Hulliger, M. (1979). Flexible balance between skeletomotor and fusimotor activity during voluntary movements in man, *Neurosci. Lett.*, Supplement 3, s103

Muscle spindle responses to rapid stretching in normal cats

A. PROCHAZKA* and P. WAND†

SUMMARY

Discharges of single muscle spindle primary afferents were recorded in normal cats during rapid, imposed limb movements which stretched the receptor-bearing muscles. The spindles discharged in bursts during the stretch, each burst preceding by about 10 msec an EMG burst in the receptor-bearing muscle. In stretches of very short duration, which only allowed sufficient time for one spindle burst, only one EMG burst was observed. This suggests that both the short latency and the longer latency EMG responses to rapid muscle stretching depend, at least in part, on the prior occurrence of bursts of spindle discharge.

INTRODUCTION

A sudden imposed extension of the forearm of a human subject results in two or more separate bursts of electromyographic (EMG) activity in the stretched muscles (Hammond *et al.*, 1956). It has long been accepted that although the first burst is probably the result of direct excitation of spinal motoneurons by muscle spindle afferents, the second burst represents a 'long-loop' or 'trans-cortical' reflex (Phillips, 1969; Tatton *et al.*, 1975). Doubt has now been cast on this explanation by the demonstration that comparable responses in normal cats are not markedly different in decerebrate and acute spinal cats (Ghez and Shinoda, 1978). We have recorded from single muscle spindle primary afferents in normal cats during imposed, rapid stretching of the receptor-bearing muscle. The spindles discharged in bursts, each burst preceding by about 10 msec an EMG burst in the muscle. Very brief stretches

* Sherrington School of Physiology, St Thomas's Hospital Medical School, London
† Max Planck Institut für Experimentelle Medizin, Göttingen

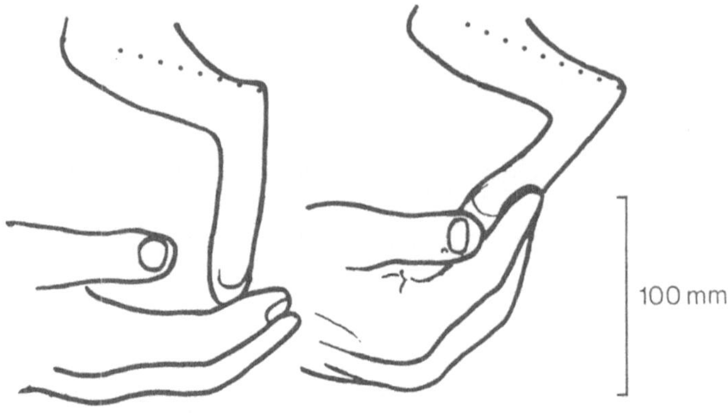

Figure 1. Tracings from a video film of the recording of figure 2A, showing the way in which rapid extension of the ankle was applied.

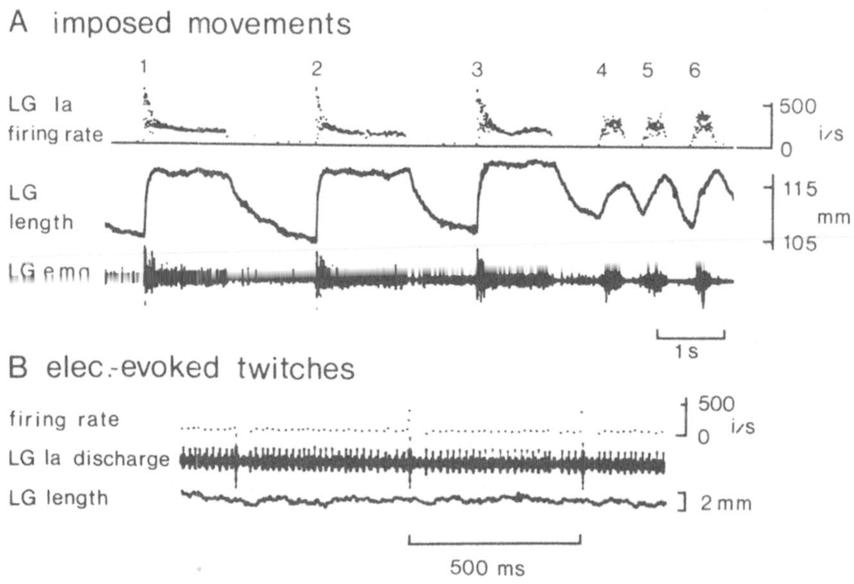

Figure 2. A, Six imposed extensions showing spindle primary (Ia) afferent firing rate, length and EMG of the ankle extensor lateral gastrocnemius (LG). B, Responses of the same afferent to three electrically evoked muscle twitches in the anaesthetised animal showing pauses characteristic of muscle spindle afferents.

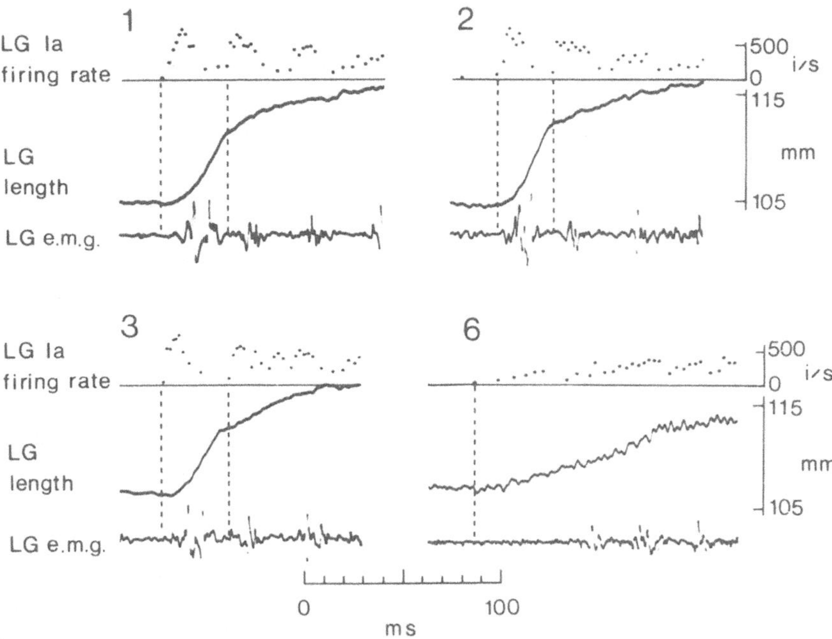

Figure 3. Three of the six imposed extensions of figure 2A, shown on an expanded time scale. Vertical dashed lines indicate onset of clearly identifiable bursts of afferent discharge. Note linking of afferent and EMG bursts in extensions 1, 2 and 3.

resulting in only one burst of spindle discharge resulted in only one EMG burst. Thus the patterns of firing of muscle spindles might, to a very large part, determine the segmented nature of the EMG response to rapid muscle stretching in mammals.

TECHNIQUE

General comments on the recording technique and afferent identification were made in our previous paper in this publication. Briefly, the discharge of a single afferent fibre is detected with the use of a free-floating microelectrode in a spinal dorsal root, and is recorded by telemetry in the freely moving animal. Afferents are identified by their responses to a series of tests applied during a brief period of deep anaesthesia. We report here observations on four spindle primary afferents of the ankle extensors in four cats.

RESULTS

Rapid stretches of the ankle extensors were applied manually as shown in figure 1. A segment of record showing the afferent and efferent responses to

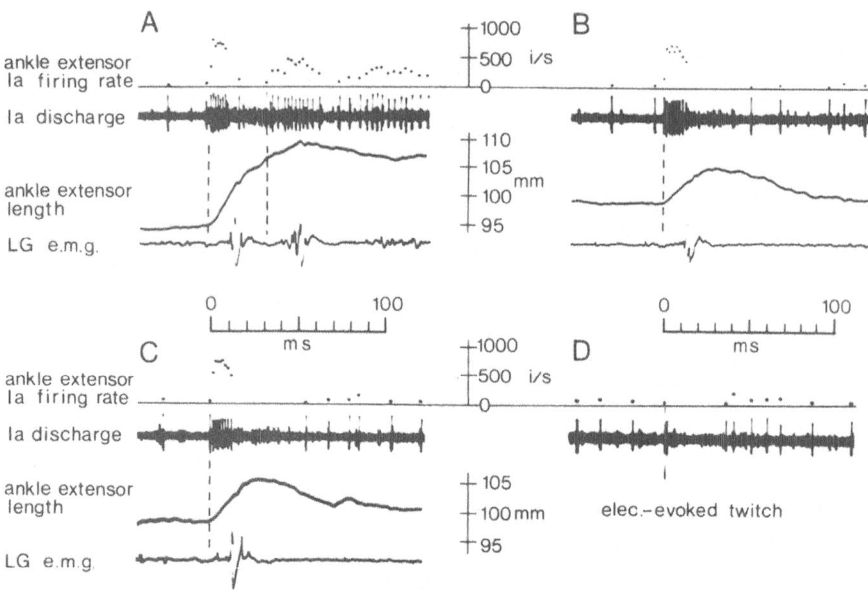

Figure 4. Responses of another spindle primary afferent. A, to rapid, maintained ankle extension. B and C, to rapid, but very brief ankle extension; note the occurrence of two main bursts of both afferent discharge and EMG in A, but only single bursts of each in B and C. D, Identification test as in figure 2B.

six such stretches, three fast and three slow, is shown in figure 2. The adapting nature of the afferent response immediately after each of the rapid stretches suggests the presence of dynamic fusimotor action on the afferent. Examination of the responses on an expanded time scale (figure 3) reveals a clear grouping of the afferent discharges, and a tendency for EMG responses to occur at latencies of about 10 msec after the onset of each afferent burst. Both the afferent response and the EMG response were greatly reduced in the slow stretch.

Figure 4 shows recordings in another animal, contrasting a maintained stretch (A) and stretches of very brief duration (B and C), obtained by tapping the plantar surface of the foot. In the maintained stretch, bursts of spindle and EMG response are evident. Due to a slightly lower overall speed of stretching, the second afferent burst occurred at a longer latency than those in figure 3 (1, 2 and 3). It is significant that the second EMG burst was similarly delayed. This would not be expected if it had been entirely the result of a 'long-loop' reflex initiated by the very first spindle response. In the very brief stretches eliciting single bursts of spindle discharge, only single bursts of EMG were observed. A similar dependence of the second burst of EMG on the duration of rapid muscle stretch has recently been reported in human subjects (Lee and Tatton, 1979).

DISCUSSION

The origin of the bursting nature of the spindle responses seen above may lie in mechanical oscillations within the muscle, fusimotor reflexes, short-range intrafusal stiffness, or a combination of these. Whatever the mechanism, the existence of the bursts does seem to determine to an important extent the pattern of EMG response. In multi-unit 'spindle-dominated' neural recordings obtained during muscle stretching in man, Burke *et al.* (1978) attributed long-latency bursts of activity to efferent responses. However, in this publication, Hagbarth describes recordings in man in which spindle afferent discharges during muscle stretch were indeed grouped at intervals of about 20 msec, a pattern similar to that described in this paper.

Our results do not disprove the existence of 'long-loop' reflexes under the conditions of our experiment. It is quite conceivable, for example, that the depolarisation of motoneurons by spindle afferents is a prerequisite for the manifestation of 'long-loop' action. However, the results do provide an explanation at the peripheral level for Ghez and Shinoda's (1978) finding that segmented EMG responses to muscle stretch may be seen even in spinalised animals. Furthermore, the observations may serve to remind researchers who use torque pulses and rapid ramp stretches as test stimuli, that the afferent responses elicited are often far from being linearly related to the time course of the stimuli.

Received on September 10th, 1980.

ACKNOWLEDGEMENTS

We thank the MRC and the Deutscheforschungsgemeinschaft SFB 33 for grant assistance, and Mr D. Earle for technical help.

REFERENCES

Burke, D., Hagbarth, K.-E. and Löfstedt, L. (1978). Muscle spindle responses in man to changes in load during accurate position maintenance, *J. Physiol., Lond.*, **276**, 159–164

Ghez, C. and Shinoda, Y. (1978). Spinal mechanisms of the functional stretch reflex, *Expl Brain Res.*, **32,** 55–68

Hammond, P. H., Merton, P. A. and Sutton, G. G. (1956). Nervous gradation of muscular contraction, *Brit. Med. Bull.*, **12,** 214—218

Lee, R. G. and Tatton, W. G. (1979). Long latency versus long loop reflexes: Dependence on the temporal characteristics of the imposed displacement, *Soc. Neurosci. Abstr.*, **5,** Abs. 1260

Phillips, C. G. (1969). Motor apparatus of the baboon's hand. The Ferrier Lecture, 1968. *Proc. R. Soc. B*, **173,** 141–174

Prochazka, A., Stephens, J. A. and Wand, P. (1979). Muscle spindle discharge in normal and obstructed movements, *J. Physiol., Lond.*, **287,** 57–66

Tatton, W. G., Forner, S. D., Gerstein, G. L., Chambers, W. W. and Liu, C. N. (1975). The effect of postcentral cortical lesions on motor responses to sudden upper limb displacements in monkeys, *Brain Res.*, **96,** 108–113

Basic patterns of muscle spindle discharge in man

ÅKE VALLBO*

INTRODUCTION

The method of recording impulses from human nerve fibres (Vallbo and Hagbarth, 1968) has provided new possibilities in gaining insight into mechanisms involved in motor control. In particular, the study of muscle spindle afferents in normal human subjects has contributed significantly to our views on proprioceptive functions. The accuracy and resolution of single unit analysis, combined with the advantages of a preparation which is truly intact and able to follow complex instructions, has clear possibilities for the study of motor functions. A large number of problems have already been explored, which illustrates the power of the method (Vallbo *et al.*, 1979). The recording technique is basically simple, although technically fairly demanding and tedious when the aim is to study well-identified single afferents.

Findings concerning spindle afferent discharge in man have not often been at variance with those obtained in nerve recordings from subhuman species and/or more indirect experiments in man. This paper considers some aspects of the function of the fusimotor system and muscle spindles as seen in the light of nerve recordings from human subjects. The emphasis will be put on basic principles, which have been presented in earlier papers, particularly with regard to the organisation of the fusimotor output in relation to the skeletomotor output. The major and simple conclusion is that there is a surprisingly intimate coactivation of the two motor systems in intact human subjects.

SKELETOMOTOR PRECEDENCE: FINDINGS AND IMPLICATIONS

As discussed in another report (Hagbarth, 1981), the available evidence indicates that fusimotor activity is absent or at least negligible when a human

* Department of Physiology, University of Umeå

subject keeps his muscles relaxed. In contrast, the fusimotor system is promptly activated at the onset of a contraction. A crucial point of relevance for some of the theories on the function of muscle spindles and the fusimotor system is whether fusimotor activity precedes the skeletomotor activity at the onset of a contraction. In neurophysiological experiments with anaesthetised, decerebrate or spinal preparations, it has been shown repeatedly that the discharge rate in spindle afferents and fusimotor neurons may start to accelerate prior to that of α-motoneurons and at lower thresholds when contractions are induced by adequate or electrical stimuli. A set of theories on the function of the fusimotor system are obviously fuelled by this finding. One is the idea that muscle contraction may be controlled exclusively or mainly via the γ-route, which would constitute an essential limb of a follow-up length servo system (Merton, 1951). Another is the idea that the fusimotor system has the function of biasing and sensitising the spindles well in advance of the contraction. Any such theory must be rejected for normal human subjects because experiments clearly demonstrate that there is no functionally significant fusimotor activation prior to contraction. This was first demonstrated in voluntary contractions under isometric conditions (figure 1A) (Vallbo, 1971). In such contractions the mechanical unloading of the sense organ would be minimal and therefore any effects of fusimotor drive would stand the greatest

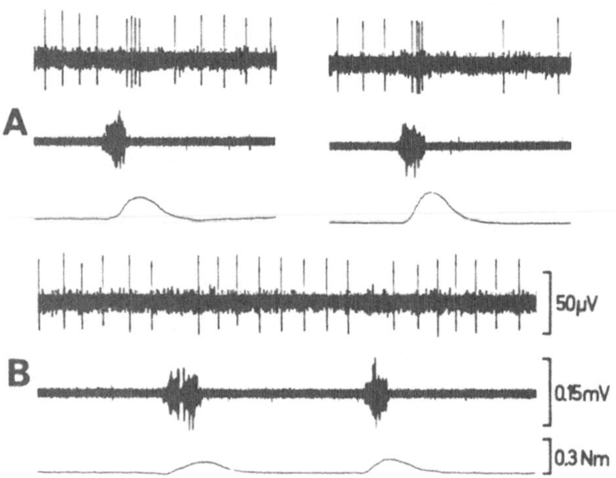

Figure 1. Discharge in a primary afferent from a muscle during isometric voluntary contractions. Top records show nerve impulses, middle records show electromyographic activity, and bottom records show the torque produced when the subject voluntarily made small contractions with his ring finger under isometric conditions. The ending was located in the flexor digitorum muscles in the forearm (ring-finger portion). Note the unloading effect in B when the contractions were smaller and slower (from Vallbo, 1971, p. 405).

chance of showing up as an acceleration in the rate of spindle discharge. Even under these conditions there was no indication of fusimotor precedence. The outcome of a large number of tests of the kind illustrated in figure 1A are collected in figure 2, where the distributions of latencies between the onset of EMG activity and spindle acceleration are shown for three units. Similar findings were subsequently reported in several kinds of experiments with various muscles—under isotonic as well as isometric conditions; with voluntary as well as reflex movements; and with test stretches applied immediately before voluntary contractions to disclose any change of spindle sensitivity (Burke *et al.*, 1980a; for further references *see* Vallbo *et al.*, 1979). Hence the outcome of a variety of experiments is totally concordant in demonstrating that there is no functionally significant fusimotor activation preceding skeletomotor activation.

It should be pointed out that spindle acceleration is not found in every test. Rather, deceleration may occur, particularly with weak contractions. An example is seen in figure 1B where the contractions are smaller and less brisk than in figure 1A. The deceleration is most reasonably interpreted as an effect of mechanical unloading of the ending due to internal shortening within the muscle, much as with an electrically induced muscle twitch. An indication of

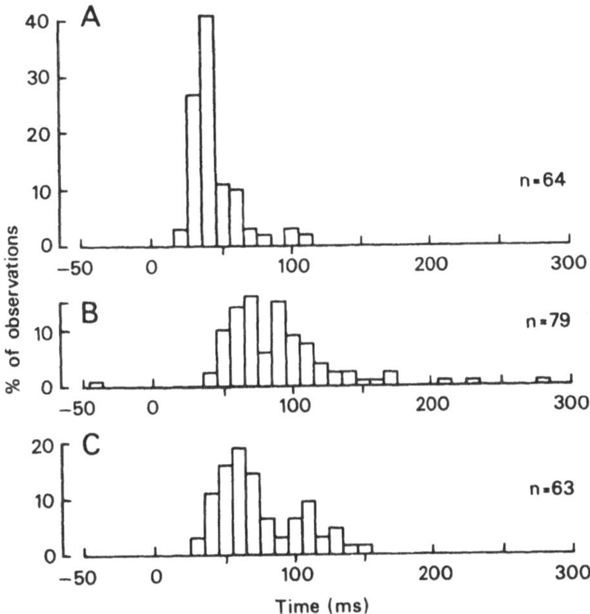

Figure 2. Histograms showing the distribution of time difference between onset of electromyographic activity and spindle acceleration with small isometric twitches as in figure 1. Data from three separate spindle afferents are shown in A, B and C. Note regular skeletomotor precedence (from Vallbo, 1971, p. 405).

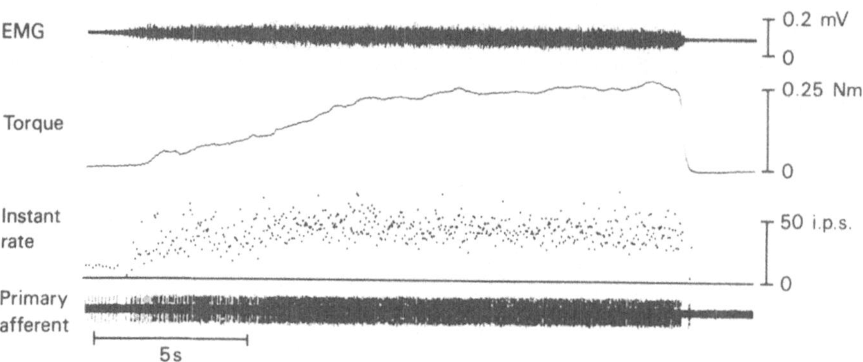

Figure 3. Discharge of a primary afferent from a muscle spindle during an isometric voluntary contraction. A minute initial unloading is followed by a sustained increase of spindle discharge rate which was crudely related to the intensity of skeletomotor contraction. The ending was located in the flexor digitorum muscles on the forearm (ring-finger portion) and the torque was produced by flexion of the ring finger alone. Same unit as in figures 5–8.

unloading is also seen initially in figure 3. It has been demonstrated that there is a variation among afferents with regard to threshold in terms of contraction force for acceleration (Burke *et al.*, 1978). As a result, some afferents do not accelerate during weak contractions, but require stronger contractions which presumably are associated with stronger fusimotor drive (see below). Hence spindle deceleration with weak isometric contractions cannot be taken as evidence of discordant changes of activity in the two systems.

From a teleological point of view, it might be argued that an earlier onset of fusimotor drive would be a great advantage in ensuring feed-back, even in the initial phase of contraction. However, in many afferents, the discharge accelerates very soon after the onset of EMG activity—the minimal latency was some 10 msec and hence before there is time for any major mechanical events in the muscle. Incidentally, the shortness of the latency suggests that the initial impulses in the fusimotor and the skeletomotor systems are simultaneously launched from the spinal cord.

The implication of a constant α-precedence is that voluntary contractions are initiated by descending impulses from supraspinal structures on to skeletomotor neurons either directly or via interneurons, but not via γ-motoneurons and muscle spindles. Whether any contractions at all are differently organised remains to be explored, but the fact that convincing γ-precedence has not been seen in any kind of motor activity studied in human subjects seems to carry considerable weight as support for the basic principle. It may be added that a rejection of the follow-up length servo hypothesis is further supported by a number of findings demonstrating that spindle afferent activity does not provide 'a major or indispensable excitatory drive to the skeletomotor neurones' (Vallbo *et al.*, 1979).

The conclusion that muscle spindles do not provide the major excitatory drive to the skeletomotor neurons is fundamental also from the point of view that this conclusion effectively limits the alternatives with regard to interpretation of experimental data. It implies that the EMG activity or the force during contractions is not mainly seen as an effect of the afferent input from the spindles, but rather as crudely reflecting the drive on to the pool of skeletomotor neurons of non-spindle origin, including supraspinal sources. The spindle discharge rate, on the other hand, may be seen as crudely reflecting fusimotor drive when the muscle length is held constant. Thus, the basic relations between the net drive on to the two kinds of motor neurons may be elucidated in the intact human subject.

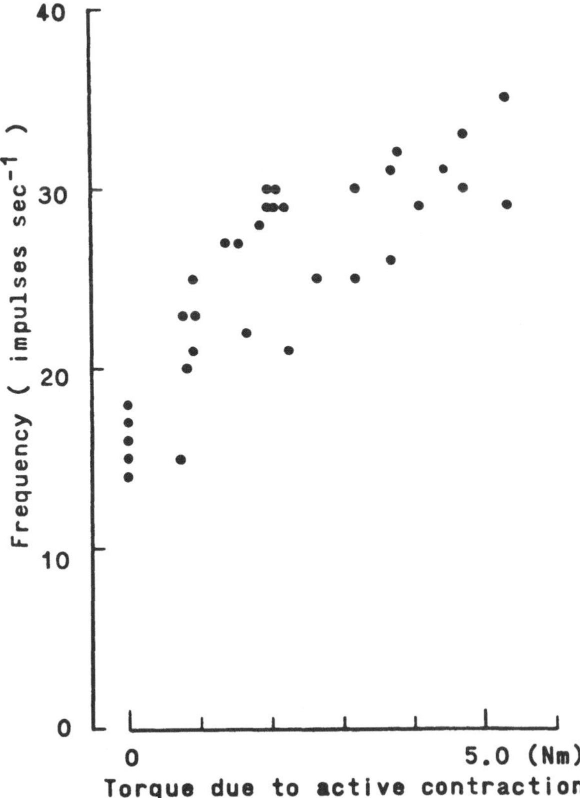

Figure 4. Plot of the discharge rate of a primary afferent from a muscle spindle against torque due to active contraction under isometric conditions. The ending was located in the flexor digitorum muscles of the forearm (ring-finger portion) and the torque was produced by flexions of the ring finger alone. Impulse frequency and average torque were measured over periods of 1 sec during semi-stationary conditions (from Vallbo, 1973, p. 251).

SPATIAL AND QUANTITATIVE ASPECTS OF FUSIMOTOR OUTFLOW

As emphasised in the preceding sections, the fusimotor system is promptly activated at the onset of a contraction. It has also been shown that the fusimotor system remains continuously active while a contraction is going on (figure 3). A study of the relation between contraction force and discharge rate in spindle afferents demonstrated that the rate increases with contraction force, indicating a parallelism between the intensity of fusimotor output and the intensity of skeletomotor output. An example is shown in figure 4. A positive relationship between force and rate—although not necessarily a linear one—has been demonstrated in a low range of contraction intensity, below approximately 20% of maximal voluntary force, whereas stronger contractions have not been tested in this respect (Vallbo, 1974). A possibility is that the parallel activation of the two motor systems constitutes one of the factors which account for the finding that the stretch reflex increases with the load on the muscle (Marsden *et al.*, 1976). However, it remains to be explored to what extent responses of human spindles to perturbations are dependent on the size of the load, as is the rate of discharge when the muscle length is held constant.

The principle of an intimate coactivation of the fusimotor and the skeletomotor systems is also evident when the spatial relations are studied. A localised voluntary contraction, for instance the flexion of a single finger, does not result in a general activation of the spindles in the muscles of the hand or the fingers, but the effect of fusimotor activation, observed as spindle acceleration, is largely confined to the contracting muscle portions (Vallbo, 1970). This basic finding has been confirmed with other muscles (for reference, *see* Vallbo *et al.*, 1979). A decrease of impulse rate is often encountered when the particular spindle studied is not located in the focus of the contraction.

FUSIMOTOR AND SKELETOMOTOR COACTIVATION

To sum up the main points of the preceding sections, a striking correspondence has been found between the skeletomotor output and muscle spindle activation in three respects—in time, in space and in quantity. The findings point to an intimate coactivation of the fusimotor and skeletomotor system in human subjects, whereas a clear and uncontroversial demonstration of an isolated activation of either the fusimotor or the skeletomotor systems has so far not been produced. Exceptions are the phasic stretch reflex and the tonic vibration reflex, which give rise to a skeletomotor activation with no indication of fusimotor activation. The statement that there is a close correspondence between skeletomotor output and spindle activation in time, in space and in quantity, should not be stretched too far, but should be seen as a first approximation indicating in outline the basic relation between the neural outputs to intrafusal and extrafusal muscle fibres. Experimental data are not available to define exactly how far the statement is valid. The spatial relations have not been analysed in much detail and the quantitative relations have been studied within a limited range of contraction intensity. Moreover, the conclusions concerning fusimotor activity are based on indirect analyses of spindle afferent activity rather than direct recordings of fusimotor impulses. Further, very little, if any, data are available on the share between dynamic and static fusimotor outflow.

SPINDLE RESPONSES TO POSITION AND VELOCITY DURING SLOW MOVEMENTS

The conclusions concerning coactivation of the fusimotor and skeletomotor systems as presented above were largely based on studies when movements were restricted as much as possible. The situation is more complex when movements are allowed. However, it seems that findings from such tests may readily be explained qualitatively on the same basis. Examples of shortening movements with a primary afferent from the finger flexor muscles are shown in figures 5–8. The muscle was working with a torque load corresponding to

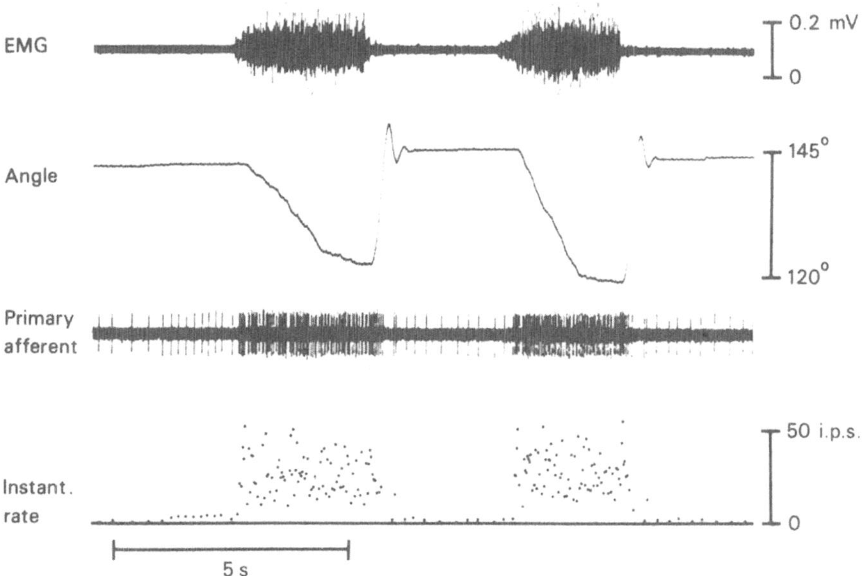

Figure 5. Discharge in a primary afferent from a muscle spindle during voluntary shortening movements; flexions at the metacarpophalangeal joint of the ring finger. The ending was located in the flexor digitorum muscles of the forearm. Same unit as in figures 3 and 6–8.

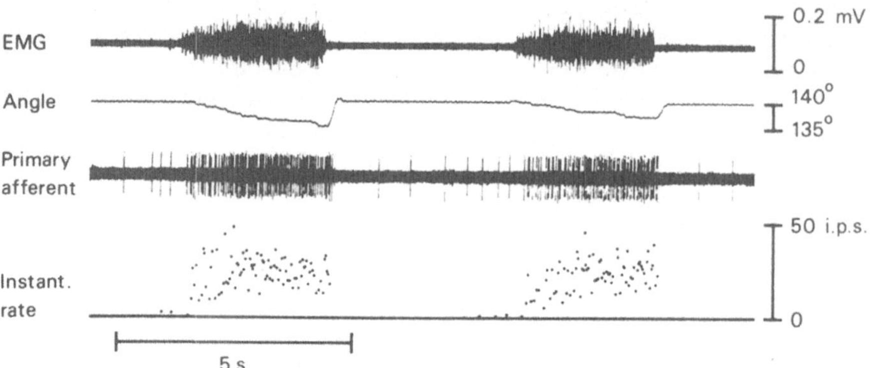

Figure 6. Discharge of a primary afferent from a muscle spindle during voluntary shortening movements; flexions at the metacarpophalangeal joint of the ring finger. The ending was located in the flexor digitorum muscles of the forearm. Same unit as in figures 3, 5, 7 and 8.

less than 3% of the maximal voluntary contraction force and with an inertial load of about ten times the inertia of the finger. Figure 5 shows shortening movements with a speed of about 3% of muscle length per second. It is striking that the afferent does not exhibit a discharge rate that is obviously related to either muscle length or velocity. The average rate is similar regardless of the muscle being long or short, the velocity being relatively high or almost at zero as in the semi-plateau phases. How should this be interpreted? Is the spindle totally insensitive to muscle length and velocity of shortening? It seems reasonable to interpret the rate as a balanced result of two opposing factors—the fusimotor drive and the changes of muscle length. During active shortening the skeletomotor activity is usually increasing continuously to

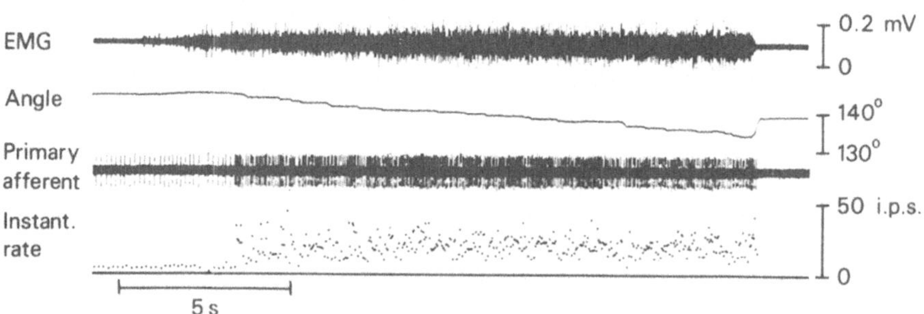

Figure 7. Discharge of a primary afferent from a muscle spindle during a slow voluntary shortening movement; flexion of the metacarpophalangeal joint of the ring finger. Note irregularities of discharge rate visible in record of impulse train and their relation to variations in speed of movement. The ending was located in the flexor digitorum muscles of the forearm. Same unit as in figures 3, 5, 6 and 8.

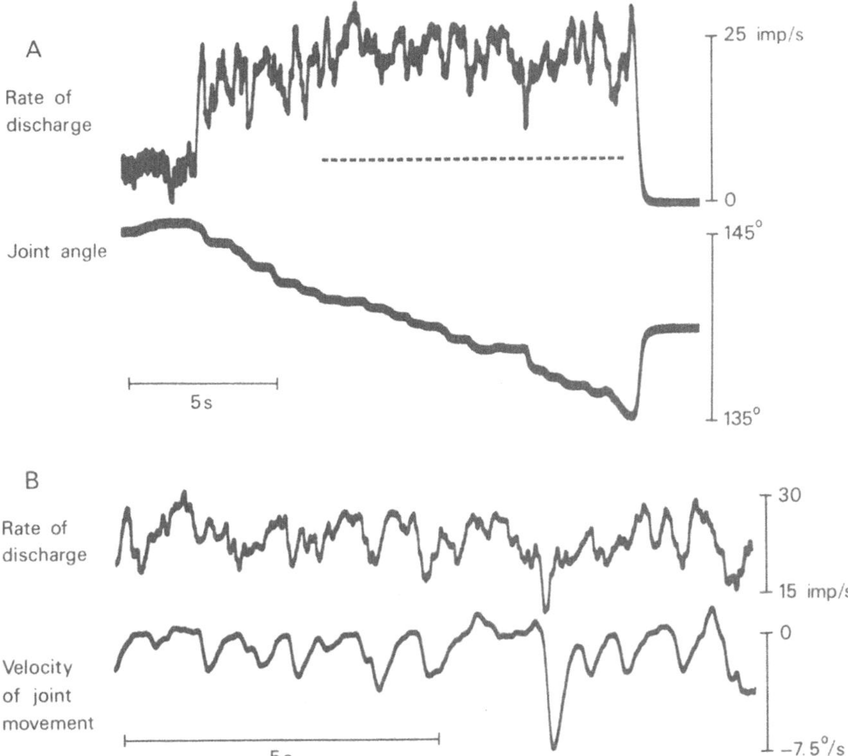

Figure 8. Discharge rate of a primary afferent from a muscle spindle related to details of a slow, voluntary shortening movement; flexion of the metacarpophalangeal joint of the ring finger. Same test as in figure 7. A shows the rate of discharge in an analogue form in the top and joint position below. The part of the sequence marked by the interrupted line in A is also illustrated in B, where the top record indicates the rate of discharge (as in A), whereas the bottom record gives the velocity of movement. Note parallel changes of rate of discharge in afferent and velocity of movement. The velocity signal was produced as the time derivative of the joint angle signal with proper filtering. (The exact time relation between impulse rate and changes of velocity of movement was slightly distorted in the processing of the original signals.) (From Vallbo, 1973, p. 251.)

overcome the intrinsic length–tension properties of the muscle. A parallel increase of fusimotor drive (see above) would tend to raise the discharge rate of the afferent, whereas the shortening would tend to decrease the rate. The net result might be a rate that remains practically independent of muscle length as in figure 5. Moreover, when movements are performed at a much slower speed, as in figure 6, the discharge rate was not much different. Hence the rate was remarkably independent of mean or desired speed of movement.

This is presumably a result of the two opposing factors increasing with the speed of movement, the net rate being similar with different speeds.

Although the average rate during the tests of figures 5 and 6 was relatively independent of the mean velocity of movement, the instantaneous discharge of primary afferents may be closely related to minute irregularities in slow movements. An example is shown in figure 7, where irregularities of discharge rate are obvious and their relations to movements may be observed. Somewhat different displays of the same test are presented in figure 8 to highlight the relationship between discharge and details of the movement. The top record of figure 8A is an analogue display of the impulse rate, produced by feeding spike-triggered standard pulses into a leaky integrator. It may be seen that the average rate of about 25 impulses/sec does not change much during the movement. On the other hand, there are variations of considerable amplitude which are closely related to the velocity of muscle shortening. In figure 8B the section of the test which is indicated by the interrupted line is illustrated on an expanded time scale. The upper record in B is the same signal as in A, whereas the lower record shows the speed of joint movement, negative values indicating muscle shortening. It may be seen that there were often parallel changes of discharge rate and velocity of movement, indicating that this spindle afferent unit provided a description of the variations in velocity. It is striking that the variations in the discharge rate associated with the small differences in velocity are considerable when compared to the rather small variations with mean speed of movement of the same spindle afferent, illustrated in figures 5 and 6.

Findings of the kind illustrated in figure 8 suggest that the afferent impulses from spindle primaries may describe the deviations from a desired velocity, and, via their central excitatory effect upon α-motoneurons, have the function of keeping oscillations and irregularities during the movement within limits (Matthews, 1964; Phillips, 1969; Vallbo, 1973; Hagbarth and Young, 1979, Young and Hagbarth, 1980). This mechanism would exert its effect regardless of which factors account for the deterioration of the movement, whether internal or external, mechanical or neural, or complex interactions between several factors. It seems that the tendency to produce irregular skeletomotor outputs during slow movement or position-holding is an inherent characteristic of the motor system. A number of factors are known to increase such irregularities, e.g. size of the load, fatigue, excitement and muscle weakness in ageing and disease (Young and Hagbarth, 1980). Thus, a mechanism for control of irregularities of movement may satisfy the need to cope with a general weakness of the motor system. This might be particularly important for muscles engaged in high-precision movements such as the small hand muscles and the neck muscles, where spindle densities are known to be high. There are indications that many primary endings are sufficiently sensitive to monitor not only irregularities of the external movement but also even internal length changes within the muscle that are too small to give rise to actual irregularities of movement (Hagbarth and Young, 1979; Å. B. Vallbo and M. Hulliger, unpublished report). Hence, it seems that the system would not only minimise irregularities during slow movements, but also even prevent the occurrence of such irregularities.

It should be pointed out, however, that many primary endings do not exhibit this kind of large response to small irregularities. Actually, it seems to be a general finding that the impulse rate and discharge pattern vary considerably from one spindle to another. A large variation has been found not only in muscles of the arm and leg in human subjects, but also in jaw muscles of cats and monkeys (Matsunami and Kubota, 1972; Taylor and Cody, 1974; Goodwin and Luschei, 1975). It remains to be assessed, however, whether this variation among spindle recordings is due to inherent differences between individual motor systems, to varying motor strategies of the subject, or to the exact spindle location in relation to the contracting muscle proportions, or whether different spindles in the same subject are differently controlled by the fusimotor system.

ROLE OF FUSIMOTOR INDEPENDENCE IN THE INTACT ORGANISM

The main thesis of this paper is that there is an intimate coactivation of the fusimotor and the skeletomotor systems in time, in space and in quantity. Available data can most simply be interpreted qualitatively on the basis of this view, although it must also be taken into account that some afferents have a higher threshold than others in terms of contraction force. Other interpretations may be possible on some points, although this would probably require that more complex relations are assumed which have so far not been demonstrated. It would be interesting to scrutinise the findings from experiments on truly intact animals to assess to what extent the same principles also hold in this case. On the other hand, in neurophysiological preparations when the normal function or integrity of the central nervous system has been interfered with by means of anaesthetics or sectioning of the neural trunk, there is a wealth of findings demonstrating fusimotor precedence rather than a more strict coactivation. Such findings are interesting because they illustrate a relative independence of the two motor systems, but it remains to be shown how this independence is utilised in the intact organism. Is it, for instance, ever used to activate the γ-motor system in isolation when the muscles are relaxed? Or is it merely used to allow a refinement of spindle control while muscles are active, adjusting spindle bias and sensitivity to meet the demands which are set by load, desired speed and precision of movement, control systems involved, etc? The findings from human subjects clearly favour the latter possibility. Positive evidence for a flexible balance between skeletomotor and fusimotor activity in human subjects has recently been reported, although some of these findings probably need confirmation (Vallbo and Hulliger, 1979; Burke *et al.*, 1980b).

CONCLUDING REMARKS

In this paper, limited aspects of muscle spindle and fusimotor function have been discussed, while other aspects which have been studied in recordings from human afferents are considered by Hagbarth (1981). A large number of problems in this area remain to be analysed, although many of them may require more quantitative approaches than have usually been undertaken so far. In particular, it seems that the definite assessment of which variable (or variables) are represented in the message from the spindles, calls for the consideration, as well as the assessment, of factors which may modify spindle sensitivity and bias. Moreover, now that some of the basic features of muscle spindle function in the intact human subject have been clarified, it seems that the time has come to study proprioceptive activity during more complex motor acts.

ACKNOWLEDGEMENTS

This work was supported by the Swedish Medical Research Council, project No. 2075

Received on June 17th, 1980.

REFERENCES

Burke, D., Hagbarth, K.-E. and Skuse, N. F. (1978). Recruitment order of human spindle endings in isometric voluntary contractions, *J. Physiol., Lond.*, **285,** 101–112

Burke, D., McKeon, B., Skuse, N. F. and Westerman, R. A. (1980a). Anticipation and fusimotor activity in preparation for a voluntary contraction, *J. Physiol., Lond.*, **306,** 337–348

Burke, D., McKeon, B. and Westerman, R. A. (1980b). Induced changes in the thresholds for voluntary activation of human spindle endings, *J. Physiol., Lond.*, **302,** 171–181

Goodwin, G. M. and Luschei, E. S. (1975). Discharge of spindle afferents from jaw-closing muscles during chewing in alert monkeys, *J. Neurophysiol.*, **38,** 560–571

Hagbarth, K.-E. (1981). Fusimotor and stretch reflex functions studied in recordings from muscle spindle afferents in man. This publication

Hagbarth, K.-E. and Young, R. R. (1979). Participation of the stretch reflex in human physiological tremor, *Brain*, **102,** 509–526

Marsden, C. D., Merton, P. A. and Morton, H. B. (1976). Servo-action in the human thumb, *J. Physiol., Lond.*, **257,** 1–44

Matsunami, K. and Kubota, K. (1972). Muscle afferents of trigeminal mesencephalic tract nucleus and mastication in chronic monkeys, *Jap. J. Physiol.*, **22,** 545–555

Matthews, P. B. C. (1964). Muscle spindles and their motor control, *Physiol. Rev.*, **44,** 219–288

Merton, P. A. (1951). The silent period in a muscle of the human hand, *J. Physiol., Lond.*, **114,** 183–198

Phillips, C. G. (1969). Motor apparatus of the baboon's hand, *Proc. R. Soc. B*, **173,** 141—174

Taylor, A. and Cody, F. W. J. (1974). Jaw muscle spindle activity in the cat during movements of eating and drinking, *Brain Res.*, **71,** 523–530

Vallbo, Å. B. (1970). Discharge patterns in human muscle spindle afferents during isometric voluntary contractions, *Acta physiol. scand.*, **80,** 552–566

Vallbo, Å. B. (1971). Muscle spindle response at the onset of isometric voluntary contractions in man. Time difference between fusimotor and skeletomotor effects, *J. Physiol., Lond.*, **218,** 405–431

Vallbo, Å. B. (1973). Muscle spindle afferent discharge from resting and contracting muscles in normal human subjects. In *New Developments in Electromyography and Clinical Neurophysiology* (ed. J. E. Desmedt), Vol. 3, Basel, Karger, 241–262

Vallbo, Å. B. (1974). Human muscle spindle discharge during isometric voluntary contractions. Amplitude relations between spindle frequency and torque, *Acta physiol. scand.*, **90,** 319–336

Vallbo, Å. B. and Hagbarth, K.-E. (1968). Activity from skin mechanoreceptors recorded percutaneously in awake human subjects, *Expl Neurol.*, **21,** 270–289

Vallbo, Å. B., Hagbarth, K.-E., Torebjörk, H. E. and Wallin, B. G. (1979). Somatosensory, proprioceptive and sympathetic activity in human peripheral nerves, *Physiol. Rev.*, **59,** 919–957

Vallbo, Å. B. and Hulliger, M. (1979). Flexible balance between skeletomotor and fusimotor activity during voluntary movements in man. *Neurosci. Lett.*, Supplement No. 3, s103

Young, R. R. and Hagbarth, K.-E. (1980). Physiological tremor enhanced by manoeuvres affecting the segmental stretch reflex, *J. Neurol. Neurosurg. Psychiat.*, **43,** 248–256

Fusimotor and stretch reflex functions studied in recordings from muscle spindle afferents in man

K.-E. HAGBARTH

As compared with other techniques used in the study of muscle stretch receptors and their role in motor control, microelectrode recordings from human nerves (microneurography) have one main, obvious advantage. They give direct information of how the muscle feedback systems operate under normal conditions in unrestrained subjects who are able to follow the instructions of the experimenter in the execution of various motor tasks. With the experimenter himself as subject in combined microneurography and EMG recordings he can easily study to what extent he, by acts of will, can control the impulse generation in his muscle spindles and alpha motoneurons, while his muscles or limbs are exposed to various (predicted or unpredicted) external stimuli.

The method is technically not as difficult as some colleagues seem to believe, even though it is often a tedious procedure to find and identify the type of nerve fibres searched for. From the motor-control point of view it is unfortunate that recordings from gamma motor fibres have not yet been accomplished, but fusimotor functions are still accessible to analysis since the technique involves a sampling bias in favour of the large, myelinated group Ia afferents of the muscle spindles. A limitation is that dislocation of the microelectrode easily occurs if the external mechanical stimuli to the extremity, or the voluntary contractions, are too brisk. Another limitation is, of course, that excitatory or inhibitory events in the skeletomotor and fusimotor systems are revealed only in terms of spike generation; sub-threshold excitability changes in the spike-generating structures are not accessible to analysis. Still, the technique represents a powerful tool for the assessment of various theories of motor control in man. The relatively large number of reports during the last decades on recordings from muscle spindle afferents in man have recently been reviewed and discussed (Vallbo *et al.*, 1979). Those with experience of the technique are not likely to agree with Houk's recent

Department of Clinical Neurophysiology, University Hospital, Uppsala

statement that 'unfortunately . . . the method is difficult to use, and, at present, only very limited observations have been reported' (Houk, 1979a).

ALPHA–GAMMA COACTIVITY AND ALPHA–GAMMA CO-SILENCE

Numerous studies in decerebrate, spinal or anaesthetised cats have shown tonic fusimotor firing in the resting state and—on initiation of muscle contraction—a rise in fusimotor activity preceding the alpha activity (*see* Murthy, 1978). In contrast, recordings from spindle afferents in human nerves provide no convincing evidence for a resting fusimotor tone in relaxed muscles, and muscle contractions are not preceded by any signs of spindle sensitisation (*see* Vallbo *et al.*, 1979; Burke, 1981a). Claims that during the Jendrassik manoeuvre, or in response to arousal stimuli, spindle sensitisation may occur without alpha engagement (Burg *et al.*, 1974) have not been verified (Hagbarth, Wallin, Burke and Löfstedt, 1975a). As recently pointed out by Granit (1979) a subject's ability to relax may well involve an act of suppression by inhibition. Still, if there were a gamma precedence one would expect to see some signs of spindle sensitisation before the appearance of the first motor unit potentials when relaxation ceases. However, both for the finger flexors, thoroughly studied by Vallbo (1971), and for leg muscles (Hagbarth, Wallin, Löfstedt, 1975b), it has been found that the extrafusal contraction, as seen in torque and EMG recordings, starts before the fusimotor-induced activation of the spindle afferents. As the subject gradually increases the contraction strength (under isometric conditions) one spindle ending after another is recruited into action, and once the recruitment threshold for a given ending has been reached its firing rate tends to increase in approximate proportion to contraction strength (Burke *et al.*, 1978c). The discrepancies between the findings in cat and man are not easily explained. It cannot certainly be excluded that there is a species difference with respect to alpha–gamma balance, but before that conclusion is drawn one would like to see convincing evidence of a tonic fusimotor tone on the spindles in EMG-silent muscles in awake intact cats.

It is often argued that a separate motor innervation of the extrafusal and intrafusal fibres would be unnecessary if the two systems are always functionally joined in coactivity and co-silence. However, the principle of alpha–gamma coactivation does not necessarily imply that the gamma system is engaged to an equal extent in all types of motor acts. Recent studies in man indicate, for instance, that vestibular stimulation can affect the recruitment threshold for individual spindle endings in slowly rising voluntary isometric contractions (Burke *et al.*, 1980b), and preliminary findings suggesting a flexible alpha–gamma balance have been reported also by Vallbo and Hulliger (1979). It should also be recalled that in cats the autogenetic excitatory effect of the spindle afferents engages the gamma motoneurons to a much lesser extent than the alpha motoneurons (Ellaway and Trott, 1978). This agrees with the findings in man that isometric tendon jerk and TVR contractions—unlike voluntary contractions—are accompanied by signs of spindle unloading rather than spindle sensitisation (Szumski *et al.*, 1974; Hagbarth *et al.*, 1975c; Burke *et al.*, 1976b).

In voluntary movements, when muscle shortening occurs during contraction, the effect of the intrafusal contraction is counteracted by the unloading effect of surrounding extrafusal fibres, and this unloading becomes more apparent with increasing speed of movement (Burke *et al.*, 1978b). Sudden obstruction of a voluntary movement causes an abrupt increase in spindle firing as the unloading effect ceases when the contraction becomes isometric. Thus, even though no stretch has occurred, it is not surprising that a motor response synonymous to a stretch reflex may appear in response to a sudden obstruction (Marsden *et al.*, 1976).

There are three main reasons why one should look with caution upon reports suggesting a lack of alpha–gamma linkage in various motor acts.

1. Spindles with a high recruitment threshold during isometric contraction may show signs of unloading before the fusimotor-induced spike acceleration starts (Burke *et al.*, 1978c).
2. Under isometric conditions, contractions in neighbouring muscles may cause unloading of the muscle in which the selected spindle is located (Vallbo, 1970; Burke *et al.*, 1980a).
3. If the receptor-bearing muscle is allowed to shorten during contraction, fusimotor-driven spindles may be silenced by the unloading effect of the shortening (Burke *et al.*, 1978b).

REFLEX EFFECTS OF SPINDLE INFLOW DURING CONTRACTION

Sustained effects

With vibratory stimulation of skeletal muscles in man it is possible to evoke a high-frequency sustained discharge in the Ia afferents, and a slowly increasing autogenetic reflex contraction (TVR) of sufficient strength to lift an arm or a leg against gravity (Burke *et al.*, 1976a; Hagbarth and Eklund, 1966). This gives a hint regarding the power of the support that the alpha motoneurons may receive across the gamma loop in an isometric voluntary contraction. As the spindle inflow increases with the strength of an isometric voluntary contraction, the tonic reflex support to the contraction is also likely to increase. It is questionable, however, whether the gamma system, even in maximal voluntary contractions, can drive the spindle afferents up to such high firing rates as can be induced by vibration (200–300 Hz). Therefore, it is also questionable whether the autogenetic reflex support via the loop ever exceeds what is required to counterbalance the force of gravity. With available data concerning the position sensitivity of spindles and their firing rate during isometric contractions, it has been estimated that the spindle inflow cannot be the main generator of the motor response that enables the subject to hold the joint position constant against increased external load (Vallbo, 1974). However, that some contribution to voluntary power is provided by the gamma loop is evidenced by the fact that part of the power loss following partial Xylocaine block of the muscle nerve can be restored by adding a vibratory stimulus (Hagbarth *et al.*, 1970).

Transient effects

Many microneurographic findings indicate that a main reflex function of the primary endings in contracting muscles is related to their ability to respond with synchronised discharges or synchronised pauses to minute mechanical perturbations such as sudden small changes in muscle length or in the speed of ongoing movement (Vallbo, 1973; Burke *et al.*, 1978a; Hagbarth and Young, 1979b; Young and Hagbarth, 1980). Irrespective of the question of whether dynamic as well as static gamma motoneurons are activated in voluntary contractions, the background discharge of the spindles during contraction provides them with a larger dynamic working range since—in contrast to silent spindles—they can respond not only in a positive direction but also in a negative direction to external load variations. In a similar way, the dynamic working range of a firing alpha motoneuron pool is larger than that of a silent pool, and, during contraction, the peaks and troughs in the spindle inflow, resulting from external perturbations, have a potent modulating effect on the skeletomotor outflow. The non-linear characteristics of the primary endings (Matthews, 1972) which enable them to respond briskly to the very initiation of a length change, and the short latency of the segmental stretch reflex, make this negative feed-back capable of providing fast motor reactions that can help to absorb impacts of mechanical disturbances. If the mechanical disturbances are due to variations in external load, the reflex can counterbalance these load changes; if the disturbances result from irregularities in skeletomotor outflow during active, graded movements, the reflex can counterbalance the unwanted changes in trajectory (Cooke, 1980) or speed of movement (Vallbo, 1973). For imposed muscle stretches with a certain speed and duration, it has been shown that the reflex can act precisely at the time when the inherent viscous resistance of the muscle yields, and it can then be said that the reflex regulates and improves the linearity of muscle stiffness, as advocated by Houk (1979b). Arguments concerning the relative importance of each of these various modes of operation are probably not very fruitful. It seems more important to emphasise that the way the reflex acts depends to a large extent on the speed and duration of the mechanical disturbance. In contracting human wrist muscles, the latency of the reflex torque in response to an abrupt wrist extension is about 60 msec, and consequently mechanical disturbances of shorter duration cannot be damped by the reflex (Hagbarth *et al.*, 1981). But even when the reflex comes after the completion of a stretch movement, it can still be functional in the sense that it tends to compensate for sudden unwanted changes in position or trajectory. For alternating flexion–extension movements, occurring at a rate of 7–9 Hz, the reflex torque resulting from each stretch movement will occur during the succeeding shortening phase and thus reinforce the mechanical oscillations. This mechanism has been invoked to explain enhanced physiological tremor (Hagbarth and Young, 1979b).

A consequence of the non-linear characteristics of primary spindle endings is that during imposed abrupt joint movements the endings are sensitive detectors of mechanical vibrations resulting from the perturbation. Thus, during muscle stretch resulting from a brisk joint displacement (generated, for instance, by a torque motor), the spindle discharges are grouped into successive volleys of impulses, time-locked to damped mechanical oscillations in the moving body parts. For suddenly applied angular wrist extensions, reaching a speed of 100–200°/sec, the vibrations have a frequency of about 50 Hz. Providing the flexor muscles are contracting at the time of the impact, the resulting successive spindle volleys give rise to reflex modulation of the skeletomotor outflow, seen in the gross EMG as successive groups of motor unit potentials, separated by intervals of about 20 msec (Hagbarth *et al.*, 1981). This implies that one has to look with considerable caution upon reports maintaining that the EMG peaks succeeding the initial one result from 'long-loop reflex' action.

PROBLEMS CONCERNING REFLEX 'GAIN' IN NORMAL AND PATHOLOGICAL STATES

As long as there is no convincing evidence that under normal conditions in man the fusimotor system can be activated prior to, or independently of, the skeletomotor system, changes in fusimotor drive cannot be held responsible for modifications of stretch reflex responsiveness of relaxed muscles. Recordings from spindle afferents combined with EMG recordings in man indicate that tendon jerk reinforcement, induced by the Jendrassik manoeuvre (Hagbarth *et al.*, 1975a), or presetting of the stretch reflex prior to a voluntary contraction (Burke, 1981a,b), may occur without involvement of the fusimotor system. Also, the voluntary suppression of the tonic vibration reflex is apparently not a fusimotor affair (Burke *et al.*, 1976b), but dependent on some central regulating mechanisms which prevent the impulses in the Ia afferents from generating any skeletomotor outflow. A simple explanation for this TVR suppression would be that voluntary commands can inhibit a motoneuron pool to such an extent that even a strong Ia excitatory drive remains a sub-threshold event. An alternative interpretation is that the Ia impulses are prevented from reaching the alpha motoneurons by inhibitory mechanisms acting at presynaptic or interneuronal sites in the reflex arc.

The term *action stretch reflex* introduced by Neilson and Lance (1978) may well be used more commonly to emphasise the contrast between the reflex in the passive and in the active states. Whereas the motoneurons of a relaxed muscle may be totally unresponsive (in terms of spike generation) even to heavy barrages of spindle impulses evoked by muscle stretch or vibration, the temporal patterning of the discharges from a firing motoneuron pool is highly susceptible to the modulating influence of even minute variations in the fusimotor-driven spindle inflow (Burke *et al.*, 1978a; Hagbarth and Young, 1979a; Young and Hagbarth, 1980). Claims that the 'gain' of the reflex is increased by background contraction (Marsden *et al.*, 1976) have been criticised by Houk (1979b), who maintains that changes in reflex responsiveness depending on variations in background excitatory drive on the motoneurons should rather be described in terms of 'resetting of the threshold of the motoservo'. It seems, however, that the increased responsiveness of the system in contracting muscles is not adequately described as either a 'gain' or a 'threshold' change. In the active state, a new dimension is added to the working range of the 'motoservo', since it is only in this state that both the spindles and the alpha motoneurons can respond in the negative direction to unloadings.

Even though the fusimotor system does not act as a 'forerunner' to the skeletomotor system in the initiation of voluntary movements, the segmental 'action stretch reflex' can act as a 'forerunner' to the longer-latency voluntary reactions to muscle stretch. Subjects instructed to resist unexpectedly applied ankle displacements generally cannot, without background contraction in the leg muscles, produce any active resistance to the displacement within the first 200–300 msec after its start. But with background contraction this time may be reduced to less than 100 msec. As judged by simultaneous nerve and EMG recordings, the main difference is that in the relaxed state the stretch-induced spindle discharges are ineffective in producing any short-latency segmental reflex responses, whereas in the active state, reflex enhancement of the skeletomotor outflow precedes the voluntary alpha–gamma-linked motor reactions (Burke *et al.*, 1978a).

The normal ability to prevent a muscle from responding to muscle stretch or vibration is lost in pathological states such as spasticity and rigidity. In none of the patients investigated so far with either of these types of hypertonus has it been possible to demonstrate any definite abnormality in the spindle responsiveness to stretch or vibration (Hagbarth *et al.*, 1973). This supports the hypothesis that the hyperreflexia in these patients is not primarily due to increased fusimotor tone, but rather to some type of central overreactivity to essentially normal spindle inputs. The hyperreflexia in spasticity may well be due either to structural changes such as sprouting of the Ia terminals, or to abnormal withdrawal of inhibitory mechanisms acting at presynaptic or postsynaptic sites in the segmental reflex arc. The hypertonus in Parkinsonian rigidity, on the other hand, is more likely to depend on faulty supraspinal handling of the proprioceptive inflow (Burke *et al.*, 1977).

Received on June 9th, 1980.

REFERENCES

Burg, D., Szumski, A. J., Struppler, A. and Velho, F. (1974). Assessment of fusimotor contribution to reflex reinforcement in humans, *J. Neurol. Neurosurg. Psychiat.*, **37**, 1012–1021

Burke, D. (1981a). The activity of human muscle spindle endings in normal motor behaviours. In: *International Review of Physiology and Neurophysiology IV* (ed. R. Porter), Baltimore, University Park Press, to be published

Burke, D. (1981b). A critical examination of the case for and against fusimotor involvement in disorders of muscle tone. In: *Progress in Clinical Neurophysiology* (ed. J. E. Desmedt), Basel, Karger, to be published

Burke, D., Hagbarth, K.-E. and Löfstedt, L. (1978a). Muscle spindle responses in man to changes in load during accurate position maintenance, *J. Physiol., Lond.*, **276**, 159–164

Burke, D., Hagbarth, K.-E. and Löfstedt, L. (1978b). Muscle spindle activity in man during shortening and lengthening contractions, *J. Physiol., Lond.*, **277**, 131–142

Burke, D., Hagbarth, K.-E., Löfstedt, L. and Wallin, B. G. (1976a). The responses of human muscle spindle endings to vibration of non-contracting muscles, *J. Physiol., Lond.*, **261**, 673–693

Burke, D., Hagbarth, K.-E., Löfstedt, L. and Wallin, B. G. (1976b). The responses of human muscle spindle endings to vibration during isometric contraction, *J. Physiol., Lond.*, **261**, 695–711

Burke, D., Hagbarth, K.-E. and Skuse, N. F. (1978c). Recruitment order of human spindle endings in isometric voluntary contractions, *J. Physiol., Lond.*,**285**, 101–112

Burke, D., Hagbarth, K.-E. and Wallin, B. G. (1977). Reflex mechanisms in Parkinsonian rigidity, *Scand. J. rehab. Med.*, **9**, 15–23

Burke, D., Hagbarth, K.-E. and Wallin, B. G. (1980a). Alpha–gamma linkage and the mechanisms of reflex reinforcement. In: *Spinal and Supraspinal Mechanisms of Voluntary Motor Control and Locomotion*, Vol. 8, *Progress in Clinical Neurophysiology* (ed. J. E. Desmedt), Basel, Karger, 170–180

Burke, D., McKeon, B. and Westerman, R. A. (1980b). Changes in the thresholds for voluntary activation of human spindle endings induced by supraspinal and segmental influences, *J. Physiol., Lond.*, to be published

Cooke, D. J. (1980). The role of stretch reflexes during active movement, *Brain Res.*, **181**, 493–497

Ellaway, P. H. and Trott, J. L. (1978). Autogenetic reflex action on to gamma motoneurones by stretch of triceps surae in the decerebrated cat, *J. Physiol., Lond.*, **276**, 49–66

Granit, R. (1979). Interpretation of supraspinal effects on the gamma system. In: *Reflex Control of Posture and Movement* (ed. R. Granit and O. Pompeiano), Elsevier, North-Holland Biomedical Press, 147–154

Hagbarth, K.-E. and Eklund, G. (1966). Motor effects of vibratory muscle stimuli in man. In: *Muscular Afferents and Motor Control. Proceedings of the First Nobel Symposium* (ed. R. Granit), Stockholm, Almqvist and Wiksell, 177–186

Hagbarth, K.-E., Hägglund, J. V., Wallin, E. U. and Young, R. R. (1981). Grouped spindle and EMG responses to abrupt wrist extension movements in man, *J. Physiol., Lond.*, to be published

Hagbarth, K.-E., Hongell, A. and Wallin, B. G. (1970). The effect of gamma fibre block on afferent muscle nerve activity during voluntary contractions, *Acta physiol. scand.*, **79**, 27A–28A

Hagbarth, K.-E., Wallin, B. G., Burke, D. and Löfstedt, L. (1975a). Effects of the Jendrassik manouvre on muscle spindle activity in man, *J. Neurol. Neurosurg. Psychiat.*, **38**, 1143–1153

Hagbarth, K.-E., Wallin, B. G. and Löfstedt, L. (1973). Muscle spindle responses to stretch in normal and spastic subjects, *Scand. J. rehab. Med.*, **5**, 156–159

Hagbarth, K.-E., Wallin, B. G. and Löfstedt, L. (1975b). Muscle spindle activity in man during voluntary fast alternating movements, *J. Neurol. Neurosurg. Psychiat.*, **38**, 625–635

Hagbarth, K.-E., Wallin, B. G., Löfstedt, L. and Aquilonius, S.-M. (1975c). Muscle spindle activity in alternating tremor of parkinsonism and in clonus, *J. Neurol. Neurosurg. Psychiat.*, **38**, 636–641

Hagbarth, K.-E. and Young, R. R. (1979a). Participation of the stretch reflex in human physiological tremor, *Brain*, **102**, 109–526

Hagbarth, K.-E. and Young, R. R. (1979b). Sensitivity of normal human muscle spindles to small spontaneous changes in muscle length during voluntary contraction, *Trans. Am. Neurol. Assoc.*, **103**, 1–4

Houk, J. C. (1979a) Motor control processes: New data concerning motoservo mechanisms and a tentative model for stimulus-response processing. In: *Posture and Movement* (eds R. E. Talbott and D. R. Humphrey), New York, Raven Press, 231–241

Houk, J. C. (1979b). Regulation of stiffness by skeletomotor reflexes, *Ann. Rev. Physiol.*, **41**, 99–114

Marsden, C. D., Merton, P. A. and Morton, H. B. (1976). Servo-action in the human thumb, *J. Physiol., Lond.*, **257**, 1–44

Matthews, P. B. C. (1972). *Mammalian Muscle Receptors and Their Central Actions*, Baltimore, Williams and Wilkins

Murthy, K. S. K. (1978). Vertebrate fusimotor neurones and their influences on motor behaviour, *Prog. Neurobiol.*, **4**, 249–307

Neilson, P. D. and Lance, J. W. (1978). Reflex transmission characteristics during voluntary activity in normal man and patients with movement disorders. In *Cerebral Motor Control: Long Loop Mechanisms*, Vol. 4, *Progress in Clinical Neurophysiology* (ed. J. E. Desmedt), Basel, Karger, 263–299

Szumski, A. J., Burg, D., Struppler, A. and Velho, F. (1974). Activity of muscle spindles during muscle twitch and clonus in normal and spastic human subjects, *Electroenceph. clin. Neurophysiol.*, **37**, 589–597

Vallbo, Å. B. (1970). Discharge patterns in human muscle spindle afferents during isometric voluntary contractions, *Acta physiol. scand.*, **80**, 552–566

Vallbo, Å. B. (1971). Muscle spindle response at the onset of isometric voluntary contractions in man. Time difference between fusimotor and skeletomotor effects, *J. Physiol., Lond.*, **218**, 405–431

Vallbo, Å. B. (1973). Impulse activity from human muscle spindles during voluntary contractions. In *New Developments in Electromyography and Clinical Neurophysiology*, Vol. 3 (ed. J. E. Desmedt), Karger, Basel, 251–262

Vallbo, Å. B. (1974). Human muscle spindle discharge during isometric voluntary contractions. Amplitude relations between spindle frequency and torque, *Acta physiol. scand.*, **96**, 319–336

Vallbo, Å. B., Hagbarth, K.-E., Torebjörk, H. E. and Wallin, B. G. (1979). Somatosensory, proprioceptive, and sympathetic activity in human peripheral nerves, *Physiol. Rev.*, **59**, 919–957

Vallbo, Å. B. and Hulliger, M. (1979). Flexible balance between skeletomotor and fusimotor activity during voluntary movements in man, *Neurosci. Letts.*, Suppl. No. 3, S103

Young, R. R. and Hagbarth, K.-E. (1980). Physiological tremor enhanced by manoeuvres affecting the segmental stretch reflex, *J. Neurol. Neurosurg. Pyschiat.*, **43**, 248–256

A critique of the papers by Vallbo and Hagbarth

A. TAYLOR*

An appraisal of the latter two papers is in effect an appraisal of the methods, the results and the interpretations of human micro-neurography.

THE METHOD

It is evident that the method works well in the hands of experienced and patient investigators. It is safe and acceptable and has yielded a great deal of interesting data. Of course, every method has its limitations, and in this case there seem to be three main problems. Firstly, difficulties of maintaining electrode stability have meant that movements studied have had to be slow and small, relative to the normal range. Secondly, most of the axons recorded from concerned with motor control have been the largest myelinated ones. It is a source of real regret to us all that gamma efferents have so far been inaccessible, but there is also a relative scarcity of recordings from fully identified spindle secondary afferents. The third problem is that in the human subject it is difficult to apply all the tests which are really required to give certain identification of the nature of each afferent recorded.

With regard to the size and speed of movements studied, our concern must be that we may be led to generalisations about muscle spindle behaviour which are not applicable to a wide range of interesting movements. In particular, Prochazka *et al.* (1979) have pointed out that by expressing velocity of muscle shortening in terms of resting lengths per second (l_r/sec), the movements studied in man are generally so slow that only a very modest constant fusimotor drive would be needed to maintain spindle afferent firing. The movements studied in alert animals (e.g. Taylor and Cody, 1974; Goodwin and Luschei, 1975; Prochazka *et al.*, 1976) commonly show reduction or silencing of spindle firing, at least in part because the rates of shortening are so

* Sherrington School of Physiology, St. Thomas's Hospital, London

much greater. In the extreme case, a number of studies of human spindle afferents have concerned isometric contractions in the expectation that, at constant muscle length, changes in spindle firing must indicate fusimotor activity changes. This may be the case, but we should be careful in concluding that the same fusimotor changes are to be expected in non-isometric conditions with the same amount of alpha motor activity. After all, the spindles, as displacement sensors, are not likely to be engaged in controlling isometric contractions in the same way as shortening contractions, if the higher functions of the controller are at all refined. We should certainly assume that they are likely to be so after such a long period of evolution accompanied by extensive elaboration of parts of the brain concerned with movement control. It is, of course, a real advantage to have the co-operation and insight of the human subject in experiments on voluntary movements, but the disadvantage is that we are pitting our own wits against the most advanced system we could choose. In the earlier stages of understanding there is merit in studying simpler systems and ones potentially more stereotyped in behaviour such as the rhythmically repeating movements of breathing, mastication and locomotion.

The shortage of spindle secondary afferent recordings by micro-neurography and the lack of gamma efferents are seen to be one of the essential limitations of the method, though it is possible that experiments with different electrode tip configurations might help. Otherwise our chief hope for direct fusimotor recordings without the use of anaesthesia must be an extension of the approach by Lund *et al.* (1979) of implanting metal microelectrodes into the motor nucleus of the fifth nerve in animals. This was technically difficult and so far has been rather unproductive, but clearly needs to be followed up.

The question of certain identification of each muscle afferent recorded is important. Because the standard tests are difficult or impossible to apply, it is vital that units should not be identified on the basis of a preconception of how they should fire in relation to alpha motor activity, because this is the subject which is to be investigated. The greatest chance for confusion exists between spindle primaries and Golgi tendon organs. The consequence of taking a tendon organ for a spindle afferent would be to demonstrate what appears to be a rigid alpha–gamma coactivation. Unfortunately, the use of succinyl choline is the most satisfactory distinguishing test and it has not been possible to use this in human experiments. The problem will be considered further below. The methods of distinction between primary and secondary spindle afferents also need to be improved. Even in animal experiments this is difficult without anaesthesia, but it is certainly worth while trying to make the distinction because the effects of static fusimotor action may be seen clearly in secondaries.

RESULTS AND INTERPRETATIONS

The data summarised in these papers have been gathered over a number of years. The main feature is the impression, as Vallbo puts it, of 'an intimate coactivation of the fusimotor and skeletomotor activity in time, in space and in quantity'. Also it appears that the alpha activity almost always precedes signs of fusimotor activity at the onset of movement.

The validity of these results rests of course in the certain identification of the afferents as from spindles rather than from tendon organs, because the latter would be expected to behave in just that way. Tendon organ afferents have essentially the same diameter as spindle primary afferents and would be equally well picked up by microneurography. Some of the units illustrated here by Vallbo are particularly worrying from this point of view. It is quite possible that a spindle could show increasing discharge during slow muscle shortening if there were a steady build up of fusimotor drive as illustrated. However, on relaxation the sudden lengthening would always be expected to produce a striking extra spindle discharge, judging from animal work. A fall in afferent frequency with relaxation is characteristic of a tendon organ afferent. It should be borne in mind that a tendon organ will tend to show a rather sudden recruitment with the onset of activity in particular muscle fibres which insert into it, and further increases in firing with recruitment of additional related units. The existence of what appears to be a rather marked velocity sensitivity of the receptor is also not definite evidence of its being a spindle primary, as tendon organs are known to be dynamically sensitive.

If we make allowances for the uncertainty of identification of some of the units, no doubt many remain which are indeed from spindles and therefore the general conclusions is valid that fusimotor activity in these particular normally organised movements does not precede alpha activity. This agrees with the conclusion of work on conscious animals and so eliminates alpha activation via the 'gamma loop' or the 'length follow-up servo' hypothesis of Merton as a possibility in these cases.

However, the demonstration of a general and intimate alpha–gamma linkage seems to be much less satisfactory. Apart from problems of unit identification already referred to, there is the question of whether it is worth continuing with such a generalisation which makes no distinction between the static and dynamic fusimotor systems. There is, after all, ample evidence in animals that separate central mechanisms exist for the two parts, and indeed some recordings in man indicate separate activation to be possible (Burg *et al.*, 1976).

The concept of alpha–gamma linkage originated 25 years ago (Granit, 1955) when static and dynamic divisions of the fusimotor system were unrecognised. Surely it is now time that we sought new hypotheses of fusimotor involvement in voluntary movement control, which embody the known differences in function between static and dynamic fusimotor systems and suggest how they may be usefully exploited?

Received on August 8th, 1980.

REFERENCES

Burg, D., Szumski, A. J., Struppler, A. and Velho, F. (1976). Influence of a voluntary innervation on human muscle spindle sensitivity. In *The Motor System* (ed. M. Shahani), Amsterdam, Elsevier, 95–110

Goodwin, G. M. and Luschei, E. S. (1975). Discharge of spindle afferents from jaw-closing muscles during chewing in alert monkeys, *J. Neurophysiol.*, **38**, 560–571

Granit, R. (1955). *Receptors and Sensory Perception*, New Haven, Yale University Press

Lund, J. P., Smith, A. M., Sessle, B. J. and Murakami, T. (1979). Activity of trigeminal α- and γ-motoneurones and muscle afferents during performance of a biting task, *J. Neurophysiol.*, **42**, 710–725

Prochazka, A., Stephens, J. A. and Wand, P. (1979). Muscle spindle discharge in normal and obstructed movements, *J. Physiol., Lond.*, **287**, 57–66

Prochazka, A., Westerman, R. A. and Ziccone, S. P. (1976). Discharges of single hindlimb afferents in the freely moving cat, *J. Neurophysiol.*, **39**, 1090–1104

Taylor, A. and Cody, F. W. J. (1974). Jaw muscle spindle activity in the cat during normal movements of eating and drinking, *Brain. Res.*, **71**, 523–530

A critique of the papers by Hagbarth and Vallbo

ALBRECHT STRUPPLER*

I should like to start by taking this opportunity to congratulate the speakers for the interesting and demanding work that they have presented. I shall add some of our own data in a discussion of some relevant and selected problems. There is no further necessity to elaborate on the principle of alpha–gamma coactivation during voluntary movement in man.

It seems to me that the main function of the muscle spindle consists of assisting compensation of movement during external disturbances. This compensation occurs very rapidly. I would like to present some findings from our own laboratory, some of which were obtained in collaboration with S. Homma, F. Erbel, F. Velho and M. Yokochi. Two introductory figures show the activity of α-motoneurons (figure 1) and spindle primary afferents (figure 2) in response to tendon taps.

A controversial aspect to be discussed is the nature of muscle spindle activity in a 'relaxed' muscle. In contrast to the findings of Hagbarth, we have been able to record spindle primary afferent impulses from a 'relaxed' muscle, in both a stretched and a mid-range position of the extremity. I would, however, agree with the opinion that muscle spindles are usually quiet in 'relaxed' human muscles. The major controversy relates to whether muscle spindles can be activated without physiological stretch of the muscles under investigation. As to the effectiveness of the Jendrassik manoeuvre, I propose that the divergence of opinion between Hagbarth's group (Hagbarth *et al.*, 1975) and my own may be reconciled by taking into account the different functions of the muscles considered. Jendrassik's original concept referred to antigravity (extensor) muscles. Different findings might well be expected in flexor muscles (figure 3).

Muscle spindles involved in a flexor reflex may also be activated in a relaxed muscle as a result of cutaneous stimuli as seen in Struppler and Velho (1976, figure 9). Figures 5 and 6 of that paper showed the discharge of two spindle

* Neurologische Klinik der Technischen Universität, München

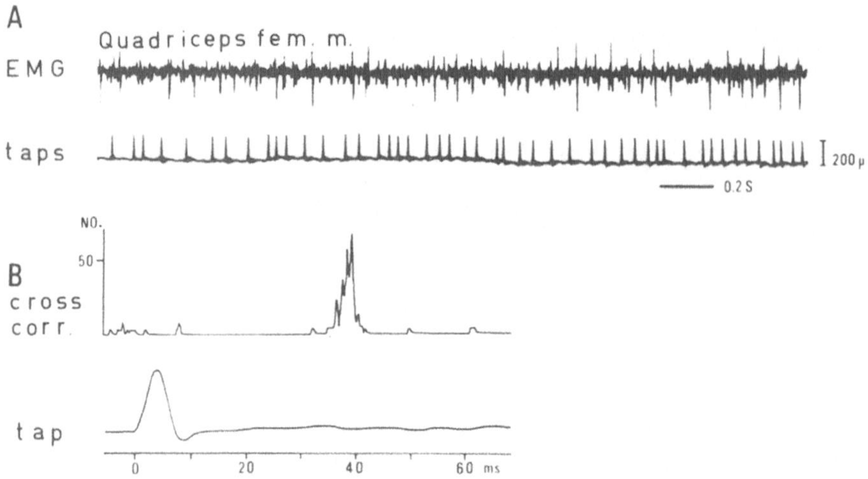

Figure 1. Electromyographic activity in voluntarily contracted quadriceps femoris muscle during application of tendon taps. A, motor unit spikes and random triangular taps (200 μm amplitude tendon taps to the patellar ligament). B, cross-correlation of motor unit spikes following tendon taps.

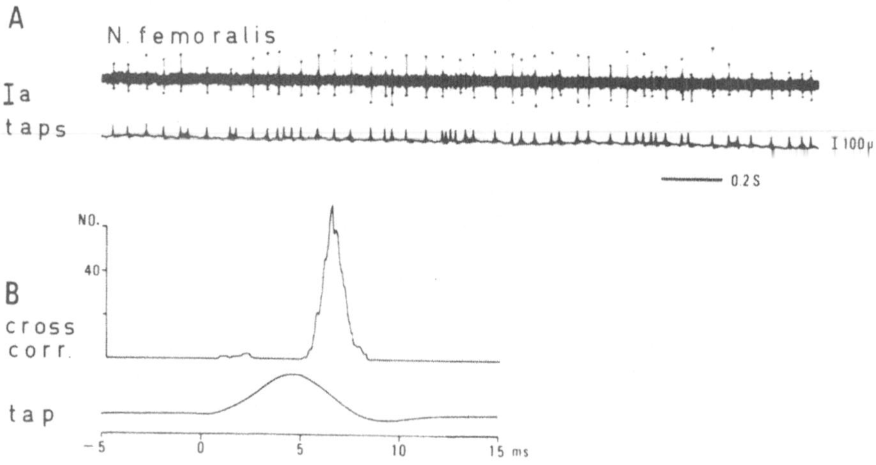

Figure 2. Activity in primary muscle spindle afferents in N. femoralis following tendon taps to the patellar ligament. Muscle at rest. A, Ia afferent activity (top; tips of spikes retouched) and triangular taps (100 μm amplitude). B, cross-correlation of Ia afferent spikes following tendon taps.

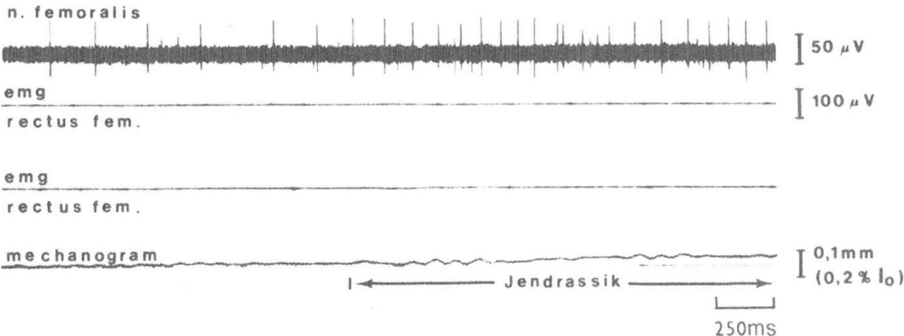

Figure 3. Muscle spindle activity during rest and reinforcement (Jendrassik manoeuvre). Note the increase in spindle discharge during reinforcement.

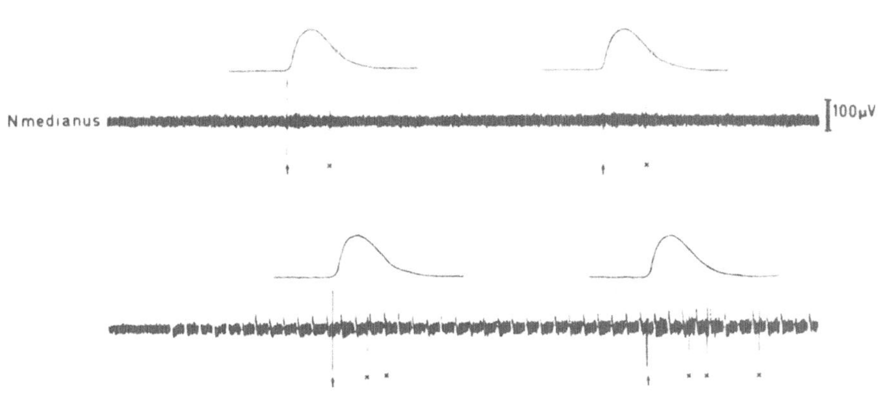

Figure 4. Effect of thalamic stimulation on muscle spindle afferent discharge elicited by electrically evoked twitch contractions (arrows indicate electrical stimuli). Upper trace: activity of a muscle spindle of a flexor muscle of the hand. Lower trace: same, during sub-threshold repetitive stimulation at the thalamic target point (see stimulation artefacts). Spindle responses have increased.

receptors, one responding to stretches in a static fashion, the other responding more dynamically.

As in animal experiments (Appelberg, 1980) spindle activity in man can be affected by stimulation of subcortical structures. Figure 4 gives an example of this (*see also* Struppler *et al.*, 1978).

Spontaneous spindle activity may arise under pathological conditions such as in spasticity. In clonus one may observe, apart from abnormal α-activity, an increased activity of spindle afferents (Szumski *et al.*, 1974, figure 2). Obviously, more recordings from clinically well-defined types of spasticity are needed.

The main interest of all of us is focused on motor control during various motor performances. As a clinician I would like to emphasise that we need more information about the sensitivity of muscle spindles under various pathological conditions, particularly in hypotonia of peripheral and central origin. Some preliminary results have been obtained and it is desirable that future investigations be carried out to clarify the underlying mechanisms of motor control and its disruption in disease.

Received August 12th, 1980.

REFERENCES

Appelberg, B. (1980). Selective midbrain control of dynamic gamma motoneurons. This publication

Hagbarth, K.-E., Wallin, G., Burke, D. and Löfstedt, L. (1975). Effects of the Jendrassik manoeuvre on muscle spindle activity in man, *J. Neurol. Neurosurg. Psychiat.*, **38,** 1143–1153

Struppler, A., Gerilowsky, L., Velho, F., Erbel, F. and Altmann, H. (1978). Mode of innervation following stereoencephalotomy, *Contemp. clin. Neurophysiol.* (EEG Suppl. No. 34), 493–500

Struppler, A. and Velho, F. (1976). Single muscle spindle afferent recordings in human flexor reflex. In *The Motor System* (ed. M. Shahani), Amsterdam, Elsevier, 197–207

Szumski, A., Burg, D., Struppler, A. and Velho, F. (1974). Activity of muscle spindles during muscle twitch and clonus in normal and spastic human subjects, *Electro-enceph. clin. Neurophysiol.*, **37,** 589–597

SECTION 4
REFLEXES MEDIATED BY MUSCLE AFFERENTS

Overview

It is very evident that however much we learn about the properties of muscle receptors, the possibilities for their employment in movement control must depend on the nature and strength of their central connections. This session was dedicated to the reflexes of muscle afferents, not from the point of view of details of synaptic arrangements, but rather trying to assess their effectiveness in various real tasks and how they may be adapted to changing needs.

The first four papers may be thought of as being inspired by the need to define a role for the segmental stretch reflexes. The lines of thought followed were in turn very much influenced by what might be termed the 'servo' approach. Generally workers in this area have been rather disappointed in the apparently limited effectiveness of the segmental reflexes initiated by muscle afferents as providing correction of muscle contraction in the face of changing loads. There has therefore been a tendency to try variations of the original theme of muscle length control to see if some other variable is more effectively controlled. The group led by Houk press the case for stiffness regulation and suggest that spindle non-linear properties play a part in giving advanced compensation for failure in stiffness during transient stretch.

The following paper presented by Hoffer examines the relationship of stiffness to the magnitude of perturbations in premammillary cats. It appeared that the gain of the tendon organ feedback was negligibly low in this preparation.

The papers by Allum and by Dietz on the effectiveness of stretch reflexes in the human leg took rather different views of the situation. Allum, using a now classical test incorporating external disturbances, found negligible correcting effects from short latency reflexes. However, Dietz, using an unexpected change in terrain during running, did show potentially significant EMG responses. The meaning of these results is well considered in the prepared critiques, but the reader may conclude that simplistic theories of linear servo control are nearing the end of their usefulness, because they encourage a patronising view of the subtleties of movement control. There is no particular reason why a linear servo control need be the best form of control and nature generally evolves mechanisms which are basically workable with a minimum of sophisticated control. The latter is added to give improvements at the extremes of performance. The effectiveness of the active neural controls may therefore be difficult to demonstrate unless our tests fully extend the system capabilities. However, survival in competition rewards only the best performance, and justifies apparently extravagant investment in quite small improvements of function.

The remaining papers presented in this session related to the ways in which reflexes may be switched during locomotion. The work of the groups led by Rossignol, Forssberg, Schomburg and Grillner shows how far we have come in modifying our views of reflex function in recent years. It is becoming difficult to distinguish between a reflex modified by the action of a central pattern generator, and a pattern generator modified by muscle afferent input.

The papers in this section were read at the symposium, with the exception of those by Proske and Walmsley and by Johnson *et al.* which appeared as posters, and that by Homma *et al.* which was presented as a video film.

Function of the spindle dynamic response in stiffness regulation—a predictive mechanism provided by non-linear feedback

J. C. HOUK*, P. E. CRAGO† AND W. Z. RYMER*

SUMMARY

Autogenetic reflex responses that act to regulate muscle stiffness are initiated too soon and are too large to be explained solely on the basis of linear feedback from muscle spindle receptors and Golgi tendon organs. In this article we present several lines of evidence that the unexpected efficacy is due to the unique non-linear response properties of primary endings. Several of these non-linear features of spindle dynamic responses are well matched to assist in the maintenance of stiffness during the transient phases of response to muscle stretch and release. As a consequence of these spindle and muscle properties, appropriate reflex actions can actually occur in advance of the failures in stiffness for which they provide compensation. This result is interpreted to mean that the system is endowed with a predictive mechanism.

INTRODUCTION

Recent studies of stretch and unloading reflexes have led to the hypothesis that the function of these autogenetic reflexes is to regulate stiffness, rather than to control muscle length or tension individually (Nichols and Houk, 1976; Crago *et al.*, 1976). Stiffness is a composite property constituted by the ratio of force change to length change, and its regulation has been viewed as a natural consequence of a balanced interplay between excitatory length feedback from muscle spindle receptors and inhibitory force feedback from Golgi tendon organs (Houk, 1978a). Figure 1 shows a simplified block diagram of the overall system that is believed to mediate stretch and unloading reflexes.

* Northwestern University, Chicago
† Case Western Reserve, Cleveland, Ohio

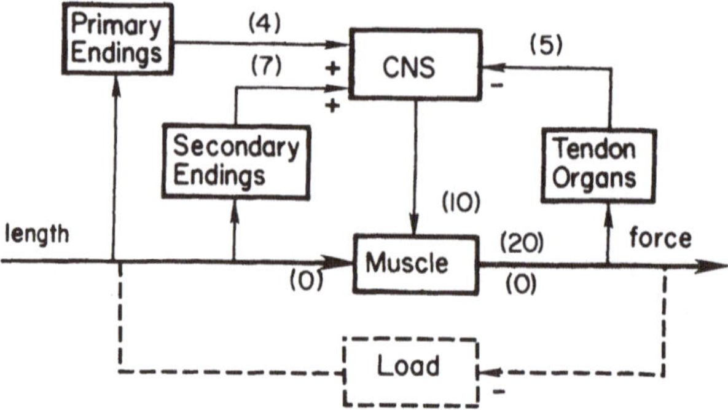

Figure 1. Block diagram of the motor servo. The numbers in parentheses give the latencies (in msec) at which different motoservo signals associated with the cat soleus muscle change in response to a rapid (100 mm/sec) ramp stretch.

This total system, which includes a muscle, its mechanical load, the muscle's proprioceptors and the associated autogenetic reflex pathways, is collectively called the *motor servo* (cf. Houk, 1978b). While there is reasonable evidence that stiffness is, in fact, well regulated by motor servo actions (cf. review of evidence by Houk, 1979), the specific mechanisms responsible for this regulation are not yet well understood.

The comparison of the reflex and muscle responses to stretch and release given in figure 2A illustrates some of the unexplained features. The outer force traces labelled REFLEX represent typical stretch and unloading reflexes obtained from the soleus muscle of the decerebrate cat when afferent and efferent pathways are intact, whereas the inner traces labelled MECHANICAL are the underlying muscle mechanical components (direct effect of length on muscle force in figure 1). The latter were derived from control records obtained by stretching and releasing the muscle while the cut ventral roots were electrically stimulated at a physiological rate (cf. Nichols and Houk, 1976). The differences between the reflex and mechanical responses, labelled REFLEX ACTION, represent the neurally mediated components of the overall reflex responses. Note that the magnitude of reflex action associated with stretch is appreciably greater than that associated with release. Nichols and Houk (1976) pointed out that this asymmetry in reflex action is appropriate to compensate for an opposite asymmetry in the muscle mechanical response, and, as a consequence, the stiffness of the overall reflex ($\Delta f/\Delta x$) is maintained approximately constant. One of the unexplained features of stiffness regulation addressed in this report is the observation that the magnitude of the asymmetry in reflex action is too large to be accounted for on the basis of a simple, linear feedback explanation of stiffness regulation (Nichols and Houk, 1976).

Although stiffness is not plotted directly in figure 2A, an instantaneous

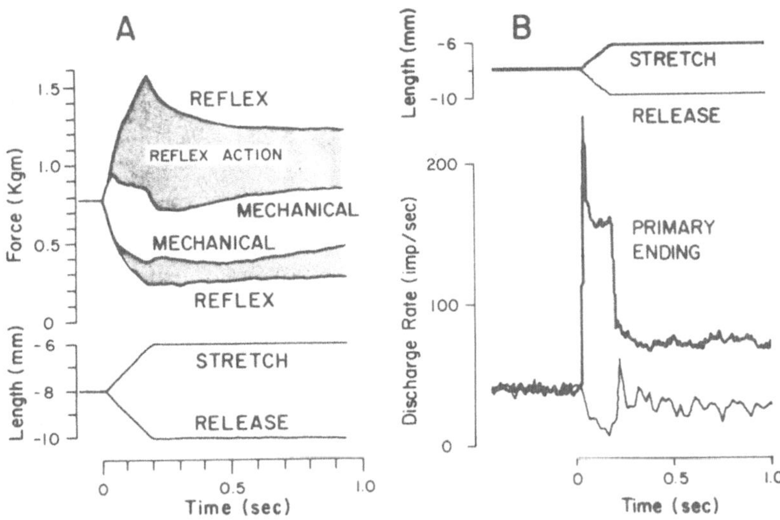

Figure 2. Asymmetry of reflex action and its origin in primary endings. The responses of several motoservo signals to stretch and release applied at 12.5 mm/sec are compared. The responses labelled MECHANICAL represent the changes in force that would occur if there were no reflexly generated changes in motor unit discharge or in the number of motor units recruited. REFLEX ACTION is the difference between the overall REFLEX and its MECHANICAL component. The greater reflex action associated with stretch, as contrasted with release, appears to be due mainly to the greater dynamic responses of primary endings to stretch, as contrasted with release.

measure of this quantity is readily estimated from the slopes of the force traces. Since length was caused to vary as a linear function of time (ramp stretch and release), the slopes represent both $\mathrm{d}f/\mathrm{d}t$ and $\mathrm{d}f/\mathrm{d}x$, the latter differential being instantaneous stiffness. Looking first at the mechanical response to stretch, it is clear that muscle mechanical stiffness is initially high but later decreases and, in the particular case illustrated, actually becomes negative. This rather abrupt failure in mechanical stiffness is called yielding, and it occurs whenever the amplitude of stretch exceeds approximately 4% of muscle fibre length (Joyce *et al.*, 1969; Nichols, 1974). Yielding is believed to result from a sudden increase in the breakdown rate of actomyosin bonds promoted by stretch beyond the limit of bond deformation. Regardless of the mechanism of yielding, it is clear that compensation is quite effective. The delayed onset of reflex action appears to coincide rather closely with the time of yield, instead of following the yield by the conduction delay around the myotatic loop. As a result, the failure in mechanical stiffness is effectively masked. This apparent match between the time of yield and the onset of reflex action is a second unexplained feature of stiffness regulation that is addressed in this chapter.

METHODS

Our experimental approach was to record the response patterns of muscle stretch receptors from small filaments of dorsal root while eliciting stretch and unloading reflexes from a functionally isolated soleus muscle in the decerebrate cat (for detailed methods, *see* Rymer *et al.*, 1979). Each muscle afferent was carefully identified on the basis of conduction velocity and response during muscle twitch. A special electrode array and dissecting plate that attached to the L7 spinous process provided stable recordings without the use of hip pins; this minimised trauma and appeared helpful in the preservation of the generally good reflex status of our animals. In spite of these precautions, there was probably some degradation of the stretch reflex that resulted from the laminectomy and exposure of dorsal roots. A vibrator attached to the moving portion of the main stretching device provided a reliable means for superimposing longitudinal tendon vibration on reflex responses to ramp stretch.

RESULTS

Origin of asymmetric reflex action

An asymmetry of reflex action that compensates for an opposite asymmetry of muscle mechanical stiffness appears to be an important and consistent feature of stiffness regulation in both the soleus muscle of the decerebrate cat (Nichols and Houk, 1976) and in the human biceps muscle (Crago *et al.*, 1976). The same pattern of asymmetry was confirmed in the present studies in the course of an investigation of the afferent mechanisms that contribute to it. Specifically, we sought to determine which categories of muscle receptor have appropriately asymmetrical responses to explain the observed asymmetry of reflex action.

For this phase of the study we used ramp stretches and releases identical to those shown in figure 2A, namely 2 mm length changes applied at 12.5 mm/sec (a typical velocity encountered during walking). About 10 trials of reflex response to both stretch and release were recorded with simultaneous responses of one to four muscle afferents. At the end of the experiment, muscle mechanical responses to the same stretches and releases were recorded while stimulating the cut muscle nerve at a physiological rate (usually 10/sec). At a later time the digitised records were scanned and sorted into appropriate groups for the construction of ensemble averages of response. The direction and degree of asymmetry was assessed visually and also by compiling stretch-to-release ratios of the averaged responses measured just prior to ramp plateau (dynamic asymmetry ratio) and 2 sec after ramp plateau (static asymmetry ratio).

The observations made upon primary and secondary endings are easiest to interpret since equal amplitudes of stretch and release were ensured by the rigidity of the stretching device. In general we found that secondary ending responses (n=24) were only slightly asymmetric, having average ratios of 0.83 and 0.74 for dynamic and static phases, respectively. In contrast, primary ending responses (n=35) were highly asymmetric in the dynamic phase (average ratio of 3.4) and mildly asymmetric in the static phase (1.4). Figure 2B shows an example which contrasts ensemble average response to stretch with that to release for one primary ending. The much larger dynamic response to stretch is a non-linear response feature, since linear systems always respond in a symmetrical fashion to symmetrical inputs.

The asymmetry of the primary ending responses shown in figure 2B is accentuated by the fact that there is no appreciable pause in discharge during release. Presumably this is attributable to a high level of gamma static activity whereas the large dynamic response to stretch is probably the result of a high level of gamma dynamic activity. While many receptors did show pauses during release, the incremental responses were nevertheless highly asymmetric. This is because the discharge during stretch was substantially more than twice the spontaneous rate present at the initial length.

In a recent study we reported that tendon organs are responsive to total muscle force, whether that force is produced by recruitment and rate modulation of motor units or by stretching recruited muscle fibres to longer lengths (Houk *et al.*, 1980). Consistent with this previous report, we found that the asymmetry ratios of tendon organs (n=12) closely paralleled the ratios observed for the overall reflex response. The dynamic asymmetry ratios for both were greater than 1.0 and the static ratios varied in the vicinity of 1.0 in all but one animal. The latter preparation had good crossed-extensor reflexes which were used to modulate the initial force, but for some unexplained reason was autogenetically areflexic (verified with emg recordings). In this animal the dynamic asymmetry ratios for both force and tendon organ responses were approximately 0.5, which is within the range of ratios observed for muscle mechanical responses (0.3–0.5).

In summary, our asymmetry observations strongly suggest that the greater reflex action during stretch as contrasted to release arises at least in part from a non-linear feature of the dynamic responsiveness of primary endings, namely the greater sensitivity to stretch as contrasted with release. The small asymmetries observed for secondary endings and for tendon organs are in a direction that would be expected to oppose the observed asymmetry of reflex action.

Timing of reflex action

In order to understand better the origin of the coincidence between the delayed onset of reflex action and the timing of muscle yield, we measured the latencies at which the various motoservo variables change following a stretch of rapid onset. We also explored how these latencies vary with stretch velocity. While some of these data were already available (Matthews, 1972; Matthews, 1975; Nichols and Houk, 1976), some were not, and, furthermore, it seemed important to collect a complete set of latencies under the same experimental conditions.

The occurrence of yield is associated mainly with a certain amplitude of stretch, although this amplitude increases somewhat at higher velocities (Nichols, 1974; Rack and Westbury, 1974). Correspondingly, one would expect the time of yield to be a decreasing function of velocity, which we found to be the case for velocities up to about 30 mm/sec. At 10 mm/sec the yield time was 36 msec whereas above 30 mm/sec it levelled off at about 18 msec. The levelling off was partly due to the acceleration limitations of our apparatus. The onset time of reflex action was measured as the first departure between records of overall reflex response and our estimates of the underlying mechanical components. This latency remained essentially constant at about 20 msec throughout the range 10–100 mm/sec. Similarly, the latencies of muscle afferent and of emg response did not vary appreciably over this range of velocities, and typical values are included in parentheses in figure 1. The latency given for tendon organs (5 msec) applied only when the muscle was initially active. When it was initially quiescent, in which case there was no initial tendon organ discharge, this latency was much longer, whereas the latencies observed for other motor servo signals were unaffected.

In conclusion, the minimum latency from the initiation of stretch to the onset of reflex action is about 20 msec, and activity at this latency appears to be dominated by primary ending discharge. The earliest increases in motor output clearly occur before muscle force yields, and therefore cannot be produced as a consquence of this failure in mechanical stiffness. Variations in velocity have relatively little effect on the latency of reflex action, even though they cause appreciable variations in the time of yield.

Disruption of compensation for yielding produced by vibration

The foregoing results suggest that the large dynamic responses of primary endings are especially important in mediating the prompt compensation for the yield in force that occurs during muscle stretch. We sought to test this hypothesis directly by using tendon vibration as a technique for blocking the dynamic responses of primary endings.

Brown *et al.* (1967) demonstrated that longitudinal tendon vibration of an appropriate amplitude (about 0.1 mm) causes the discharge of most primary endings to become phase-locked to the vibratory frequency without influencing the discharge of most secondary endings. Vibration not only turns on primary ending discharge, but it can also clamp this discharge at the vibratory frequency such that primary endings no longer respond to stretch. Matthews

(1969) and others have used the latter effect to block the participation of primary endings in the tonic stretch reflex, and thus to obtain evidence that secondary endings contribute excitation. So far vibration has rarely been used to study the transient phase of the stretch reflex (*but see* Westbury, 1972), presumably due to uncertainty as to the efficacy of phase-locking under these conditions. Our experimental procedures provided an opportunity to study the effects of vibration on reflex activity while simultaneously testing its effects on the muscle afferent population.

Vibration at amplitudes of 0.1–0.2 mm and frequencies of 150–200 Hz had consistent, parallel effects on primary ending discharge and on yielding in all 12 animals that were studied. There was little or no effect on secondary ending discharge. Primary endings became phase-locked to the vibration and showed only a few extra spikes above the vibratory frequency during the period of stretch. Thus, the dynamic response was effectively blocked, and, correspondingly, there was a substantial reduction in the transient phase of compensation for yielding.

The effects of vibration on tendon organ discharge and on the tonic component of the stretch reflex were more variable. Tendon organs were sometimes phase-locked (usually at a sub-harmonic of the vibratory frequency) throughout the observation period before, during and after the stretch. In other cases they became phase-locked only as a consequence of stretch, and in yet others they were unaffected by vibration. Our population (n=7) is too small to make useful conclusions about the relative frequency of the various effects.

The tonic component of the stretch reflex was completely blocked by vibration in several cases, leaving a response that was indistinguishable from the purely mechanical response of the muscle. At the other extreme, vibration sometimes had no effect on the tonic stretch reflex, causing only a disruption of the early transient phase of the reflex. A good example of the latter situation is shown in figure 3. There were also intermediate cases in which vibration partially blocked the tonic reflex. We do not understand the reasons for these variations.

DISCUSSION

The present results strongly support the notion that dynamic responses of primary endings are largely responsible for the transient phase of compensation for stretch-induced failures in muscle stiffness. Three lines of evidence were provided. (1) Primary ending responses are large during stretch and small during release (figure 2B), which is precisely the pattern required to produce an appropriate asymmetry of reflex action (figure 2A). (2) The timing of primary ending responses and of the reflex effects they provoke are consistent with the observed latencies of reflex action (figure 1). (3)

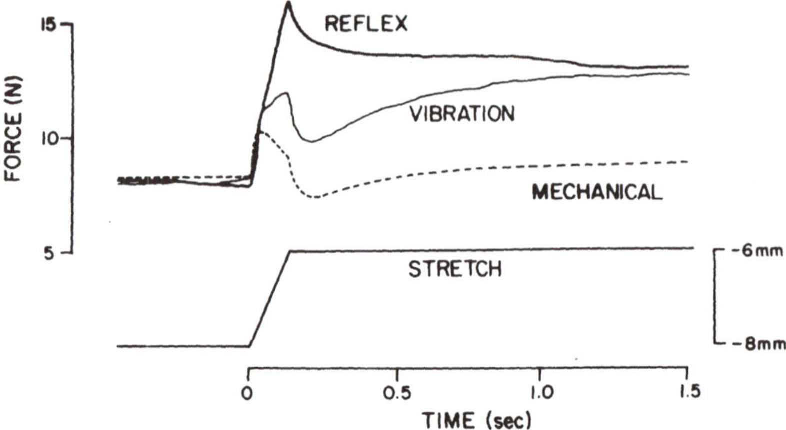

Figure 3. Attenuation of compensation for yielding produced by vibration. Longitudinal tendon vibration was used to block the stretch responses of primary endings, in order to demonstrate the importance of the dynamic response in the transient phase of stiffness regulation. The initial force of the normal REFLEX response was augmented to the initial force developed in the presence of VIBRATION with a crossed-extensor reflex. The difference between REFLEX and VIBRATION traces is attributed mainly to primary endings, whereas the difference between VIBRATION and MECHANICAL traces is attributed mainly to secondary endings, although tendon organ inhibition may limit the magnitude of this component.

Longitudinal tendon vibration of sufficient amplitude to clamp primary ending discharge at the vibratory frequency, which effectively prevents dynamic responses to stretch, consistently causes an appreciable attenuation of compensation for muscle yielding (figure 3).

Figure 4 outlines the manner in which compensation for yielding appears to evolve. At the onset of stretch, muscle force rises fairly steeply, due to the short-range elasticity of actomyosin bonds. Although muscle force will not yield until 18 msec later, the reflex response that will compensate for the yield begins to evolve almost immediately. Primary ending discharge rate rises steeply, due to the enhanced sensitivity of these receptors in the small-signal region, and then goes through an initial burst, following which it levels off to a response of reduced slope that persists for the remainder of the constant-velocity phase of stretch. The primary ending response in figure 4 shows these features and also includes the 4 msec delay associated with conduction to the spinal cord in Ia afferent fibres. After a central delay and an efferent conduction time an emg signal is initiated in the muscle. The similarity between the time course of the illustrated emg and that of primary ending discharge is based on the findings by Houk *et al.* (1977), whereas the delay to onset is based on the results presented here. The 'active state' is assumed to be a delayed and slightly slowed version of the emg signal. This signal then

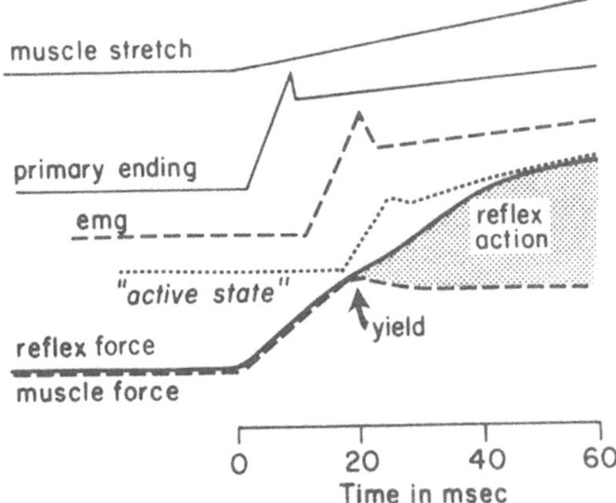

Figure 4. Origin and timing of compensation for yielding. The large dynamic response of the primary ending develops well in advance of yielding. It is then delayed during conduction around the myotatic loop, but at this particular velocity (50 mm/sec) it arrives at about the right time to mask the failure in muscle stiffness associated with yielding. The fact that the compensatory change in motor output (emg) occurs in advance of yielding is interpreted as a predictive mechanism built into the motor servo as a consequence of the non-linear properties of primary spindle receptors.

undergoes further lag as it gets translated into the neurally mediated component of force called *reflex action*.

The analysis provided in the previous paragraph suggests that compensation for yielding represents the action of a predictive mechanism, since the compensatory emg signal is generated before the time at which an error in stiffness actually occurs. In a more conventional type of feedback regulator the compensatory signal would be produced by the error in stiffness, and thus would occur only after the yield in muscle force had been detected by tendon organs. The production of a compensatory signal in advance of the error instead is more typical of the actions of a feedforward regulator (cf. Houk, 1980, for a discussion of the contrasting properties of feedback and feedforward regulators).

The reflex pathway from spindle receptors does in fact operate in a feedforward mode under the conditions imposed in most experimental studies of functionally isolated muscles, including the present ones. This is because muscle length is rigidly controlled by the stretching apparatus, which procedure serves to open the length feedback loop. Muscle force is not allowed to influence its own length as it normally does by way of its actions on the mechanical load (dashed portion of the length feedback loop in figure 1). Thus, under certain experimental conditions, spindle receptors function as feedforward sensors, whereas under most natural conditions they function as feedback sensors. It is likely that the predictive mechanism is operative in both cases. Correspondingly, it seems unwise to use the feedforward nomenclature, since this would stress operation under restricted experimental conditions, rather than operation under more natural circumstances.

Questions of semantics aside, one can attribute several of the predictive features of the motor servo to non-linearities in the response properties of primary endings. The fact that dynamic responses to stretch are large and those to release are small (asymmetry) is well matched to the probability that stretch will cause a failure in stiffness whereas release will not. The fact that stretch responses rise rapidly to plateau values (differential sensitivity to small and large stretch) ensures a large, quick response well matched to the probability that there will soon be an abrupt failure in stiffness. The fact that the dynamic response is relatively independent of velocity (motion detection feature discussed elsewhere in this book (Houk *et al.*, 1981)) seems well matched to the probability that about the same magnitude of compensation is required relatively independent of velocity.

Other properties of the primary ending pathway appear not to be especially well matched to the compensatory task, as is sometimes the case with predictive designs. One is that the precise timing of the compensatory response does not change appropriately with velocity in order to match the latency of muscle yielding. Another is that the amplitudes of primary ending responses fail to scale in proportion to the amplitudes of length change (Hasan and Houk, 1975); they are disproportionately large for small changes in length and then tend to saturate. While this tendency is also seen in reflexes (Matthews, 1969), the latter do scale reasonably well with amplitude (Nichols and Houk, 1976). Thus, amplitude scaling, which seems to be both available and useful, is not well accounted for by the properties of primary endings.

The emphasis of this paper has been upon the transient phases of stretch and unloading reflexes. Primary endings and predictive mechanisms may be much less important than secondary endings, tendon organs and more conventional linear feedback mechanisms in the regulation of the tonic phase.

ACKNOWLEDGEMENTS

This research was supported in part by NIH grants NS06828, NS07226 and NS14703.

Received on June 9th, 1980.

REFERENCES

Brown, M. C., Engberg, I. E. and Matthews, P. B. C. (1967). The relative sensitivity to vibration of muscle receptors of the cat, *J. Physiol., Lond.*, **192**, 773–800

Crago, P. E., Houk, J. C. and Hasan, Z. (1976). Regulatory actions of the human stretch reflex, *J. Neurophysiol.*, **30**, 925–935

Hasan, Z. and Houk, J. C. (1975). The transition in the sensitivity of spindle receptors that occurs when the muscle is stretched more than a fraction of a millimeter, *J. Neurophysiol.*, **38**, 673–689

Houk, J. C. (1978a). A two-stage model of neural processes controlling motor output. In *Cybernetics 1977* (edited by G. Hauske and E. Butenandt), Munich and Vienna, Oldenbourg Verlag, 35–46

Houk, J. C. (1978b). Participation of reflex mechanisms and reaction-time processes in the compensatory adjustments to mechanical disturbances. In *Progress in Clinical Neurophysiology, Vol. 4: Cerebral Motor Control in Man: Long Loop Mechanisms* (edited by J. E. Desmedt), Basel, Karger, 193–215

Houk, J. C. (1979). Regulation of stiffness by skeletomotor reflexes, *A. Rev. Physiol.*, **41**, 99–114

Houk, J. C. (1980). Homeostasis and control principles. In *Medical Physiology, Fourteenth Edition* (edited by V. B. Mountcastle), Vol. 2, St. Louis, Mosby, 246–270

Houk, J. C., Crago, P. E. and Rymer, W. Z. (1980). Functional properties of the Golgi tendon organs. In *Prog. Clin. Neurophysiol., Vol. 8: Spinal and Supraspinal Mechanisms of Voluntary Motor Control and Locomotion* (edited by J. E. Desmedt) Basel, Karger, 33–43

Houk, J. C., Rymer, W. Z. and Crago, P. E. (1977). Complex velocity dependence of the electromyographic component of the stretch reflex, *Proc. XXVII Int. Cong. Physiol. Sci.*, Paris

Houk, J. C., Rymer, W. Z. and Crago, P. E. (1981). Nature of the dynamic response and its relation to the high sensitivity of muscle spindles to small changes in length, this publication

Joyce, C. G., Rack, P. M. H. and Westbury, D. R. (1969). The mechanical properties of cat soleus muscle during controlled lengthening and shortening movements, *J. Physiol., Lond.*, **204**, 461–474

Matthews, P. B. C. (1969). Evidence that the secondary as well as the primary endings of the muscle spindles may be responsible for the tonic stretch reflex of the decerebrate cat, *J. Physiol., Lond.*, **204**, 365–393

Matthews, P. B. C. (1972). *Mammalian Muscle Receptors and Their Central Actions*, London, Arnold

Matthews, P. B. C. (1975). The relative unimportance of the temporal pattern of the primary afferent input in determining the mean level of motor firing in the tonic vibration reflex, *J. Physiol., Lond.*, **251**, 333–361

Nichols, T. R. (1974). Soleus muscle stiffness and its reflex control (Ph.D. thesis), Cambridge, Harvard University

Nichols, T. R. and Houk, J. C. (1976). The improvement in linearity and the regulation of stiffness that results from the actions of the stretch reflex, *J. Neurophysiol.*, **39**, 119–142

Rack, P M. H. and Westbury, D. R. (1974). The short range stiffness of active mammalian muscle and its effect on mechanical properties, *J. Physiol., Lond.*, **240**, 331–350

Rymer, W. Z., Houk, J. C. and Crago, P. E. (1979). Mechanisms of the clasp-knife reflex studied in an animal model, *Brain Res.*, **37**, 93–113

Westbury, D. R. (1972). A study of stretch and vibration reflexes of the cat by intracellular recordings from motoneurones, *J. Physiol., Lond.*, **226**, 37–56

Limitations in the servo-regulation of soleus muscle stiffness in premammillary cats

J. A. HOFFER*† and S. ANDREASSEN*‡

SUMMARY

It has recently been proposed that proprioceptive reflexes may regulate muscle stiffness. To test this hypothesis, we have studied the effects of the amplitude, shape and velocity of length perturbations, as well as the reflex response to muscle potentiation, in the soleus muscle of cats decerebrated at the premammillary level. A twenty-fold increase in stretch amplitude (from 0.1 to 2.0 mm) caused a two-fold decrease in the reflex-mediated incremental stiffness. Although the intrinsic muscle stiffness doubled for triangular ramp stretches (0.5–5.0 mm/sec) compared to rectangular stretches of the same amplitude (1.0 mm), the net reflex component of stiffness was unaltered. The only singularity was observed using 10 Hz perturbations, a frequency that was presumed to cause resonance around the segmental reflex loop. The intrinsic stiffness of soleus potentiated by 32–66% during prolonged muscle activation, whereas the reflex component was essentially unchanged. These data suggest that in the premammillary decerebrate the gain of Golgi tendon organ feedback is negligibly low; the observed reflex effects are attributed to spindle group Ia and II feedback.

INTRODUCTION

Evidence obtained in recent years has led to the suggestion that segmental reflexes subserved by muscle proprioceptors have a role in the regulation of muscle stiffness. Nichols and Houk (1976) observed that reflex action can

* Department of Physiology, University of Alberta

† Present address: Laboratory of Neural Control, NINCDS, Building 36, Room 5A 29, National Institutes of Health, Bethesda, Md 20205, USA

‡ Present address: Institut for Elektroniske Systemer, Aalborg Universitetscenter, DK-9100 Aalborg, Denmark

compensate effectively for the yielding response demonstrated by areflexic, active muscle when it is stetched beyond a fraction of a millimetre. In addition to being an important departure from more traditional views on what is the controlled variable of proprioceptors, the concept of servo-regulation of muscle stiffness by segmental pathways is the cornerstone of a recent modelling effort that offers a simple, general interpretation of reflex action. Houk (1977, 1978, 1979a, 1979b) proposed a model for the processing of sensorimotor information that includes two levels: (1) segmental pathways involved in the automatic regulation of muscle stiffness, and (2) a central 'stimulus-response processor', triggerable by afferent inputs. According to this model, the action of a segmental (and perhaps also suprasegmental) 'motor servo' would cause muscles to respond to perturbations more or less like well-behaved springs, having reasonably constant stiffness at different operating lengths and forces. If stiffness were indeed a regulated variable, the central nervous system would need to concern itself only with dictating the level of muscle activation, since the operating point of the muscle (its length and force) and its threshold to reflex effects (viz. Matthews, 1959a,b; Feldman and Orlovsky, 1972) would then be specified.

In previous work (Hoffer and Andreassen, 1981) we tested the validity of the assumptions in this model, and found that the stretch reflex indeed makes the stiffness of the soleus muscle more uniform over the physiological range of operating lengths and forces. Further, we determined that in cats decerebrated at the premammillary level, any intrinsic changes in the gain of the stretch reflex depend strictly on the operating point of the muscle. In this work we have investigated some potential limitations in the regulation of muscle stiffness, revealed by the response of soleus to length perturbations of different shapes, amplitudes and velocities, and under circumstances when the intrinsic stiffness of the muscle itself was made to potentiate during prolonged activation. We chose the premammillary cat preparation so as to remove the voluntary component of motor control while still retaining a wide behavioural repertoire which allowed us to study reflex effects against a background of different levels and modes of muscle activation.

Some of these findings were reported in abstract form (Hoffer and Andreassen, 1978).

METHODS

Experiments were performed in cats decerebrated at the premammillary level. Surgical procedure and experimental approach have been described in detail elsewhere (Hoffer and Andreassen, 1981). The length of the soleus muscle was perturbed by a servo-controlled torque motor following periodic rectangular or triangular signals of 0.1–2.0 mm in amplitude and 0.1–2.0 sec in duration. Incremental muscle stiffness was determined from the ratio of the force change measured at the end of a stretch, divided by the imposed length change. Stiffness was measured against a background of operating forces covering the physiological range of the muscle, both in the presence and in the absence of segmental reflexes. When reflexes were present, soleus activation was triggered by exteroceptive stimuli such as blowing on the face or stroking the fur. With reflex effects extinguished by deep Halothane anaesthesia, the intrinsic stiffness of areflexic muscle was evaluated against operating forces generated by supramaximal stimulation of the soleus nerve for 3–6 sec at several frequencies in the range 10–100 Hz. The soleus nerve was cut at the end of the experiment and stimulation was repeated, giving the same results as during anaesthesia. Potentiation of the intrinsic stiffness of soleus was achieved by stimulating the nerve for longer periods (10–20 sec), or by stimulating electrically the contralateral mesencephalic tegmentum with reflexes present, which causes tonic activation of soleus (Maffei and Pompeiano, 1962; Hoffer and Andreassen, 1981).

RESULTS

Stiffness dependence on amplitude of perturbation

Figure 1(a) shows the response of soleus muscle to periodic rectangular stretches lasting 200 msec, with reflexes present. Stretch amplitudes were set at 0.1, 0.2, 0.4, 1.0 and 2.0 mm in consecutive runs. Perturbations were presented once per second against a background of varying operating forces. Baseline muscle length was held at L_{max} – 7 mm (7 mm short of maximum physiological length) by the torque motor. Data points represent measurements of incremental force, determined at the end of individual 200 msec stretches, as shown in figure 1(b). The lines shown were fitted by eye. Figure 1(c) shows the dependence of the average force increment on the amplitude of stretch, at two reference operating forces (500 g and 1000 g). Note that force increments were not linearly dependent on length increments. Figure 1(d) shows the dependence of the incremental stiffness of the muscle (obtained by dividing the incremental force change by the incremental length change) on the amplitude of stretch, at the same operating forces. Incremental stiffness values were relatively higher for smaller amplitude stretches, as was observed by Matthews (1969).

The data in figure 1(d) indicate also a moderate dependence of stiffness on operating force; stiffness values obtained at 1000 g operating force were uniformly higher than at 500 g. Such dependence was observed earlier for the low range of operating forces (Houk *et al.*, 1970; Rosenthal *et al.*, 1970; Hoffer and Andreassen, 1981). However, for intermediate and high forces the incremental stiffness reaches a plateau value and remains constant within ± 15% (Hoffer and Andreassen, 1981).

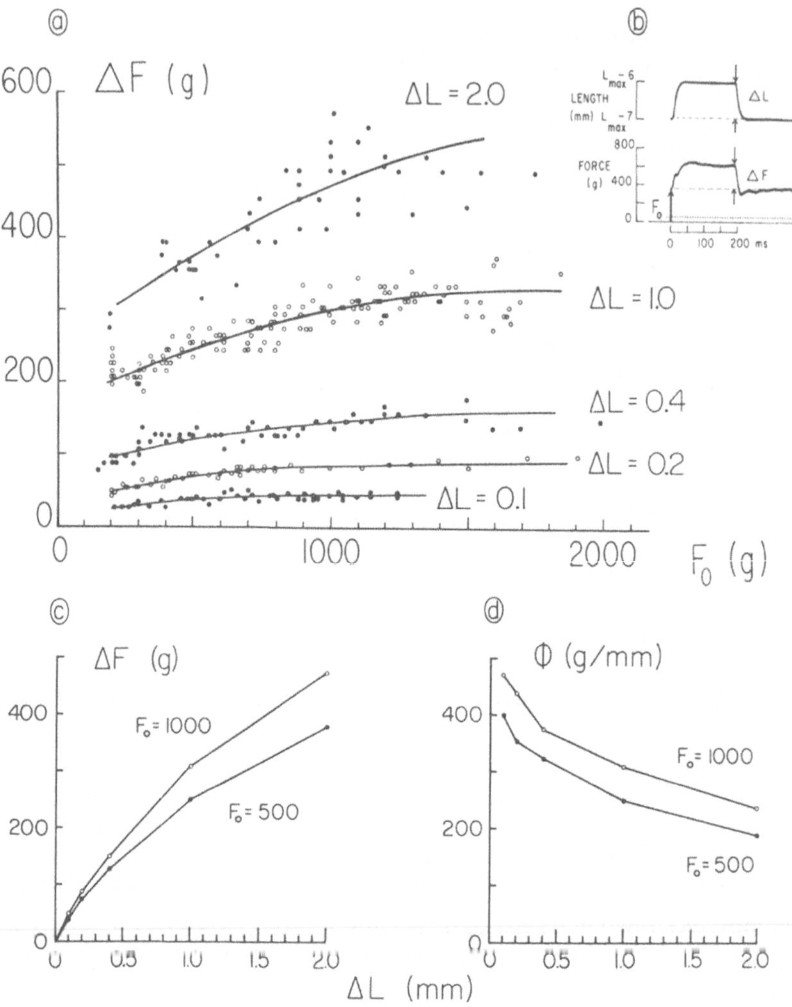

Figure 1. Effects of perturbation amplitude on soleus muscle stiffness. Rectangular stretches lasting 200 msec were presented once per second against varying background forces. Reference muscle length was $L_{max} - 7$ mm. As shown in (b), incremental force ΔF was measured at the end of each stretch and plotted against operating force F_0 in (a), for five stretch amplitudes, ΔL, from 0.1 mm to 2.0 mm. Lines were fitted by eye. Dependence of incremental force on incremental stretch amplitude is shown in (c) for two reference operating forces—500 g and 1000 g. Dependence of incremental stiffness ($\Phi = \Delta F/\Delta L$) on ΔL is shown in (d) for the same two operating forces. Cat 14.

Stiffness dependence on shape of perturbation

Effects of the shape of the perturbation on the incremental stiffness of soleus are shown in figure 2. Data correspond to triangular perturbations 1 mm in amplitude, lasting 500 msec. The incremental stiffness was defined by the

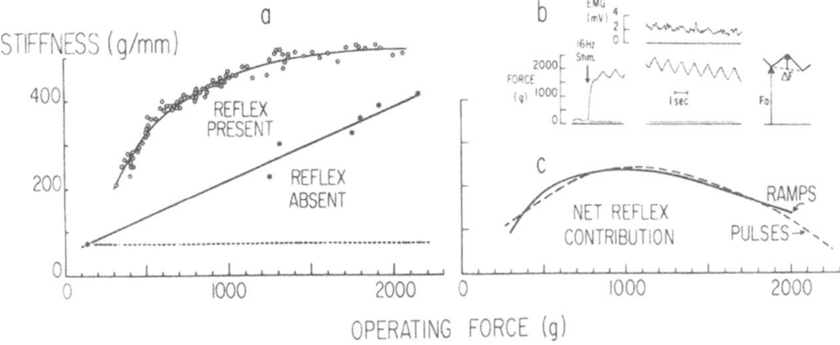

Figure 2. Dependence of incremental stiffness on operating force for 1 mm × 500 msec triangular perturbations. (a) Stiffness with reflex present (O) is compared with intrinsic muscle stiffness (∗). Dotted line indicates passive muscle stiffness. Operating forces in absence of reflex were generated by soleus nerve stimulation at 10, 12, 14, 16, 19 and 25 Hz. Parameters of linear fit are in last line of table 1. (b) Examples of original records of force obtained in absence of reflex, during 16 Hz stimulation, and in presence of reflex, during ongoing force modulations. Determination of operating force F_0 and force increment ΔF is shown schematically at right. (c) Net reflex contribution to incremental stiffness (the difference between the two fitted curves in (a)) is compared with the net reflex contribution determined for rectangular stretches of similar amplitude and duration (from Hoffer and Andreassen, 1981, figure 3d). Cat 17, $L_{max} - 5$ mm.

ratio of change in force to change in length at the end of each ramp stretch, shown in figure 2(b). Measurements were made in the presence of reflex, and were repeated minutes later in the deeply anaesthetised preparation, against background forces generated by electrical stimulation of the soleus nerve. Individual measurements obtained over the physiological range of operating forces are shown in figure 2(a), both for reflex present (circles) and in the anaesthetised preparation (asterisks).

Note that the intrinsic stiffness of areflexive muscle is linearly related to the operating force. Parameters of the best line fit are presented in table 1 for this and other preparations. A linear relation is predicted from cross-bridge theory for the short-range component of muscle stiffness (Flitney and Hirst, 1978; Huxley, 1974), which applies in cat soleus for stretches smaller than 0.4 mm (Morgan, 1977; Rack and Westbury, 1974). A linear relation was also found to apply for 1 mm rectangular stretches, when the force was allowed to recover for 200 msec or longer (Hoffer and Andreassen, 1981). However, the slope of the intrinsic stiffness/force relation for triangular perturbations shown in figure 2(a) is approximately twice that found for rectangular perturbations of identical amplitude and duration (*see* data for rectangular and triangular perturbations, cat 17, in table 1).

With reflexes present, muscle stiffness was larger for all operating forces

Table 1. Potentiation of muscle stiffness

				Unpotentiated line fit				*Potentiated line fit*				*Ratio of slopes*
Cat	L_0 (mm)	Δt (ms)	Stretch	Slope (mm^{-1})	Intercept (g/mm)	S.E.	ρ	Slope (mm^{-1})	Intercept (g/mm)	S.E.	ρ	
14	−7	200	rectangular	0.088	17	0.013	0.96	0.133	28	0.003	0.99	1.51
15	−7	500	rectangular	0.062	37	0.009	0.96	0.081	32	0.012	0.95	1.32
17	−5	500	rectangular	0.085	58	0.004	0.99	0.141	49	0.012	0.98	1.66
17	−5	500	triangular	0.171	48	0.012	0.98	0.234	62	0.008	0.99	1.37

L_0 = operating length with respect to L_{max}, Δt = duration of perturbation and S.E. = standard error of slope

(circles, figure 2(a)). Reflex effects are greatest at low and intermediate operating forces; at high forces the intrinsic stiffness of actively contracting muscle tends to dominate. The net additional contribution of the stretch reflex can be evaluated by subtracting the intrinsic component (fitted line in figure 2(a)) from the total response (fitted curve). This net reflex contribution is shown by the solid curve in figure 2(c). For comparison, the net reflex contribution obtained in the same preparation using rectangular instead of triangular perturbations is shown by the dashed curve in figure 2(c) (from Hoffer and Andreassen, 1980, figure 3(d)). The net reflex stiffness response to stretches of similar amplitude (1 mm) appears to be independent of the shape of the stretch. The curves obtained with rectangular and with triangular perturbations essentially overlap for all operating forces, in contrast to the corresponding intrinsic stiffness line fits, which differ by a factor of two (table 1). We will return to this point in the discussion.

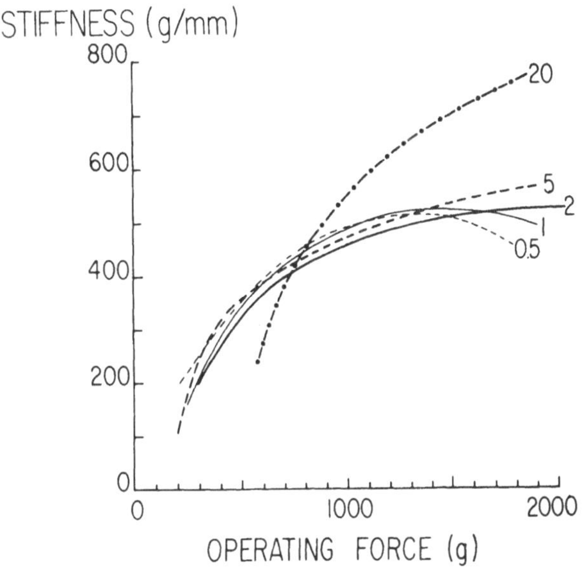

Figure 3. Effect of ramp velocity on incremental stiffness. Triangular perturbations of identical amplitude (1 mm) were presented at 0.5, 1, 2, 5 and 20 mm/sec. Note agreement over most of force range for ramps between 0.5 mm/sec and 5 mm/sec, whereas at 20 mm/sec the curve shifted markedly. Cat 17, L_{max} − 5 mm.

Stiffness dependence on velocity of perturbation

The finding that the reflex contribution to muscle stiffness depends on the amplitude of the length perturbation but not on its shape prompted us to test the effect of risetime by perturbing the muscle with 1 mm ramps of different velocities. The data of figure 2 were obtained using triangular perturbations 1 mm in amplitude and lasting 500 msec; therefore the rate of stretch (and of release) was 2 mm/sec. Figure 3 shows results obtained using ramps of similar amplitude, but of velocities ranging betwen 0.5 mm/sec and 20 mm/sec. Only the total stiffness with reflex present is shown here; the intrinsic muscle stiffness was not tested in each case. The preparation was the same as in figure 2.

Stiffness curves for ramp velocities ranging between 0.5 mm/sec and 5.0 mm/sec were essentially coincident over most of the operating force range. The tendency for the curves to diverge at high operating forces may be attributed to the ongoing decline in the force baseline, following an exponential time course (*see* Hoffer and Andreassen, 1981, figure 2) which became more noticeable for slow ramps. In contrast, ramps delivered at 20 mm/sec caused an altered response in the muscle; stiffness values were lower at low operating forces and markedly increased at higher forces. However, it should be noted that ramp velocities of 20 mm/sec with 1 mm amplitude impose a 10 Hz oscillation on the muscle, close to the natural frequency of reflexly mediated oscillations in soleus (8–12 Hz, Stein and Oğuztöreli, 1976; 12–14 Hz, Nichols *et al.*, 1978). It is therefore likely that triangular perturbations imposed at 10 Hz caused resonance around the spinal reflex loop. From a methodological standpoint, subjecting the muscle to repeated stretches and releases 10 times per second violated our own protocol of making measurements of stiffness only after the full effects of the reflex had time to occur (150 msec or longer in these preparations; Hoffer and Andreassen, 1981, figure 1). Using 10 Hz ramps, measurements of stiffness had to be made only 100 msec after onset of each ramp stretch, and thus the full effect of the reflex could not have been expressed. In addition, at high forces the tail end of the decay phase in the intrinsic muscle response was still observed at 100 msec, causing higher readings of stiffness than would have occurred at later times. At low operating forces partial cancellation of the transient component of the reflex due to resonance around the spinal loop probably caused the muscle to appear more compliant (Joyce and Rack, 1974; Nichols *et al.*, 1978).

It should be noted that the 10 Hz behaviour was the only frequency-dependent singularity observed in these preparations. Perturbations with higher frequency components (rectangular pulses) or lower components (0.25–2.5 Hz ramps) exhibited comparable values of net reflex stiffness (e.g. fig 2(c)).

Stiffness dependence on prior activation history of muscle

Finally, we studied the effect on the reflex-mediated muscle stiffness of changes in the mechanical behaviour of the muscle itself under conditions of intense activation for prolonged periods of time. As was described earlier under 'Methods' we observed that prolonged electrical activation of soleus caused gradually increasing values of stiffness at matched operating forces and lengths. Examples are shown in table 1. The increase in the slope of the intrinsic stiffness versus operating force relation is shown in the last column of the table. Stiffness values in the potentiated state increased by 32–66% in different preparations. This increase was observed uniformly throughout the force range; the correlation coefficients of the corresponding line fits were comparable. Potentiation was observed with both rectangular pulse stretches and ramp stretches. The potentiation effect was reversible; a return to normal values was observed after several minutes in each case.

For comparison, figure 4 (top) shows data obtained with reflexes intact during prolonged stimulation of the mesencephalic tegmentum using trains at 45 Hz. Maintained high levels of activation of soleus could be elicited in this manner. During such prolonged activation, total stiffness values were observed to rise and exceed control values measured in the same preparation just minutes before onset of stimulation. The graph in figure 4 shows stiffness values collected 50–100 sec after onset of continued brainstem stimulation (open circles) and for up to 60 sec after stimulation was discontinued (filled circles). Stiffness values increased by 60–120 g/mm at matched operating forces. These measurements fell well outside control values determined for the same preparation just before stimulation (dashed curve; from figure 7, Hoffer and Andreassen, 1981). The increase in total stiffness during brainstem stimulation with reflexes present matched closely the increase in intrinsic stiffness caused by potentiating the muscle with electrical stimulation in the absence of reflexes (solid line vs. dashed line in figure 4), suggesting that the effect was due primarily to an alteration of the intrinsic properties of muscle as a consequence of prolonged activation, rather than to changes in reflex gain, particularly since stiffness remained high for many seconds *after* stimulation was discontinued. However, from these data it cannot be known whether some form of potentiation was also taking place along central and/or segmental pathways as a consequence of brainstem stimulation.

The net contribution to total stiffness in the potentiated case, given by the difference between the two solid lines in figure 4, was about 30 g larger than for the rested muscle case (given by the difference between the two dashed lines). The discrepancy could signify that synchronous stimulation of the soleus nerve and brainstem stimulation using a needle electrode were not equally effective in bringing about potentiation in muscle. Table 1 shows that even when comparable electrical stimulation was used, the degree of stiffness potentiation obtained in different muscles could vary within a factor of two. Nevertheless, it is of interest that the net reflex addition to muscle stiffness was relatively unchanged even though the intrinsic stiffness was markedly increased due to potentiation. Figure 4 and figure 2 both show that the reflex contribution to the total stiffness of soleus muscle in the premammillary cat is relatively insensitive to changes in the intrinsic behaviour of the muscle,

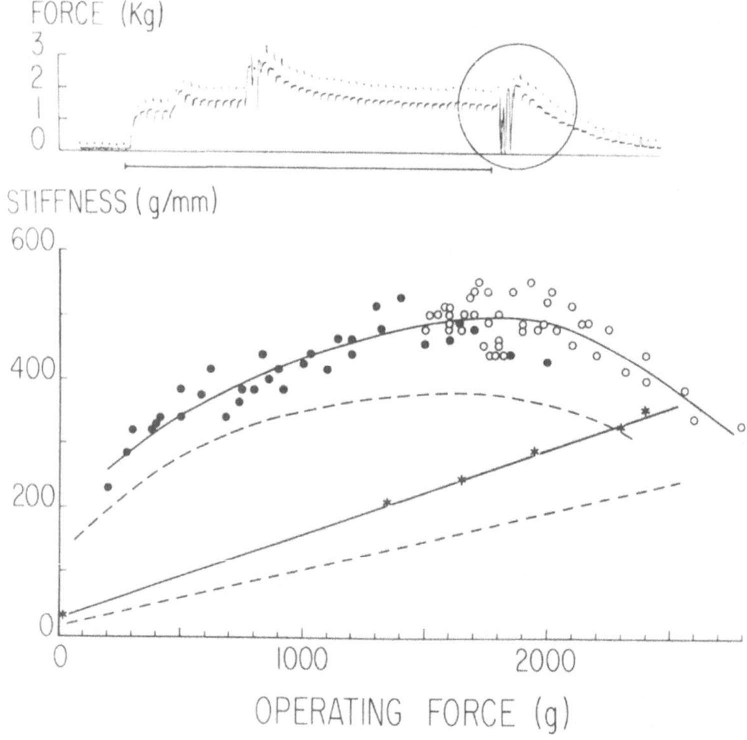

Figure 4. Effect of prolonged muscle activation on incremental stiffness. Top: Record of force obtained from soleus muscle held at L_{max} − 7 mm by torque motor. Perturbations 1 mm in amplitude and 200 msec in duration were presented once per second. Brainstem stimulation at 45 Hz, maintained for about 1 min (bar) caused high levels of force in soleus. Force decayed after stimulation ended. Readings of potentiated stiffness were obtained during circled period, in this and similar runs.

Bottom: Incremental stiffness measurements obtained in presence of reflex (circles) and from areflexic muscle activated by nerve stimulation at several frequencies (asterisks). Leftmost point indicates passive muscle stiffness. Data in presence of reflex were obtained late in period of brainstem stimulation (○) or just after stimulation ended (●). Note that curve fitted to data points is shifted from control curve (dotted; obtained in same preparation minutes earlier). Similarly, line fitted to intrinsic muscle stiffness data has larger slope than control line (dotted). Parameters of line fits in table 1. Cat 14.

whether revealed by using perturbations of different shape or by prolonged activation causing potentiation.

DISCUSSION

In this paper we have extended our previous study on the intrinsic and reflex components in the regulation of soleus muscle stiffness in premammillary cats, by investigating the effects of the amplitude, shape and velocity of length perturbations, as well as of the activation history of the muscle under circumstances that bring about potentiation of the intrinsic stiffness.

In agreement with earlier observations (Matthews, 1969) we observed a moderate dependence of the incremental stiffness on the amplitude of the imposed stretch. This dependence was not very pronounced, however, since a twenty-fold decrease in the amplitude of perturbation was accompanied by only a two-fold increase in the incremental stiffness. An enhanced small-signal spindle sensitivity (Hasan and Houk, 1975) may contribute to greater operating point stability (Matthews and Stein, 1969) and perhaps also to reduction of amplitude of tremor (Goodwin *et al.*, 1978). Thus, although reflex-regulated muscle does not behave quite like a 'perfect' spring, the observed non-linearity for small-amplitude perturbations may in fact be a desirable feature. Interestingly, areflexive muscle can also show marked stiffness non-linearities for small stretches (e.g. Rack and Westbury, 1974, figure 5). The increased stiffness attributed to non-linear spindle firing behaviour should also be attributed in part to intrinsic muscle stiffness non-linearities.

Changing the shape of perturbations revealed differences between the intrinsic muscle response and the total response with reflexes present. Whereas the areflexic muscle subjected to triangular perturbations was approximately twice as stiff as when rectangular perturbations of the same amplitude were used (table 1), the *additional* stiffness contributed by the stretch reflex was virtually identical in the two cases, since the two net reflex curves coincided within 20 g over most of the operating range (figure 2). This suggests that in the premammillary decerebrate preparation the reflex-mediated response is essentially completely determined by the amplitude of stretch, and is more or less impervious to changes in muscle properties. Modelling of reflex action indicates that the gain of Golgi tendon organ feedback is negligibly small (Andreassen and Hoffer, unpublished), complementing observations from other types of experiments done previously in decerebrate cats (Eccles and Lundberg, 1959; Houk *et al.*, 1970).

The insensitivity of the reflex to perturbation shape may appear surprising. However, it must be kept in mind that our measurements were made when the full reflex effects had time to be expressed (200–500 msec after onset of perturbation). It is believed that the dynamic phase of the muscle response is mediated primarily by short-latency Ia pathways, while the steady-state response is mediated principally by spindle group II, polysynaptic Ia and Golgi tendon organ pathways (Houk, 1979b; Houk *et al.*, 1977; Kanda and Rymer, 1977; Matthews, 1969, 1970), and it is this steady-state incremental response that was measured in these experiments. Since spindle secondaries and Golgi tendon organ receptors monitor chiefly the amplitude of the stimulus rather than its velocity (Houk and Henneman, 1967; Matthews, 1969; Rymer *et al.*, 1977), the shape of the perturbation was essentially unimportant in our experiments.

With regard to the effects of velocity of stretch on our measurements of incremental stiffness, it must be noted that whenever measurements were made 200 msec or later after onset of 1 mm perturbations, reflex stiffness contributions at matched operating forces were essentially similar, independently of stretch velocity. The discrepancy found for 10 Hz vibration derived from making measurements 100 msec after onset of stretch, thus not allowing the development of the full reflex-mediated response. Consequently we were looking at transient aspects of the response, rather than at the steady-state incremental stiffness.

Under conditions when the intrinsic stiffness of soleus was allowed to potentiate through continued activation, the net reflex contribution to the total response was essentially unchanged even though the intrinsic muscle response was substantially increased. A similar finding stemmed from comparison between ramps and rectangular pulses of similar amplitude. This dissociation between intrinsic properties and reflex effects indicates the predominant role of length detectors in determining the total reflex response. The observed shortcomings in reflex compensation for variations in muscle stiffness in premammillary cats lend themselves to a common interpretation: force feedback is largely ineffectual in this preparation. This conclusion can be substantiated rigorously using a modelling approach (Andreassen and Hoffer, unpublished) but the argument can be made in a qualitative way from each of the paradigms presented in this paper. Increased Golgi tendon organ feedback gain would have been expected to provide better compensation for increases in the intrinsic muscle stiffness observed for small-amplitude stretches, using triangular instead of rectangular perturbations, and during muscle potentiation.

The servo-regulation of muscle stiffness as proposed by Houk (1978, 1979a, 1979b) requires the participation of both length and force feedback. Ironically, a remarkable resemblance between Golgi tendon organ firing patterns and operating force is observed in decerebrate preparations (Houk *et al.*, 1978). It appears that in spite of such faithful monitoring of muscle force, Golgi tendon organ afferent pathways operate at very low gain in decerebrate cats (Eccles and Lundberg, 1959; Houk *et al.*, 1970). Three implications stem from these observations:

1. Adequate reflex regulation of muscle stiffness cannot be expected to occur in decerebrate cats, due to inadequately low force feedback gain. This observation holds for premammillary levels as well as lower levels of decerebration.
2. There is evidence, however, that the total gain of the stretch reflex can be reduced in the decerebrate cat by stimulating brainstem areas (Hoffer and Andreassen, 1981), an observation consistent with possible enhancement of force feedback gain.
3. Whether reflexes manage to regulate muscle stiffness effectively in intact

systems remains to be shown. Since no reliable data on the magnitude of the gain of force feedback are available at this time, this is clearly an important point to be resolved.

ACKNOWLEDGEMENTS

The authors wish to thank Dr T. R. Nichols for invaluable guidance during the early phases of this work, Dr R. B. Stein for offering the use of his laboratory, and Drs M. J. Pinter and C. A. Pratt for commenting on the manuscript.

J. A. Hoffer was a Muscular Dystrophy Association of Canada Post-Doctoral Fellow. S. Andreassen was a Technical Research Council of Denmark Visiting Fellow.

Received on June 30th, 1980.

REFERENCES

Eccles, R. M. and Lundberg, A. (1959). Supraspinal control of interneurones mediating spinal reflexes, *J. Physiol., Lond.*, **147**, 565–584

Feldman, A. G. and Orlovsky, G. N. (1972). The influence of different descending systems on the tonic stretch reflex in the cat, *Expl Neurol.*, **37**, 481–494

Flitney, F. W. and Hirst, D. G. (1978). Cross-bridge detachment and sarcomere 'give' during stretch of active frog's muscle, *J. Physiol., Lond.*, **276**, 449–465

Goodwin, G. M, Hoffman, D. and Luschei, E. S. (1978). The strength of the reflex response to sinusoidal stretch of monkey jaw closing muscles during contraction, *J. Physiol., Lond.*, **279**, 81–111

Hasan, Z. and Houk, J. C. (1975). Transition in sensitivity of spindle receptors that occurs when muscle is stretched more than a fraction of a millimetre, *J. Neurophysiol.*, **38**, 673–689

Hoffer, J. A. and Andreassen, S. (1978). Factors affecting the gain of the stretch reflex and soleus muscle stiffness in premammillary cats, *Neurosci. Abstr.*, **4**, 935

Hoffer, J. A. and Andreassen, S. (1981). Regulation of soleus muscle stiffness in premammillary cats: intrinsic and reflex components, *J. Neurophysiol.*, to be published

Houk, J. C. (1977). An assessment of stretch reflex function, *Prog. Brain Res.*, **44**, 303–314

Houk, J. C. (1978). Participation of reflex mechanisms and reaction-time processes in the compensatory adjustments to mechanical disturbances, *Prog. clin. Neurophysiol.*, **4**, 193–215

Houk, J. C. (1979a). Regulation of stiffness by skeletomotor reflexes, *A. Rev. Physiol.*, **41**, 99–114

Houk, J. C. (1979b). Motor control processes: new data concerning motoservo mechanisms and a tentative model for stimulus-reponse processing. In *Posture and Movement* (eds R. E. Talbot and D. R. Humphrey), New York, Raven Press, 231–241

Houk, J. C., Crago, P. E. and Rymer, W. Z. (1978). Functional properties of golgi tendon organs. In *Segmental Motor Control in Man* (ed. J. E. Desmedt), *Prog. clin. Neurophysiol.*, Basel, Karger

Houk, J. C. and Henneman, E. (1967). Responses of Golgi tendon organs to active contractions of the soleus muscle of the cat, *J. Neurophysiol.*, **30**, 466–481

Houk, J. C., Rymer, W. Z. and Crago, P. E. (1977). Complex velocity dependence of the electromyographic component of the stretch reflex, *Proc. XXVII Congr. Physiol. Sci.*, Paris, Abstr. No. 981

Houk, J. C., Singer, J. J. and Goldman, M. R. (1970). An evaluation of length and force feedback to soleus muscles of decerebrate cats, *J. Neurophysiol.*, **33**, 784–811

Huxley, A. F. (1974). Muscular contraction, *J. Physiol., Lond.*, **243**, 1–43

Joyce, G. C. and Rack, P. M. H. (1974). The effects of load and force on tremor at the normal human elbow joint, *J. Physiol., Lond.*, **240**, 375–396

Kanda, K. and Rymer, W. Z. (1977). An estimate of the secondary spindle receptor afferent contribution to the stretch reflex in extensor muscles of the decerebrate cat, *J. Physiol., Lond.*, **264**, 63–87

Maffei, L. and Pompeiano, O. (1962). Effects of stimulation of the mesencephalic tegmentum following interruption of the rubrospinal tract, *Arch. Ital. Biol.*, **100**, 510–525

Matthews, P. B. C. (1959a). The dependence of tension upon extension in the stretch reflex of the soleus muscle of the decerebrate cat, *J. Physiol., Lond.*, **147**, 521–546

Matthews, P. B. C. (1959b). A study of certain factors influencing the stretch reflex of the decerebrate cat, *J. Physiol., Lond.*, **147**, 547–564

Matthews, P. B. C. (1969). Evidence that the secondary as well as the primary endings of the muscle spindles may be responsible for the tonic stretch reflex of the decerebrate cat, *J. Physiol., Lond.*, **204**, 365–393

Matthews, P. B. C. (1970). A reply to the criticism of the hypothesis that the group II afferents contribute excitation to the stretch reflex, *Acta physiol. scand.*, **79**, 431–433

Matthews, P. B. C. and Stein, R. B. (1969). The sensitivity of muscle spindle afferents to small sinusoidal changes in length, *J. Physiol., Lond.*, **200**, 723–743

Morgan, D. L. (1977). Separation of active and passive components of short-range stiffness of muscle, *Am. J. Physiol.*, **232**, 45–49

Nichols, T. R. and Houk, J. C. (1976). Improvement in linearity and regulation of stiffness that results from actions of stretch reflex, *J. Neurophysiol.*, **39**, 119–142

Nichols, T. R., Stein, R. B. and Bawa, P. (1978). Spinal reflexes as a basis for tremor in the premammillary cat, *Can. J. Physiol., Pharmacol.*, **56**, 375–383

Rack, P. M. H. and Westbury, D. R. (1974). The short range stiffness of active mammalian muscle and its effect on mechanical properties, *J. Physiol., Lond.*, **240**, 331–350

Rosenthal, N. P., McKean, T. A., Roberts, W. J. and Terzuolo, C. A. (1970). Frequency analysis of the stretch reflex and its main subsystems in triceps surae muscles of the cat, *J. Neurophysiol.*, **33**, 713–749

Rymer, W. Z., Houk, J. C. and Crago, P. E. (1977). The relation between dynamic response and velocity sensitivity for muscle spindle receptors, *Proc XXVII Congr. Physiol. Sci.*, Paris

Stein, R. B. and Oğuztöreli, M. N. (1976). Tremor and other oscillations in neuromuscular systems, *Biol. Cybern.*, **22**, 147–157

Observations on the control of human ankle position by stretch reflexes

J. H. J. ALLUM*

SUMMARY

Load compensation in human ankle muscles was investigated by applying disturbances which rotated the foot while the subject was endeavouring to maintain a constant position against a pre-existing force. Three separate stages of the force response were distinguished. *First*, for the initial 100 msec an 18% increase in force (expressed as a percentage of the total force required to correct the disturbance) which was attributed to the elastic resistance of muscles active prior to the disturbance. A short latency (SL) EMG response at 33 msec occurred in the soleus (SOL) muscle only when the foot was dorsiflexed, but no SL EMG response was observed in the tibialis anterior (TA) muscle following plantar flexion. It was difficult to attribute a part of the first increase in force to SL EMG activity in SOL. *Second*, after approximately 130–150 msec a medium latency (ML) increase in force which was preceded by an increase in EMG activity. After a dorsiflexion disturbance the ML EMG response at 120 msec in SOL was followed by an 18% increase in force. But, after a plantar flexion disturbance, the ML EMG response in TA at 100 msec was followed by ML EMG response in SOL at 130 msec, and a 2% increase in force. *Third*, a long latency increase in force accompanied EMG activity at 230 msec and was compatible with voluntary action which restored the foot to its original position. These results are consistent with previous results obtained in shoulder muscles (Allum, 1975) and support the hypothesis that the ML response is a test signal designed to inform the CNS of the current loading on the muscle.

*Brain Research Institute, University of Zürich

INTRODUCTION

The term *servo-regulation* or *servo-assistance* used to describe the actions of stretch reflexes in response to a load disturbance is generally accepted to mean that these reflexes cause a graded servo-control of limb position. Three major premises underlie this concept. Firstly, the input–output properties of the reflex systems which give rise to electromyographic (EMG) responses in the perturbed muscles should be linear. Secondly, the visco-elastic responses of muscles resisting and assisting the disturbance should be linearly related to disturbance amplitude. Thirdly, the forces generated by these mechanisms should be powerful enough to return the limb to its original position.

With respect to the first premise, recent investigations have demonstrated servo-like properties for the EMG responses in the leg (Gottlieb and Agarwal, 1979, 1980) and arm (Marsden *et al.*, 1976) even when single motor unit responses were examined (Tatton and Bawa, 1979). On the second premise, however, the visco-elastic responses of active muscle are known to be extremely non-linear (Joyce *et al.*, 1969; Nichols and Houk, 1976, figure 13). This result has prompted the suggestion that the forces produced by reflex EMG responses compensate for the muscle non-linearities and not for the perturbed limb position (Nichols and Houk, 1976; Crago *et al.*, 1976; Kwan *et al.*, 1979).

To date, few studies have approached the question of whether the forces generated by reflex EMG responses and muscle visco-elasticity are large enough to correct disturbances. From indirect evidence it may be recognised that in the neck muscles of monkeys (Bizzi *et al.*, 1978) and in the human ankle muscles (Melvill Jones and Watt, 1971) reflex mechanisms are unable to regulate limb position. More direct testing of the servo-regulatory force capability in shoulder muscles led to the conclusion that only a newly initiated response of suprasegmental structures could provide effective position compensation (Allum, 1975). The implication here was that the reflex systems act as sensors to help set up the new programme which eventually corrects for the disturbance. This present study examines whether this principle is applicable to ankle muscles. The ankle muscles were chosen for this investigation since the longer reflex transmission delays allow a satisfactory determination of the time delay after which active muscle contraction is added to the muscle's visco-elastic stretch response.

METHODS

For the present experiments subjects were seated with the knee flexed 110° and the bare right foot placed on a platform which could be rotated about an axis co-linear with the ankle joints. The angle of the platform was visually displayed to the subject together with a reference position of 90° ankle flexion which the subject was asked to maintain. The subject was instructed, should a disturbance occur, to return his foot and the platform to the reference position as quickly as possible with minimum overshoot, i.e. to adopt a minimum settling time strategy. A servo-controlled torque motor regulated the position of the platform. The force applied on the platform by a subject's ankle muscles was measured with a strain gauge transducer and fed back to the torque motor controller. The feedback gain ensured that the load of the platform experienced by the subject was a stiff spring (normally 4 Nm ankle torque per degree of ankle rotation).

A PDP 11/03 digital computer issued torque motor command signals for the bias force at the reference position, i.e. for the spring-load equilibrium position, and for terminated ramp position disturbances away from reference. A computer program randomly varied the interval between each disturbance and its direction (ankle flexion or extension). Disturbance magnitudes could take on one of 16 equally spaced values between two prefixed limits. Each magnitude occurred once in a sequence of 16 stimuli, but its ordering within the sequence was randomly selected. The total duration of the disturbance and the duration of its rise time were preselected at 4 sec and 70 msec, respectively.

The transducers used to measure foot force and platform acceleration had damped natural frequencies of 100 and 450 Hz, respectively. Their signals were low-pass filtered at 80 Hz. Surface EMG electrodes to record ankle extensor activity were placed over the lower medial part of the triceps surae muscle so that recorded responses were dominated by SOL activity. Transducer signals, force, angle and acceleration, as well as SOL and tibialis anterior surface EMG recordings, were stored on FM magnetic tape. The EMG signals were recorded differentially with pairs of silver–silver chloride electrodes at a bandwidth of 75 Hz and 1.4 kHz. Off-line data analysis consisted of averaging the parameters of force, acceleration, angle and surface EMG with a time resolution of 1.0 msec. For improved visualisation the EMG signals were full-wave rectified and smoothed prior to averaging.

RESULTS

Responses of ankle muscles to dorsiflexing load disturbances

Figure 1 illustrates the force and EMG responses obtained when a subject resists a 2.0° disturbance dorsiflexing the ankle and thereby stretching the SOL muscle. Prior to the disturbance the subject maintained a plantar flexion torque of 12 Nm by contracting triceps surae muscles while tibialis anterior (TA) was electrically silent. The disturbance produced a strong reflex EMG activity in SOL which commences 33 msec after the onset of platform

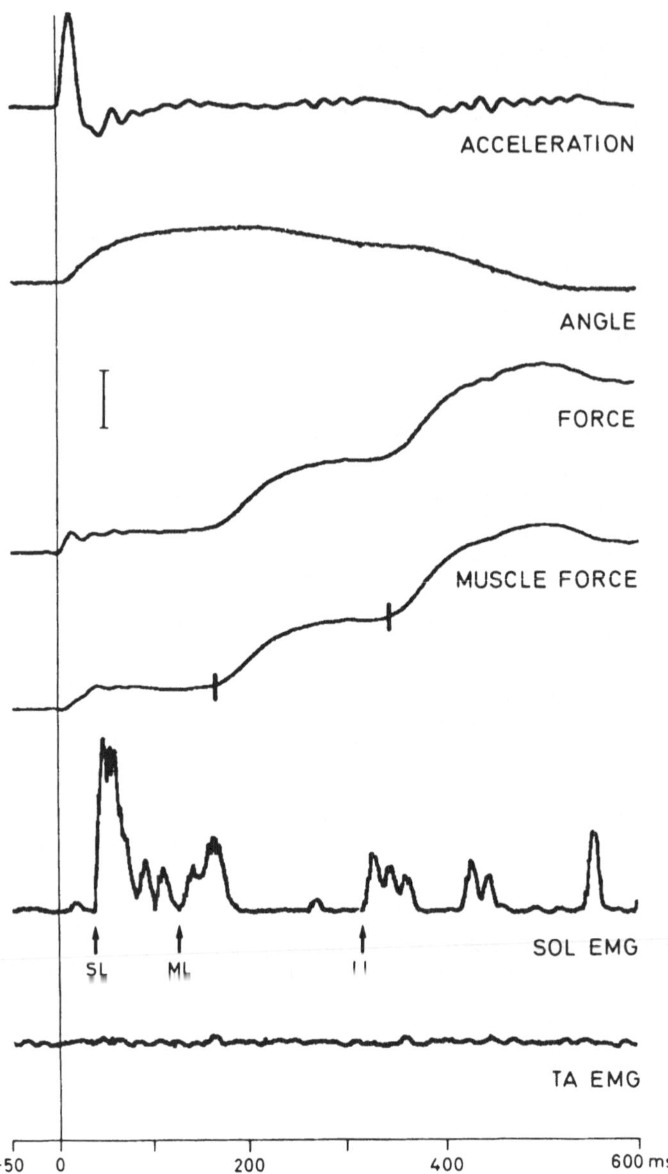

Figure 1. Force and EMG responses to a forced dorsiflexion of the foot with the subject instructed to resist and return the foot to the reference position. Prior plantar flexion force 12 Nm. The records are aligned with the onset of disturbance acceleration. EMG responses have been subdivided into short, medium and long latency components whose onsets are marked by arrows SL, ML and LL, respectively. Muscle force has been computed as the recorded force minus the inertial reaction force. Two vertical lines on the muscle force record bound the ML force response. Vertical calibration, acceleration 1340°/sec^2, angle 2°, force and muscle force 2.25 Nm, SOL and TA EMG 48 μV.

acceleration. The rapid rise in acceleration should be an effective stimulus for Ia muscle spindle afferents which project monosynaptically to alpha motoneurons and probably transmit the signals responsible for the initial short latency (SL) reflex activity. The onset SL activity is marked in figure 1.

Subsequent peaks in the SOL EMG activity are shown in figure 1. The peak at 120 msec corresponds to activity termed a *functional stretch reflex* by Melvill-Jones and Watt (1971). To avoid prejudging the function of this activity the neutral descriptor medium latency (ML) response will be employed. Following a period of electrical silence SOL is again active at 300 msec. The ensuing EMG activity is termed long latency (LL) activity. Correction of the disturbance requires an increased contraction by the triceps surae muscles, thus it is not surprising that the TA electrodes recorded no activity during the 500 or so milliseconds that the subject required to reposition the platform at its reference position.

When responses to disturbance sizes ranging from 1.5° to 2.0° are averaged together as shown in figure 2 the SL EMG response is easily identifiable since its latency is relatively constant. ML and LL responses recorded with surface electrodes have more variable latencies, are less securely driven with the small ankle rotations used in the present study (Gottlieb and Agarwal, 1980), and are not as well defined as intramuscular recordings (Bawa and Tatton, 1979). Hence ML and LL responses are smoothed out by averaging in figure 2. Nevertheless, the activity pattern shown in figure 1 may be recognised in figure 2. Apart from minor differences in the latency measurements, the pattern of SL and ML EMG activity is identical to that observed by Gottlieb and Agarwal (1980) under similar experimental conditions. The slightly shorter latencies obtained in the present study result from using acceleration onset as the starting point for latency measurements and possibly from differences in transmission lengths between subjects.

Force evoked by the disturbance has four components whose sum is recorded by the platform force transducer. These components consist of, firstly, initial reaction forces proportional to the disturbance acceleration; secondly, the visco-elastic forces of the ankle flexor and extensor muscles; thirdly, the active forces generated by these muscles following an increase in EMG activity; and fourthly, changes in force following a movement of the foot's centre-of-gravity projection on to the platform. The latter gravitational forces can be neglected as the angle of platform rotation is less than 5°. Initial reaction forces may be removed from the measured forces by subtracting the product of a constant, proportional to the moment of inertia of the foot, and the acceleration. Force remaining after this subtraction is then equal to the visco-elastic and active muscle force. Such force is labelled MUSCLE FORCE in the figures, and for brevity this term will be used in the text.

The recordings of figure 1 were chosen for illustration because the three phases of muscle force generation which act to correct a load disturbance are distinct. Simplest to describe is the force associated with the LL EMG response which provides the largest fraction of the force required to correct the disturbance (hereafter termed 'correcting force'). The onset of the LL force response is marked by a vertical line on the muscle force record in figure 1, and preceded by the onset of LL EMG activity. The ML force response in figure 1 was assumed to commence at the point marked by a vertical line some 35 msec after the onset of the ML EMG response. Somewhat exceptionally, this section of the force response provided 43% of the correcting force. None of the ML force response can be considered a visco-elastic force since the disturbance terminated some 50 msec earlier. From the onset of the disturbance until the onset of the ML force response the muscle force recorded resembles the time course of the platform displacement suggesting that this section of the force response is due to the visco-elastic response of the stretched and released ankle muscles. Despite the well-defined SL EMG response it is not followed by an observable increase in muscle force. In this respect the SL EMG activity differs from ML and LL activity.

Figure 2 shows the relation between the average integrated EMG activity and muscle force. Determination of the onset times of ML and LL muscle force responses is more difficult for average responses than for single responses (e.g. figure 1). However, over the section 150–250 msec (i.e. between the vertical marks on the muscle force record of figure 2) muscle force can be attributed to ML EMG activity. Average muscle force increases in this section amount to 18% of the correcting force. During the first 150 msec the average muscle force also increases by 18% in a manner which closely parallels the time course of the disturbance. Figure 2 confirms the finding, shown in figure 1, that no increase in muscle force is observed consistent with the occurrence of SL EMG activity.

Responses of ankle muscles to plantar flexing load disturbances

A plantar flexion displacement of the platform stretches the TA muscles and releases the SOL muscle provided subjects maintain a pre-existing plantar flexion force. The response shown in figure 3 was elicited by a 1.7° plantar flexion with a pre-existing force of 12 Nm plantar flexion. Averaged responses are shown in figure 4. From these figures it is apparent that plantar flexion responses are fundamentally different from dorsiflexion responses in three aspects. These are: the lack of an SL EMG response in the stretched TA muscles; the earlier occurrence of the ML response in TA (by 20 msec); and the presence of an ML response in the released SOL muscle. With similar experimental procedures Gottlieb and Agarwal (1979, figure 10) obtained a similar difference between plantar flexion and dorsiflexion responses. Activity in SOL prior to its plantar flexion ML response can be associated with muscle activity producing the pre-disturbance bias force. The lack of the SL response in TA is probably due to the state of excitation of the TA motoneuron pool. A plantar flexion load disturbance imposed when TA is already stretched causes an SL EMG response in TA (Allum and Büdingen, unpublished observations). Interestingly, the double pattern of ML EMG responses in TA and SOL is also observed when both feet of a standing subject are given a sudden plantar flexion (Allum and Büdingen, 1979). Although it is beyond the scope of the present study, it may not be coincidental

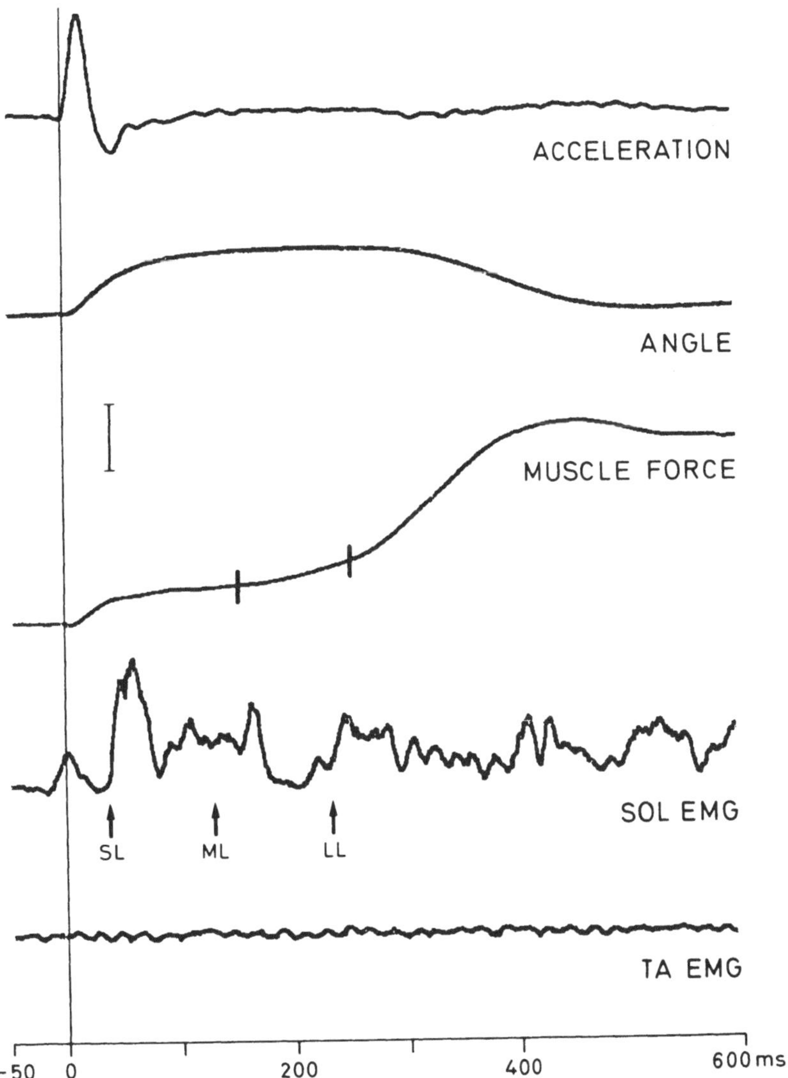

Figure 2. Averaged force and EMG responses to eight forced dorsiflexions ranging from 1.5° to 2.0°. Alignment and vertical calibration as figure 1 except EMG records 24 μV.

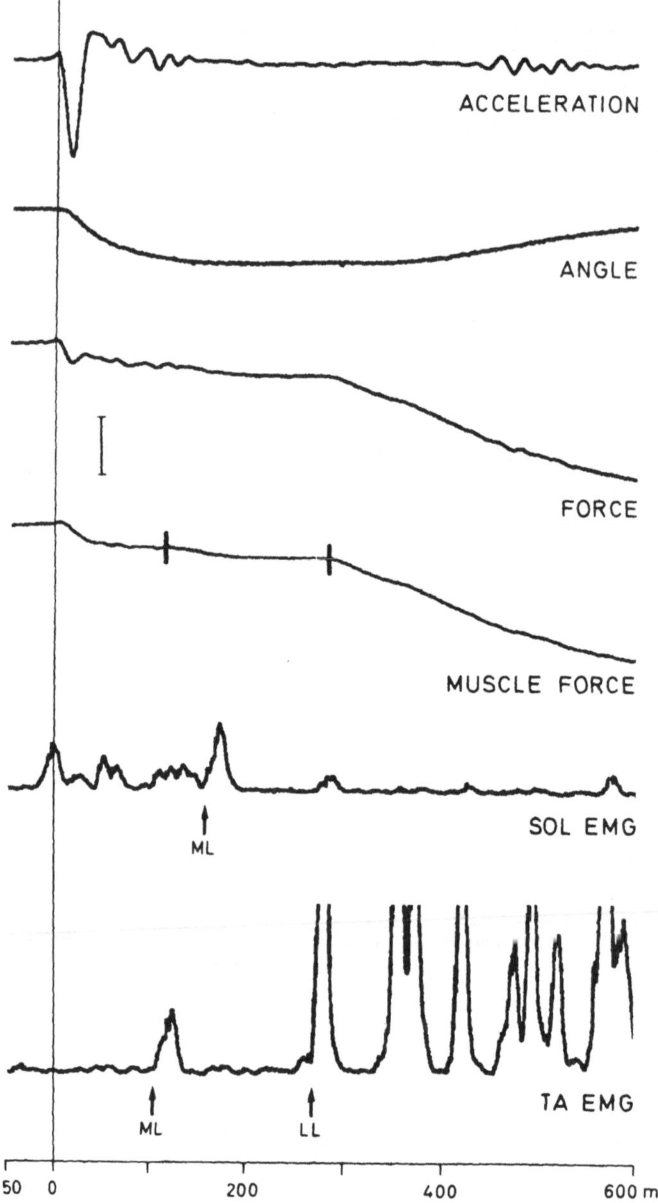

Figure 3. Force and EMG responses to a forced plantar flexion of the foot which assists the prior plantar flexion of 12 Nm produced by the subject. The amplitude of the disturbance was 1.7°. A contraction of the TA muscle was required to return the foot to the reference position. EMG responses have been marked with the onsets of ML and LL components. Note the absence of SL in TA and the presence of ML in SOL. The ML force response is bounded by vertical lines on the muscle force record. Alignment and vertical calibration as figure 1.

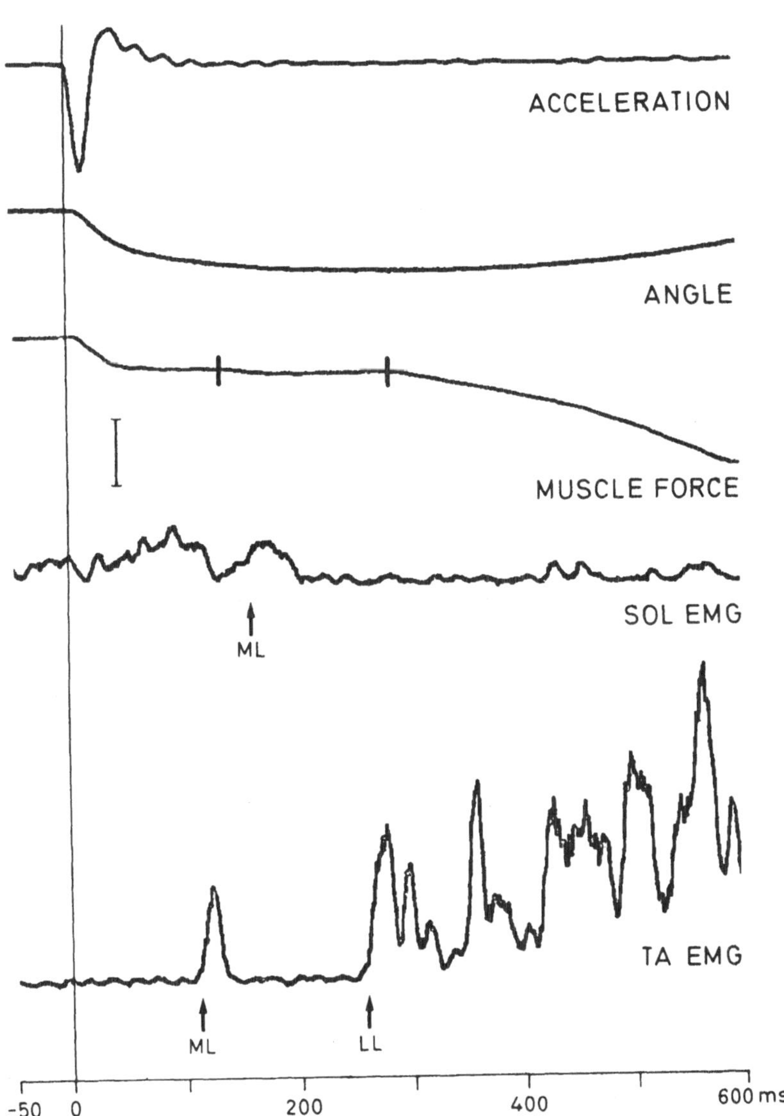

Figure 4. Averaged force and EMG responses to eight forced dorsiflexions ranging from 1.5° to 2.0°. Alignment and vertical calibration as figure 1 except EMG records 24 μV.

that the time interval between the two ML responses in TA and SOL is approximately equal to the latency of the SL EMG response in SOL following a dorsiflexion disturbance (compare EMG responses in figures 1 and 3).

Both figures 3 and 4 clearly demonstrate that only the muscle force associated with the LL EMG activity at approximately 250 msec has the capability to correct the disturbance. The increase in muscle force over the interval 100–280 msec can be associated with ML EMG activity. This interval is bounded by vertical marks on the muscle force records in figures 3 and 4, and the increase in force within the interval is quite small, 4% and 2% of the correcting force in figures 3 and 4, respectively. It may be noted, however, that the almost simultaneous occurrence of ML responses in the two antagonistic muscles (TA and SOL) probably accounts for the minor force increase.

In view of the lack of an SL EMG response in TA it is somewhat surprising that the increase of muscle force in the first 100 msec is practically identical to that occurring during a dorsiflexion disturbance. The force increase in figure 4 over this section is 17.3% of the correcting force compared with 18% in figure 2. Both force increases appear to be related to the time course of the displacement and not its velocity, suggesting that the observed force primarily represents the elastic properties of the preactivated ankle muscles. If this suggestion is correct, then a question must be raised about the role of SL EMG response in SOL during load compensation, since the symmetry of the early part of the force responses stands in contrast to the asymmetry of SL EMG responses.

DISCUSSION

The results reported in this study are not in accord with the proposition that ML responses in ankle muscles function as part of a load-compensating servo-system. The concept of servo regulation of load disturbances is built on the premise that the forces generated by ML EMG responses should be powerful enough to return the limb to its original position. ML responses contributed 18% of the force required to correct a dorsiflexion disturbance and 2% of the correcting force for a plantar flexion disturbance. The lower percentage appears to result from the coupled nature of ML EMG responses in TA and SOL following a plantar flexion disturbance. Similar coupled EMG responses are observed for both dorsiflexion and plantar flexion disturbances imposed on standing subjects (Allum and Büdingen, 1979). Thus, it is probable that the 2% value has a greater behavioural significance than the 18% value. If it is accepted that the 18% value represents the amount of force produced by the triceps surae muscles when no opposing ML response occurs in the ankle flexion muscles, it still is insufficient to correct for the load disturbance. Furthermore, it is within a few per cent of the amount of force produced by ML action in human shoulder muscles (Allum, 1975) and primate neck muscles (Bizzi *et al.*, 1978). Both these studies reported that ML responses lacked sufficient force to correct for load disturbances.

The force observed during the first 100 msec provided 18% of the correcting force. This section of the force response was identical for dorsiflexion and plantar flexion disturbances and was a predominantly spring-like increase in force as it covaried with the platform angle. The symmetry of this early force response for the two directions of platform movement contrasts with the marked asymmetry of the SL EMG response. An SL EMG response was recorded in SOL, and was also observed in other triceps surae muscles (Gottlieb and Agarwal, 1979), but was not present in TA when it was stretched. In addition it was not possible to correlate an increase in muscle force with the SL activity in SOL.

The size principle of motor unit recruitment (Henneman *et al.*, 1965) could underlie the absence of an observable SL force response in SOL. According to this principle, motor units recruited during the SL response will be smaller than those recruited during ML or LL responses because the muscle force is lower. In this case SL-recruited units will be less apparent in our force records, and the early force response is simply dominated by the visco-elastic response of the ankle muscles. To explain the symmetry of the early dorsiflexing and plantar flexing force responses, in contrast to the highly asymmetric visco-elastic responses of cat soleus muscle to stretch and release (Joyce *et al.*, 1969; Nichols and Houk, 1976), a smoothing action by the antagonist ankle muscles must be assumed.

An alternative explanation for the apparent absence of an SL force response was suggested by Tatton and Bawa (1979), namely, that the force contributions of SL and ML coalesce together. They based this assumption on the tendency of slow-twitch motor units to be recruited during SL EMG responses and fast twitch units during ML responses. While this explanation might explain an unobserved SL force response in monkey arm muscles, the argument cannot be extended to human ankle muscles. Based on the twitch properties of human gastrocnemius motor units (Garnett *et al.*, 1978), the majority of triceps surae motor units recruited during the SL EMG response should have their peak contraction some 30 msec prior to the ML EMG response.

The symmetrical early force responses, despite an asymmetry in SL EMG responses, could be taken as perfect stiffness compensation of the type studied by Nichols and Houk (1976). They would reason that the balance of muscle spindle length and Golgi tendon organ force feedback exactly compensates for the asymmetrical visco-elastic responses at the time point when muscle yielding occurs. Since the visco-elastic responses are asymmetric, an asymmetric compensating mechanism is required (cf. Crago *et al.*, 1976). If this assumption is to be accepted then it should be possible to test experimentally whether or not the force response of a muscle stretch which terminates prior to the SL EMG response is fundamentally different from one that terminates after the SL EMG. In the former case, reflex transmission delays prevent afferent feedback from compensating for muscle yielding.

The early force response provided 18% of the correcting force, thus it cannot function as an effective load compensation mechanism. However, the response was symmetrical for stretch and release of triceps surae muscles and equally powerful to the ML force response. Both responses do provide some degree of servo action, albeit insufficient to correct for the disturbance.

What, then, is the function of the ML response if it is unable to compensate for load disturbances? One explanation that has been proposed is an extension of the stiffness regulation hypothesis of Nichols and Houk (1976). Kwan *et al.* (1979) proposed that the ML response provides a background of constant muscle stiffness upon which the correcting LL response is added. An alternative explanation considers the information content of afferent signals more important than their possible segmental and supraspinal reflex actions, and delegates the immediate resistance to the disturbance to the action of inherent muscle elasticity (Allum, 1975). According to this 'servo-information' hypothesis, the ML response is released following the receipt of afferent information indicating a disturbance, to act against the unexpected load. The force required to move the load a given distance is registered by the muscle mechanoreceptors and this information is then used by the CNS to select an LL response appropriate for the situation. The SL EMG response cannot be used for this purpose because of its absence when stretch velocities are low (Gottlieb and Agarwal, 1979), or when the muscle has insufficient prior stretch (Allum and Büdingen, 1979), even though, in the latter circumstances, muscle spindle afferents are known to respond (Burke and Eklund, 1977). In addition, this study has indicated that the SL activity does not produce a distinguishable force response. The stiffness regulation and the servo-information explanations for the function of ML responses need not be considered as mutually exclusive alternatives. It is clear that further experimental testing using additional disturbances during the ML response is required to define the role of ML responses during load compensation.

ACKNOWLEDGEMENTS

The assistance of the technical staff of the Institute for Brain Research is gratefully acknowledged. It is a pleasure to thank Werner Hanselman, Francis Horber and Rudi Kaegi for their valuable help, and Ingrid Roth for typing the manuscript. Valuable comments on the manuscript were provided by Dr Manuel Hulliger.

This research was supported by the Swiss National Science Foundation grants 3.079.76 and 3.585.79, the Dr Eric Slack Gyr Foundation and the Sandoz Foundation in Zurich.

Received on July 1st, 1980.

REFERENCES

Allum, J. H. J. (1975). Responses to load disturbances in human shoulder muscles: The hypothesis that one component is a pulse test information signal, *Expl Brain Res.*, **22**, 307–326

Allum, J. H. J. and Büdingen, H. J. (1979). Coupled stretch reflexes in ankle muscles: An evaluation of the contributions of active muscle mechanisms to human posture stability. In *Reflex Control of Posture and Movement* (eds R. Granit and O. Pompeiano), Elsevier (*Prog. Brain Res.*, **50**, 185–195)

Bawa, P. and Tatton, W. G. (1979). Motor unit responses in muscles stretched by imposed displacements of the monkey wrist, *Expl Brain Res.*, **37**, 417–437

Bizzi, E., Dev, P., Morasso, P. and Polit, A. (1978). Effects of load disturbances during centrally initiated movements, *J. Neurophysiol.*, **41**, 542–556

Burke, D. and Eklund, G. (1977). Muscle spindle activity in man during standing, *Acta physiol. scand.*, **100**, 187–199

Crago, P. E., Houk, J. and Hasan, Z. (1976). Regulatory actions of human stretch reflex, *J. Neurophysiol.*, **39**, 925–935

Garnett, R. A. F., O'Donovan, M. J., Stephens, J. A. and Taylor, A. (1978). Motor unit organization of human medial gastrocnemius, *J. Physiol., Lond.*, **287**, 33–43

Gottlieb, G. L. and Agarwal, G. C. (1979). Response to sudden torques about ankle in man. I: Myotatic reflex, *J. Neurophysiol.*, **42**, 91–106

Gottlieb, G. L. and Agarwal, G. C. (1980). Response to sudden torques about ankle in man. II: Postmyotatic reactions, *J. Neurophysiol.*, **43**, 86–101

Henneman, E., Somjen, G. and Carpenter, D. O. (1965). Functional significance of cell size in spinal motoneurons, *J. Neurophysiol.*, **28**, 560–580

Joyce, G. C., Rack, P. M. H. and Westbury, D. R. (1969). The mechanical properties of cat soleus muscle during controlled lengthening and shortening movements, *J. Physiol., Lond.*, **204**, 461–474

Kwan, H. C., Murphy, J. T. and Repeck, M. W. (1979). Control of stiffness by the medium latency electromyographic response to limb perturbation, *Can. J. Physiol. Pharmacol.*, **57**, 277–285

Marsden, C. D., Merton, P. A. and Morton, H. B. (1976). Stretch reflex and servo action in a variety of human muscles, *J. Physiol., Lond.*, **259**, 531–560

Melvill-Jones, G. and Watt, D. G. D. (1971). Observations on the control of stepping and hopping movements in man, *J. Physiol., Lond.*, **219**, 709–727

Nichols, T. R. and Houk, J. C. (1976). Improvement in linearity and regulation of stiffness that results from actions of the stretch reflex, *J. Neurophysiol.*, **39**, 119–142

Tatton, W. G. and Bawa, P. (1979). Input–output properties of motor unit responses in muscles stretched by imposed displacements of the monkey wrist, *Expl Brain Res.*, **37**, 439–457

Contribution of spinal stretch reflexes to the activity of leg muscles in running

VOLKER DIETZ*

INTRODUCTION

How are stereotyped, complex movements, like those in running, controlled, and which parts of the nervous system are involved in their regulation? Most neurophysical investigations in this field have been performed in the cat (e.g. Engberg and Lundberg, 1969; Grillner, 1975). There are, of course, large differences between the bipedal gait of man and the quadrapedal gait of cats (Dietz *et al.*, 1978). In studies of complex motor tasks like hopping (Melvill Jones and Watt, 1971) and landing from falls (Greenwood and Hopkins, 1976) conflicting observations regarding the contribution of spinal stretch reflexes to the activation of leg muscles in human subjects have been reported. The purpose of this study was therefore to evaluate the relative contributions of spinal stretch reflexes and central 'programmes' to the control of leg muscles during running in man. Recent investigations by our group have supported the idea that segmental stretch reflexes do indeed contribute to gastrocnemius EMG in running subjects (Dietz *et al.*, 1979). The strength of this reflex activity was thought likely to depend on the velocity of muscle stretch, on the basis of observations in decerebrate cats (Nichols and Houk, 1976). We have examined this possibility in man under different conditions of running.

METHODS

The techniques used in these experiments have been extensively described in previous papers (Dietz and Noth, 1978; Dietz *et al.*, 1979, 1980).

*Department of Neurophysiology, University of Freiburg

Gastrocnemius EMG was recorded by telemetry. The rectified EMG was averaged over a number of locomotory cycles, using foot contact to align segments. The resulting averaged EMG records were integrated over the time period 40–100 msec after foot-fall. This integral was then compared with the angular velocity at the ankle, monitored with a goniometer.

In one experimental arrangement, the height of the supporting surface was unexpectedly varied during on-the-spot running. Under these circumstances, changes in the stretch velocities and EMG of gastrocnemius were observed (cf. Noth and Dietz, 1979). In another set of experiments, the athletes were running at different speeds (6, 9 and 12 km/h) on a treadmill. The strength of gastrocnemius EMG after ground contact was compared with the stretch velocity of this muscle for each speed of running.

RESULTS

It was demonstrated in an earlier paper (Dietz *et al.*, 1979) that spinal reflexes contribute to gastrocnemius EMG in the stance phase of running. This spinal stretch reflex activity evidently enhances the activity of the leg extensor muscles for a quick pushing off of the leg and body at the end of the stance phase in running. This assumption is based on three observations (Dietz *et al.*, 1979). Firstly, during running, the electrical activity of the gastrocnemius muscle increased sharply 35–45 msec after ground contact and reached its maximum at the end of muscle stretch. This delay is consistent with the time needed for the segmental stretch reflex to become active. Secondly, the peak level of gastrocnemius EMG in the stance phase of a sprint was normally much higher than the activity during maximum voluntary muscle contraction. Thirdly, the increase in gastrocnemius EMG, 35–45 msec after ground contact, was markedly reduced after a partial blockade of group I afferents by ischaemia. It was shown that impulse conduction of the efferent nerve fibres was apparently unaffected. By these observations, the contribution of spinal stretch reflex activity to the gastrocnemius EMG was qualitatively demonstrated.

The next step in the further investigation of the significance of stretch reflexes during running was to obtain quantitative data about the relationship between pre-programmed and stretch-induced EMG activity of leg extensor muscles in different running conditions. To this end, the activity pattern was analysed for running at different speeds on a treadmill. In another set of trials, the moment of ground contact was unexpectedly varied by varying the height of the ground support during on-the-spot running. Under these conditions different stretch velocities of the contracting gastrocnemius muscle could be obtained.

The ground support was unexpectedly heightened or lowered during on-the-spot running by randomly adding or withdrawing a step of 80 mm height.

A Running even plateau

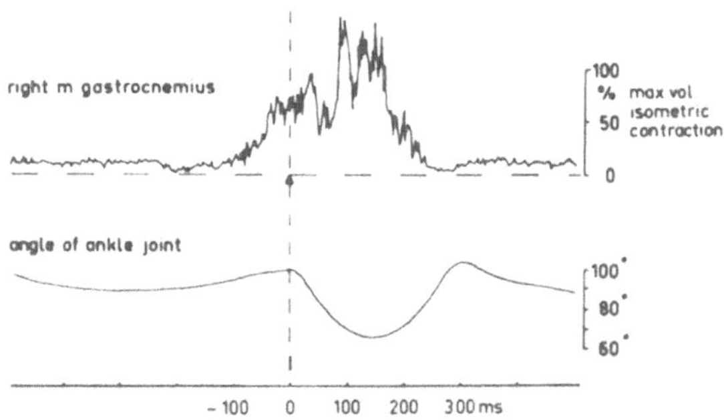

B Randomly added step

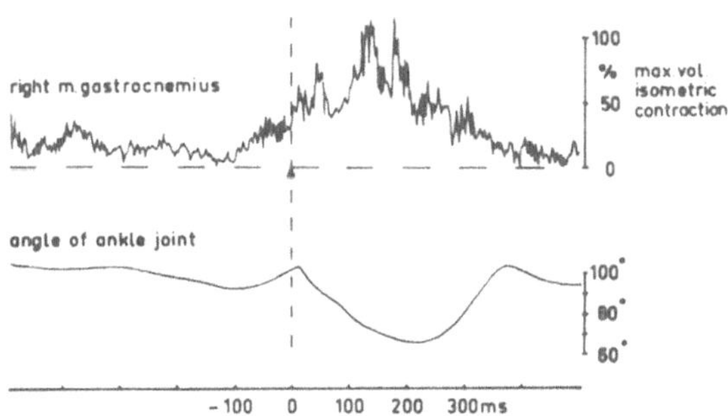

C Randomly withdrawn step

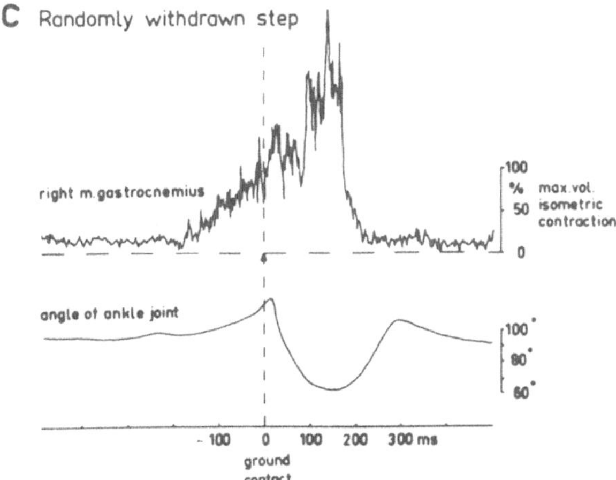

Figure 1. Rectified and averaged (n = 30) gastrocnemius EMG together with the goniometer signal of the ankle joint during on-the-spot running. During running on an even surface (A) the ground contact under the right leg was randomly lifted (B), or lowered (C), by adding or withdrawing a pedestal of 80 mm in height.

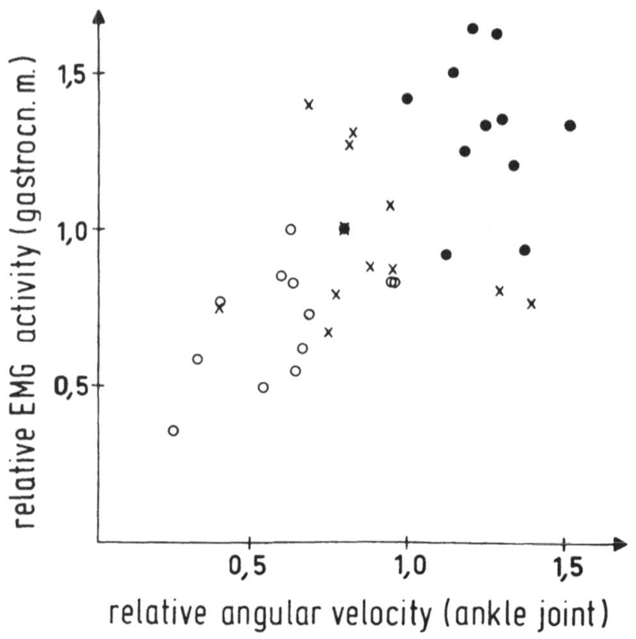

Figure 2. Correlation between the integrated (40–100 msec after onset of stretch), rectified and averaged (n = 30) gastrocnemius EMG and the peak angular velocity of the ankle joint. The results of 12 normal subjects are summarised. The symbols characterise the different running conditions. ×, normal height; ○, unexpectedly higher than ground level; and ● unexpectedly lower than ground level. Values are normalised relative to the reflex strength and the angular velocity of the ankle joint during normal running.

Visual or acoustic information about these changes was excluded. In figure 1 a typical example of the rectified and averaged gastrocnemius EMG and the goniometer signal of the right ankle joint (n = 30) is shown. In A, ground contact for the right foot was at the expected height, in B it was higher (80 mm) and in C lower (80 mm). Because the moment of ground contact was earlier in B and later in C when compared to that in A, the durations of EMG activity prior to landing were shorter in B and longer in C when compared to that in A. The EMG responses in gastrocnemius in all three conditions, however, began about 40 msec after the onset of stretch, i.e. at the onset of dorsiflexion of the foot. Due to differences in mechanical loading in the three conditions, different stretch velocities of the gastrocnemius muscle, and therefore corresponding spinal stretch responses, were obtained. When ground contact occurred later than expected (C), dorsiflexion of the foot was faster than that in A, and this was associated with a larger stretch reflex response. In B, dorsiflexion was slower than that in A, and there was a correspondingly smaller reflex response. This dependency of the stretch

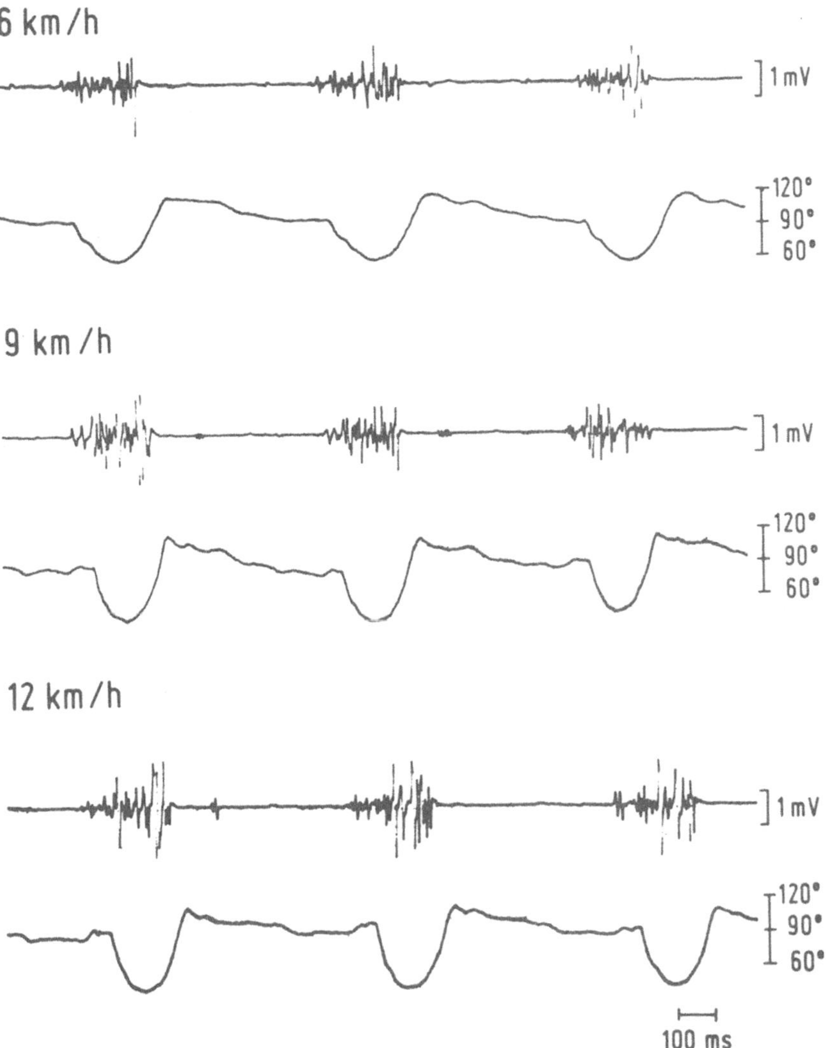

Figure 3. Raw gastrocnemius EMG together with the goniometer signal of ankle angle during running on a treadmill. From top to bottom, running speed was 6, 9 and 12 km/h, respectively.

reflex response on the stretch velocity is summarised in figure 2 on the basis of the results of 12 normal subjects. All coefficients shown in this figure are given relative to the reflex strength (integrated gastrocnemius EMG from 40 to 100 msec after onset of stretch) and the angular velocity of the ankle joint during normal free running. The histogram shows a nearly linear relationship between spinal stretch reflex response and stretch velocity of the gastrocnemius muscle.

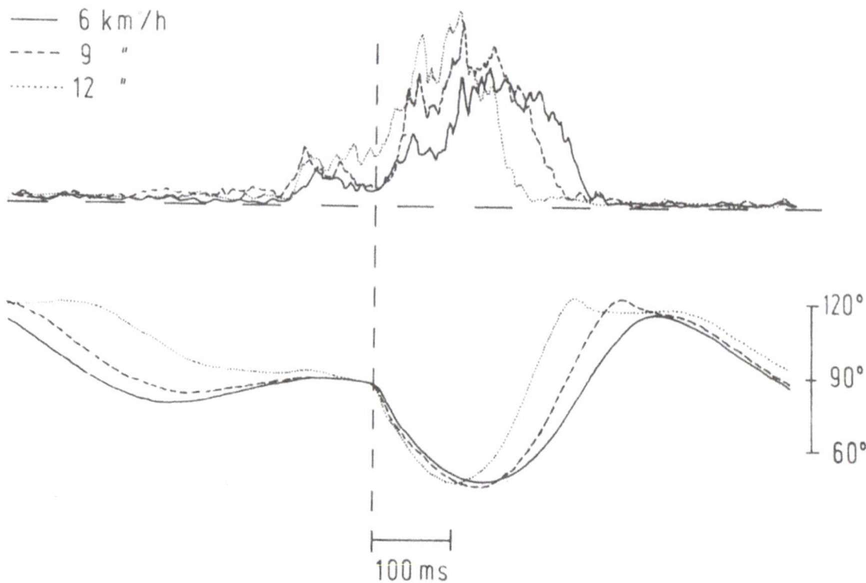

Figure 4. Rectified and averaged (n = 30) gastrocnemius EMG together with the averaged goniometer signal of the running condition described in figure 3. The dashed vertical line indicates the moment of ground contact.

Is there also a correlation between the size of the EMG reflex and locomotion speed? For these experiments the gastrocnemius EMGs of nine trained athletes were analysed during running on a treadmill at 6, 9 and 12 km/h over 3 min. Each period of running was followed by a 3-min rest period. In such trials, the gastrocnemius EMG of 30 running cycles was rectified, averaged and analysed in the usual way. Figure 3 shows a recording of the raw gastrocnemius EMG together with the goniometer signal of ankle joint angle during running at different speeds. In these recordings there are no obvious differences in the EMG patterns, or in the goniometer recordings at the different running speeds. However, when the recordings are averaged (figure 4), systematic differences do indeed emerge. While the amplitude of ankle angle change after ground contact is similar in all three conditions, the peak velocity of change increases from 295°/sec (at 6 km/h) to 340°/sec (at 9 km/h), and to 410°/sec (at 12 km/h). Parallel to this the gastrocnemius EMG, integrated between 40 msec and 100 msec after ground contact, is 1.7 times greater at 9 km/h, and 2.2 times greater at 12 km/h, than that at 6 km/h, although there are only small differences in the amplitudes of EMG activity prior to foot contact. A similar relationship was found in all nine volunteers tested. It can also be seen from figure 4 that even the duration of the enhanced gastrocnemius activity parallels that of the stretch, i.e. it is longer at lower locomotion speed and shorter at higher speed. These results show a similar linear relationship between the stretch velocity of triceps surae and stretch-induced gastrocnemius EMG, as seen in the experiments with added or withdrawn steps.

CONCLUSION

In both experimental conditions, on-the-spot running with a randomly changed ground support and running at different speeds on a treadmill, a clear positive correlation was found between the velocity of foot dorsiflexion and the stretch-induced gastrocnemius EMG activity at the beginning of the stance phase of the running cycle. A similar relationship has been reported in experiments in which dorsiflexion of the foot was applied by a torque motor (Gottlieb and Agarwal, 1979). The tendency for a loss of balance to occur when the ground level was unexpectedly changed was presumably compensated for by the stretch-modulated gastrocnemius activity (among other responses). This rapid response can only be mediated by a segmental pathway, because a long-loop stretch reflex would come too late to become effective in the same running cycle. Therefore it can be concluded that the stretch reflex contribution to the total EMG activity of leg extensors is essential for optimum, ground-adapted muscle activation in running.

The results obtained during running at different speeds demonstrate an increasing contribution of the stretch reflex to the activation of gastrocnemius for increasing speeds. In this situation, too, the stretch reflex has the advantage of optimally regulating the timing and strength of muscle activation. Such fine control for the leg thrust would neither be possible by pre-programming nor by long-loop reflexes or voluntary command. However, the contribution of the spinal stretch reflex is only valuable in the context of the pre-programmed activation pattern (first described in man by Melvill Jones and Watt, 1971) and the mechanical properties of the activated muscle fibres. The significance of the spinal stretch reflex described here is restricted to the activation of the leg muscles during stereotyped movements like running. In other motor performances, in which voluntary control of muscle activity is more important, as for instance in arm muscles, relatively smaller segmental reflex responses were recorded after rapid muscle stretch (Dietz *et al.*, 1980), and indeed in hand muscles, spinal stretch reflexes are often undetectable (Marsden *et al.*, 1972).

SUMMARY

The spinal stretch reflex contribution to the activity of leg muscles under different conditions of running was evaluated. In one experiment, the level of ground contact was unexpectedly varied in height during on-the-spot running.

In another experiment, different speeds of treadmill running were investigated. In both conditions a linear relationship between the speed of dorsiflexion at the beginning of the stance phase and the strength of the reflex response evoked by this stretch, was found. It was concluded that the spinal stretch reflex activity contributed significantly, both in its timing and its strength, to leg extensor EMG during running. This would enable fast adjustments to specific running conditions to be made by the central nervous system.

ACKNOWLEDGEMENT

This work was supported by the Deutsche Forschungsgemeinschaft (SFB 70—Hirnforschung und Sinnesphysiologie).

Received on June 6th, 1980.

REFERENCES

Dietz, V. and Noth, J. (1978). Pre-innervation and stretch responses of triceps bracchii in man falling with and without visual control, *Brain Res.*, **142,** 576–579

Dietz, V., Noth, J. and Schmidtbleicher, D. (1980). Interaction between pre-activity and stretch reflex in human triceps bracchii during landing from forward falls, *J. Physiol., Lond.*, to be published

Dietz, V., Schmidtbleicher, D., Ledig, T. and Noth, J. (1978). Timing of stance and swing phases and occurrence of phasic stretch reflex in running man, *Eur. J. Physiol.*, **373,** Suppl. R71

Dietz, V., Schmidtbleicher, D. and Noth, J. (1979). Neuronal mechanisms of human locomotion, *J. Neurophysiol.*, **42,** 1212–1222

Engberg, I. and Lundberg, A. (1969). An electromyographic analysis of muscular activity in the hindlimb of the cat during unrestrained locomotion. *Acta physiol. scand.*, **75,** 614–630

Gottlieb, G. L. and Agarwal, G. C. (1979). Response to sudden torques about ankle in man: Myotatic reflex, *J. Neurophysiol.*, **42,** 91–106

Greenwood, R. and Hopkins, A. (1976). Landing from an unexpected fall and a voluntary step, *Brain*, **99,** 375–386

Grillner, S. (1975). Locomotion in vertebrates: Central mechanisms and reflex interaction, *Physiol. Rev.*, **55,** 247–304

Marsden, C. D., Merton, P. A. and Morton, H. B. (1972). Servo action in human voluntary movement, *Nature, Lond.*, **238,** 140–143

Melvill Jones, G. and Watt, D. G. D. (1971). Observations on the control of stepping and hopping movements in man, *J. Physiol., Lond.*, **219,** 709–727

Nichols, T. R. and Houk, J. C. (1976). Improvements in linearity and regulation of stiffness that results from actions of stretch reflex, *J. Neurophysiol.*, **39,** 119–142

Noth, J. and Dietz, V. (1979). Spinal stretch reflexes in self-initiated falls and in running movements, *Agressologie*, **20B,** 159–160

A critique of the papers by Houk, Crago and Rymer; Hoffer and Andreassen; Allum; and Dietz

PETER M. H. RACK*

The stretch reflex has often been regarded as part of a system for the servo-control of limb position, through which a discrepancy between the actual muscle length and an intended length generates activity that minimises this error. If it is to work effectively, a servo-control system of this type must have a high gain around the feedback loop, so that a small error leads to a powerful correction; all the available measurements and calculations of this loop gain do, however, suggest that it is rather disappointingly low (Matthews, 1972; Vallbo, 1974) and this apparent shortcoming has sometimes puzzled physiologists.

Although some slightly different mechanisms of servo-control have been proposed, and various different neural pathways may be employed (Hammond, 1960; Phillips, 1969; Evarts, 1973), none of these can be expected to maintain an effective control of limb position if they generate only a relatively feeble correcting force. We are left therefore in a rather unsatisfactory position; there is ample anatomical and physiological evidence that the stretch reflex pathways exist, but a quantitative examination seems to indicate that they cannot do the job that we have frequently attributed to them.

Of the four contributors to this publication who address the problem, Allum (1981) concludes that the stretch reflex responses do little to correct a displacement and he suggests that they have another quite different function. Dietz (1981) suggests that the stretch reflex may perform a useful function when it operates in the course of an ongoing movement. The other two (Houk *et al.*, 1981; Hoffer and Andreassen, 1981) examine the possibility that the stretch reflex may in fact be functioning effectively, but as a servo-controller of stiffness rather than of muscle length.

* The Medical School, The University of Birmingham

SERVO-CONTROL OF STIFFNESS?

If Houk is correct in his suggestion that muscle stiffness is the controlled variable (Crago *et al.*, 1976; Nichols and Houk, 1976; Houk, 1979), then we have indeed been looking for the wrong thing, and the problem of the apparently low reflex gain may disappear since a sudden muscle extension or limb displacement would only be expected to generate a sufficient reflex resistance to keep the stiffness (force/displacement) constant. The gain of the reflex should then be measured by the stiffness that the reflex contributes in response to a departure from the 'intended' muscle stiffness.

It was suggested (Rack, 1970) that the short-range stiffness of an active muscle might often provide an effective resistance to sudden extensions until such time as the stretch reflex could activate the muscle to provide a further resistance; the extreme sensitivity of the muscle spindles to the first part of a movement would ensure that the reflex activity began as early as possible, and would to some extent mitigate the effects of the reflex delay (Joyce and Rack, 1974). The experimental results reported by Houk *et al.* (1981) indicate that in the soleus muscle of the decerebrate cat there is indeed a smooth transition from the inherent muscle resistance to a reflex resistance, and during the ramp extension that they used, the combination of these two effects did indeed keep the muscle stiffness at an approximately constant level.

Can we then accept that the concept of the stretch reflex as a stiffness controller explains the measurements and calculations that were hitherto taken to indicate so low a reflex gain? I do not think we can, because we cannot safely generalise from results obtained during ramp extensions of decerebrate cat muscles to other animals under other circumstances. In Allum's experiments (Allum, 1981) the first rather small reflex force occurred at a long interval after displacement of the human ankle, at least 170 msec after the movement; this result is compatible with the EMG responses recorded by other workers (Melvill Jones and Watt, 1971; Chan *et al.*, 1979). This interval is very different from the 20 msec recorded in the decerebrate cat soleus (Houk *et al.*, 1981) and it is no surprise that at the human ankle there is a considerable interval after the initial muscle resistance before the reflex force begins. Although the movements generated in Allum's experiments were very small ones, it is difficult to see how a reflex force which occurs so late could maintain a constant stiffness through the course of a longer movement.

THE CONSEQUENCES OF A REFLEX DELAY

When one looks at the possible consequences of a long reflex delay, it ceases to be a surprise that the gain is small. Consider the behaviour of a (theoretical) reflex pathway in which a large loop gain is combined with a long delay—any displacement would be met by a powerful correcting force, but this would occur late, at some interval after the displacement had begun and it might continue to act for some time after the displacement had been corrected. The supposed correction would thus be very likely to become an over-correction with a movement through the starting position in the opposite direction which would in turn initiate another reflex correction. The system would then be in danger of breaking into an oscillation in a way that is all too familiar to control engineers.

The factors that determine the stability of systems of this sort are fully dealt with in engineering textbooks, but for our purposes it is sufficient to make two (somewhat simplified) points: (1) when the delay is long, oscillations are likely to occur at lower frequencies than when the delay is short; (2) continuous or increasing oscillations occur when the reflex pathway has a high gain for signals at these potentially oscillatory frequencies. The truth of these two statements may be seen when a human subject exerts a high force against a suitably yielding load (Joyce and Rack, 1974). The increase in force is accompanied by an increase in the stretch reflex gain (Marsden *et al.*, 1972) and this may lead to a spontaneous oscillation or clonus. At the elbow joint, which is largely controlled by a fast spinal reflex, this clonus has a frequency of 8–12 Hz (Joyce and Rack, 1974), whereas the thumb interphalangeal joint which is controlled mainly by a slower reflex pathway (Marsden *et al.*, 1976) undergoes spontaneous clonus at 4–6 Hz (Brown *et al.*, 1977).

Engineers employ a variety of methods (differentiation, phase-advance) to raise the natural frequencies of their feedback system above the range in which they intend it to operate; they then use selective filtering to reduce the gain at these higher, potentially oscillatory frequencies, while retaining the lower frequency responses that they need. We see evidence that animals employ similar methods; the muscle spindles act as differentiators (*see* Matthews, 1972) and the reflex signal is further phase-advanced as it traverses the spinal cord (Jansen and Rack, 1966; Westbury, 1971). Often, however, the delays are so long that a natural frequency of the stretch reflex still remains within the frequency range where we regularly make spontaneous movements (Rack, 1981) and although the muscles do act as low-pass filters (Marshall and Walsh, 1956; Bawa and Stein, 1976) there can be no possibility of retaining a high feedback gain throughout the working range without precipitating a spontaneous clonus. The rather low gains of the stretch reflex are therefore a consequence of the long delays in their pathways. If the gains were not low, the reflexes would not be stable.

Although it may at first have seemed a disappointment that animals do not use the simpler forms of high-gain feedback control with all their attendant advantages, it makes more sense to put the facts the other way round and say that relatively large animals with their long reflex delays have increased their speed and facility of movement beyond the point at which they could operate a simple high-gain servo-mechanism and keep it stable.

These long reflex delays impose the same restrictions on the gain of a servo-mechanism whether it is for the control of muscle length, stiffness or force. In larger animals, therefore, the stretch reflex could not provide a very effective correction for 'errors' in muscle stiffness and it is not surprising that Hoffer and Andreassen (1981) found their different movements were met by differing muscle stiffnesses.

THE SERVO-INFORMATION HYPOTHESIS

In the experiments reported by Allum, subjects exerted a steady flexing force at the ankle until the sudden additional disturbing force was applied. The immediate reflex resistance to the disturbance was quite small, a result that confirms the low gain of the stretch reflex. Allum (1975, 1981) suggests that the stretch reflex response should be looked upon as a test signal rather than one that can by itself correct a disturbance. He postulates that the nervous system uses the afferent response to the efferent reflex burst as a source of information about the state of the limb and its muscles.

Allum's suggestion is in two parts which can be taken separately. One can hardly disagree with the suggestion that the nervous system will probably make use of any information that is obtained as a result of the stretch reflex action. Indeed, one would expect the nervous system to make use of information from all available sources in computing its future motor activities, but do we have to abandon any idea that the stretch reflex may do something effective in its own right and accept that its main function is to provide this test signal?

The long delays in the reflex pathway seem to present a serious obstacle to the servo-information hypothesis; does the animal whose limb is subjected to a sudden disturbance defer any effective correction until it has despatched a test signal to the muscles and waited for the response before computing and sending out a formal correcting signal? This would involve a doubling of the delays in the peripheral nerves and muscles. In Allum's experiments the subject maintained a steady force before the disturbances so that some of the motoneurons must already have been activated. Could not the nervous system use the response to this continuing activation as a source of information about the state of the muscles? Why does it need to wait for the response to the particular burst of stretch reflex activity?

Returning to the question of whether the stretch reflex can itself provide a useful response to a limb displacement—in the experiments of Allum and of most other workers, the limb is held in some steady state until the disturbance occurs; the stretch reflex does not then have, and cannot have, sufficient gain to provide an effective resistance. There is, however, the possibility that the stretch reflex could provide a more useful response to perturbations if they occurred during the course of some other planned movement.

STRETCH REFLEXES DURING THE COURSE OF A MOVEMENT

Although a stretch reflex that had its gain fixed at a high level would be likely to break into oscillations, this potential complication need not be so serious if the gain of the stretch reflex is increased to high levels for short periods only, since there would then be little opportunity for oscillations to build up. It seems appropriate therefore to look for evidence of an enhanced stretch reflex during brief periods and perhaps during particular activities. The increases in H reflexes that sometimes accompany or precede voluntary movements (Gurfinkel and Pal'tsev, 1965; Gottlieb *et al.*, 1970) strengthen this suggestion and imply that the gain increase may be brought about by an alteration in the motoneuron excitability.

Dietz (1981) describes results obtained from human subjects during running; his results show that unexpected changes in the surface beneath the feet are followed by changes in the calf muscle EMG. These changes follow extension of the calf muscles with a timing that indicates a spinal stretch reflex. This is an encouraging result so far as it goes, but unfortunately it is very difficult, and perhaps impossible, to estimate from the EMG signals how much force can be attributed to the reflex action (Calvert and Chapman, 1977). During an impulsive movement, or during the response to an impulsive disturbance, the electrical responses of a number of near synchronous motor units may add together in ways that produce larger surface EMG signals than one would see if they were activated independently.

One cannot therefore draw any very clear quantitative conclusions from the work of Dietz, but it is tempting to speculate that a main function of the stretch reflex is to provide a pathway through which the animal can, during chosen short periods, generate an effective and rapid resistance to a disturbance. If this is so, then experiments which merely examine the reflex responses of a limb that is held in a steady state may be eliciting something that is only a feeble attenuated version of a reflex that under other circumstances has powerful and important functions.

If, as seems possible, the effectiveness of the stretch reflex is being continuously changed in anticipation of possible disturbances, then the term 'gain' ceases to be as useful as it is in some of the simpler forms of servo-control theory, and indeed, descriptions of the stretch reflex in terms of the simpler engineering systems may be too restrictive to be helpful.

Received on August 7th, 1980.

REFERENCES

Allum, J. H. J. (1975). Responses to load disturbances in human shoulder muscles: The hypothesis that one component is a pulse test information signal, *Expl Brain Res.*, **22**, 307–326

Allum, J. H. J. (1981). The control of human ankle position by stretch reflexes. This publication

Bawa, P. and Stein, R. B. (1976). Frequency response of human soleus muscle, *J. Neurophysiol.*, **39**, 788–793

Brown, T. I. H., Rack, P. M. H. and Ross, H. F. (1977). Tremor in the human thumb, *J. Physiol.*, **269**, 3P–4P

Calvert, T. W. and Chapman, A. E. (1977). The relationship between the surface e.m.g and force transients in muscle: Simulation and experimental studies, *Proc. Instn electl electron Engrs*, **65**, 682–689

Chan, C. W. Y., Melvill Jones, G., Kearney, R. E. and Watt, D. G. D. (1979). The 'late' electromyographic response to limb displacement in man. 1: Evidence for a supraspinal contribution, *Electroenceph. clin. Neurophysiol.*, **46**, 173–181

Crago, P. E. J., Houk, J. C. and Hasan, Z. (1976). Regulatory actions of human stretch reflex, *J. Neurophysiol.*, **39**, 925–935

Dietz, V. (1981). Contribution of spinal stretch reflexes to the activity of leg muscles in running. This publication

Evarts, E. V. (1973). Motor cortex reflexes associated with learned movements, *Science, N.Y.*, **179**, 501–503

Gottlieb, G. L., Agarwal, G. C. and Stark, L. (1970). Interactions between voluntary and postural mechanisms of the human motor system, *J. Neurophysiol.*, **33**, 365–381

Gurfinkel, V. S. and Pal'tsev, Ye. I. (1965). Effects of the state of the segmental apparatus of the spinal cord on the execution of a simple motor reaction, *Biophys.*, **10**, 944–951

Hammond, P. H. (1960). An experimental study of servo action in human muscular control. In *Proc. 3rd Conf. Med. Electron., London*, 190–199

Hoffer, J. A. and Andreassen, S. (1981). Limitation in the servo-regulation of soleus muscle stiffness in premamillary cats. This publication

Houk, J. C. (1979). Regulation of stiffness by skeletomotor reflexes, *A. Rev. Physiol.*, **41**, 99–114

Houk, J. C., Crago, P. E. and Rymer, W. Z. (1981). Function of the spindle dynamic response in stiffness regulation—a predictive mechanism provided by non-linear feedback. This publication

Jansen, J. K. S. and Rack, P. M. H. (1966). The reflex response to sinusoidal stretching of soleus in the decerebrate cat, *J. Physiol., Lond.*, **183**, 15–36

Joyce, G. C. and Rack, P. M. H. (1974). The effects of load and force on tremor at the normal human elbow joint, *J. Physiol., Lond.*, **240**, 375–396

Marsden, C. D., Merton, P. A. and Morton, H. B. (1972). Servo action in human voluntary movement, *Nature, Lond.*, **238**, 140–143

Marsden, C. D., Merton, P. A. and Morton, H. B. (1976). Servo action in the human thumb, *J. Physiol., Lond.*, **257**, 1–44

Marshall, J. and Walsh, E. G. (1956). Physiological tremor, *J. Neurol. Neurosurg. Psychiat.*, **19**, 260–267

Matthews, P. B. C. (1972). *Mammalian Muscle Receptors and Their Central Actions*, London, Arnold

Melvill Jones, G. and Watt, D. G. D. (1971). Observations on the control of stepping and hopping movements in man, *J. Physiol., Lond.*, **219**, 709–727

Nichols, T. R. and Houk, J. C. (1976). Improvement in linearity and regulation of stiffness that result from actions of stretch reflex, *J. Neurophysiol.*, **39**, 119–142

Phillips, C. G. (1969). Motor apparatus of the baboon's hand, *Proc. R. Soc. B*, **173**, 141–174

Rack, P. M. H. (1970). The significance of mechanical properties of muscle in the reflex control of posture. In *Excitatory Synaptic Mechanisms* (eds P. Andersen and J. K. S. Jansen), Oslo, Universitetsforlaget, 317–321

Rack, P. M. H. (1981). Limitations of somatosensory feedback in control of posture and movement. In *Handbook of Physiology. The Nervous System*, Vol. II: *The Motor System*, Washington D.C., Am. Physiol. Soc.

Vallbo, A. B. (1974). Human muscle spindle discharge during isometric voluntary contractions. Amplitude relations between spindle frequency and torque, *Acta. physiol. scand.*, **90**, 319–336

Westbury, D. R. (1971). The response of motoneurones of the cat to sinusoidal movements of the muscles they innervate, *Brain Res.*, **25**, 75–86

REPLY TO RACK'S CRITIQUE BY HOUK, CRAGO AND RYMER

Dr Rack suggests that reflexes could not be fast enough and powerful enough to maintain either muscle length or stiffness, due to the limitations on high gain in a servo loop that has time delays. While this logical deducation is interesting, it does not address the direct experimental evidence that *stiffness is well regulated* under many conditions (Nichols and Houk, 1976). Also, Rack's discussion of feedback gain appears not to consider that phase lag and instability problems may be less severe in the tendon organ loop, since lags associated with mechanical loads will be bypassed. Finally, the data contained in our paper suggest that *non-linear feedback* may be one of the most important mechanisms for alleviating the problems of long delays raised by Rack.

Received on November 13th, 1980.

Reference

Nichols, T. R. and Houk, J. C. (1976). *J. Neurosphysiol.*, **39**, 119–141

REPLY TO RACK'S CRITIQUE BY ALLUM

Dr Rack should be complimented on the extremely clear application of control engineering principles to current thinking on the significance of reflexes mediated by muscle afferents which he used as a framework for his critique. Briefly, I wish to answer a question he raised, and to convey a point of information. First, why, under the servo-information hypothesis, does the CNS need to wait for afferent activity following the medium latency (ML) muscle activity? Simply, because the CNS cannot estimate, separate from the force of impact, the parameters (stiffness, mass, etc.) of the new load accompanying a disturbance without imposing a counteracting force—however small. Second, Rack asks if it is appropriate for the stretch reflex gain to be increased during a movement. Interestingly, Gottlieb and Agarwal (1980) recently showed that EMG gain increases during voluntary plantar flexion, consistent with the gain change expected from the equivalent tonic contraction. For 100 to 200 msec after an imposed stretch, the gain decreases below the level obtained from the relaxed stationary triceps surae (TS) muscles.

Received on November 4th, 1980.

Reference

Gottlieb, G. L. and Agarwal, G. C. (1980). *J. Neurophysiol.*, **44**, 233–246

A critique of the papers by Houk, Crago and Rymer; Hoffer and Andreassen; Allum; and Dietz

J. A. STEPHENS*

This critique is addressed primarily to the papers by Dietz and Allum. Both are concerned with finding functions for the stretch reflex in the lower limb in man.

In the first paper, Dietz concludes that spinal stretch reflex activity contributes significantly both in its timing and its strength to leg extensor muscle activity during running. Furthermore this activity is modulated in amplitude according to the rate and extent of muscle lengthening and in a way entirely appropriate for regulating the rate and extent of changes in joint angle. Recordings made during unexpected footfall show stretch-related reflex responses to be acting in a compensatory manner.

In the second paper, Allum concludes that the simple short latency stretch reflex does not produce a useful force output for load compensation in human ankle muscles.

At this point I am reminded of words of advice given to me by Doug Stuart when we were struggling over the question of how many different types of motor units there were in cat medial gastrocnemius—'avoid false controversies'. I will do my best, but I have to state my prejudice, which is that the mechanisms producing the short latency spinal stretch reflex I demonstrate to the students each year in the class must have some functional significance.

There will, of course, be those that argue that the ankle jerk, or knee jerk, or for that matter any jerk, are contrivances of the physiology laboratory. That may be true, but the reflex contractions so produced are powerful and generated, we assume, through a fast monosynaptic pathway from spindle endings to motoneurons innervating the receptor-bearing muscle. The surprising fact remains that the function of this, the best known and most extensively studied of all reflex pathways in movement control, is still a matter of debate. The level of our ignorance is made the more notable when we recall

* Sherrington School of Physiology, St Thomas's Hospital Medical School

that the fast monosynaptic pathway from spindle ending to motoneuron is probably the only monosynaptic reflex in the body. Simple anatomy does not lead to simple physiology.

Now of course muscle stretch does not only produce activity in spindle endings. Indeed the afferent input generated by muscle stretch is hard to predict and depends on the amount of active tension in the muscle when it is stretched, and on the rate and extent of stretch. In the relaxed subject one can suppose that the rapid short-duration muscle stretch caused by a tendon tap produces a relatively pure volley of impulses in primary and secondary spindle afferent fibres, producing a sharp EPSP capable of bringing many motoneurons to firing and hence produce a good-sized reflex contraction.

The situation may be more complicated when a muscle is active when stretched. Nevertheless, even when the ankle extensors are active, a sharp tap to the Achilles tendon will still elicit a good-sized reflex synchronous muscle action potential in these muscles. Under these circumstances quick muscle stretch can be expected to result not only in the excitation of spindle endings but also in the excitation of tendon organs responding to the sharp increase in muscle force following sudden muscle stretch. The fact that a stretch reflex can still be recorded shows that the net reflex effect of the mixed afferent volley is still excitatory. Continuing this line of thought one could imagine that there might be a situation in which the mixed afferent volley is dominated by tendon organ discharge. In this case muscle stretch might produce a brief period of muscle relaxation.

All this may be summarised by the simple statement that the reflex response of a muscle to stretch depends on the nature of the mixed afferent cocktail generated by that stretch. This simple fact is often overlooked by those who study the reflex effects of muscle stretch and is at the root of a false controversy which surrounds this subject.

Continuing this line of thought a little further we should also remember that the reflex response of a muscle to stretch also depends very much on the temporal pattern of afferent discharge generated by that stretch. Often EMG recordings show that the reflex response of a muscle to stretch consists of a series of little bursts of activity, often referred to as 'the 3Ms', in honour, perhaps, of the three authors most closely associated with this type of work. Traditionally these different latency components have been attributed to activity in different central pathways. While there may be some truth in this, such interpretations must now be more critically received. This fact was brought home to me most forcibly earlier this year when Prochazka showed me a record of the firing of a muscle spindle primary ending in a conscious cat obtained while the receptor-bearing muscle was subject to varying rates and extents of ramp and hold stretch. It was clear from these recordings that the afferent ending produced one or more separate bursts of impulses depending on the rate and extent of stretch. Significantly, each of these bursts was followed by a burst of EMG at short latency in the receptor-bearing muscle.

The pattern of EMG bursts was remarkably reminiscent of the 3M type of response. Also in this symposium Hagbarth showed a beautifully clear recording from a spindle ending in man showing discrete bursts of firing following ramp and hold stretch and in a temporal pattern that would again be appropriate to account for each of the bursts of reflex EMG in a typical 3M type of response. Proper deduction of the mechanism underlying a reflex response demands that the nature and temporal sequence of the afferent input should be properly defined. In the absence of such information, theories built simply on consideration of the different phases of an EMG response are built on flimsy foundations.

I find little to comment about in Dietz's paper. The only point concerns his conclusions about the pathways involved which produce the increase in muscle electrical activity in gastrocnemius that follows foot contact. It is true that the response has a latency similar to the latency of the tendon jerk for this muscle and that its magnitude depends on the velocity of muscle stretch. The problem in my mind is that it is implied that because the response has the same latency as the tendon jerk, it is generated along the monosynaptic spindle pathway. This might be true for the very earliest part of the increase in EMG, but other spinal and supraspinal mechanisms presumably join in and make their contribution as time goes on and muscle stretch continues. This is not simply a semantic point, as the use of imprecise nomenclature leads to woolly thinking.

In his discussion, Dietz states that activity in longer latency pathways would arrive too late to become effective in the same running cycle. To my mind the EMG burst shown in Figure 1 is long enough to be due in part to longer latency pathways, possibly involving supraspinal structures. The fact that a response begins at spinal latency is not evidence that it is entirely of spinal origin.

Another semantic point lies in the use of the well-worn phrase, 'long loop stretch reflex activity'. This should read 'long latency stretch evoked activity'. The words 'loop' and 'reflex' carry with them implications beyond the scope of observation. A long latency response need not have a long distance pathway, and 'reflex' is a word normally reserved for a more or less stereotyped reaction to a particular stimulus. The long latency responses that follow muscle stretch are far from stereotyped and would be better described by the neutral word 'response'.

Dietz ends his discussion by comparing his present results in the leg during running with those obtained by himself and others in the upper limb, and concludes that stretch reflexes might be less important for muscles taking part in voluntary rather than postural or locomotor tasks. This generalisation is not new and is based on two main experimental findings. Firstly, finger jerks are relatively difficult to elicit in the relaxed subject and this would indicate that the strength of the monosynaptic spindle afferent motoneuron pathway is relatively weaker for the finger flexor and extensor muscles than in the lower limb. This difference may, however, be more apparent than real. For a tendon jerk to be elicited, the incoming excitatory volley set up by the transient rapid muscle stretch must be sufficiently large to cause the resting membrane potential of some motoneurons to cross their threshold for firing. Assuming, for example, that 10 mV depolarisation is required then a volley producing 9 mV depolarisation will not produce any detectable reflex response. In such a case a powerful reflex connection remains completely undetected and its potential importance seriously underestimated. The power of the reflex remains hidden until the motoneurons are firing, as for example during a voluntary contraction. Under these circumstances and assuming a linear membrane potential trajectory between spikes, the level of motoneuron depolarisation is within 9 mV of threshold nine-tenths of the time. Unlike in the relaxed subject, the true strength of the excitatory effect of the afferent volley set up by the tendon tap is now properly revealed. Tendon jerks can, in fact, quite readily be elicited in the finger flexor and extensor muscles and for that matter in the small intrinsic muscles of the hand, the only requirements being that the muscle be active and a stimulus relatively potent for Ia endings be used, such as, for example, a short-duration tap applied to the muscle belly.

The second piece of evidence which is used to deny the importance of the stretch reflex for the control of finger movements is the fact that, during voluntary muscle contraction, sudden changes in muscle length produced by sudden changes in external load are not accompanied by obvious short latency changes in the EMG of the contracting muscles. Those changes in muscle electrical activity that do take place occur 20–30 msec later than would be expected for a simple spinal reflex. Conventional thinking would therefore imply that the spinal stretch reflex is not important for the control of these movements and plays no part in any load-compensating mechanisms. To my mind this conclusion is potentially misleading. A more correct interpretation of events would be that the summed reflex effects of the mixed afferent input on motoneuron firing generated as a result of the peripheral disturbance is neutral. The stretch reflex is not a single entity in the sense that it is mediated along a single afferent and efferent pathway, as is, for example, the corneal reflex, rather it is the final expression of the action and interaction of a number of different individual reflexes.

The title of this section of the symposium was 'Reflexes Mediated by Muscle Afferents'. For me this can only be deduced by a careful study of the separate reflex effects of the different afferents using carefully controlled stimuli to produce a more or less well-defined and known afferent input. It seems to me unreasonable to suppose that we will understand the function of muscle receptor reflexes when we test the system using procedures which produce an input consisting of an indeterminate mixture of spindle and tendon organ and cutaneous input, and with no knowledge of the temporal pattern of firing during the test. The black box approach to the study of stretch reflexes and motor control made popular by Merton in relation to his servo hypothesis has had an important place in the history of our subject, but the continuing repetition of those same experiments does not guarantee their importance in contemporary physiology.

Allum's approach to the study of the significance of reflexes mediated by muscle afferents is straightforward and familiar enough. Subjects are required to maintain a constant angle at the ankle joint against a torque motor arranged with feedback to behave like a stiff spring-type load. From time to time the current in the motor is increased, thereby changing the angle of the ankle joint. Surface EMG recordings are made from the skin overlying tibialis anterior and soleus muscles. Other transducers measure the force exerted by the subject on the foot platform and platform acceleration. In figure 1 we see the results of a single trial in which the torque motor is driven in such a way as to decrease the ankle angle by about 2°. The ankle angle then remains fairly constant in the face of the continuing motor torque. Finally, after about 200 msec the ankle angle begins to return to the desired position and overshoots at about time 500 msec.

The 'muscle force' recording shows three phases. The first increase in force is attributed to the visco-elastic response of the stretched and relaxed ankle muscles. This conclusion is reached on the basis that from the onset of the disturbance to the onset of the second phase of the mechanical response, the 'muscle force' recording resembles the time course of the platform displacement. To my mind this conclusion, denying any reflex component, is insufficiently supported by experiment to be acceptable. The author himself seems puzzled, commenting on the fact that the well-defined short latency reflex response in soleus is not followed by an observable increase in 'muscle force'. The mystery deepens when smaller bursts of EMG activity from the same electrodes are followed by big changes in 'muscle force' later in the same record. Similar problems exist for the understanding of figure 2 in which large changes in 'muscle force' take place in the absence of changes in soleus EMG.

The reader will find a number of explanations for these findings in the discussion section of the paper but to my mind they are most easily explained by the fact that the force exerted by the foot on the platform is a complicated function of the different force outputs of all the muscles acting around the ankle joint. Central to the theme of the paper is the idea of trying to separate intrinsic from reflex muscle forces as they take part in load compensation. This can only be properly achieved if force and EMG are recorded in isolation from the same muscle. The author attaches great significance to the fact that he finds no increase in 'muscle force' consistent with the occurrence of short latency soleus EMG activity. This finding and the lack of association between muscle electrical activity in soleus and force in figure 2 suggest to me that most of the force exerted on the foot plate in these experiments is generated by activity in muscles other than soleus. Under these circumstances changes in soleus EMG cannot be used safely to interpret the mechanical records.

Interpretation is made more difficult by the lack of a zero on the averaged EMG records. The significance of fluctuations in the averaged EMG signal are hard to judge in the absence of this reference.

Another point to emerge from figure 2 is the fact that averaging a few trials leads to a rather less segmented pattern of force and EMG responses. The triphasic picture shown in figure 1 becomes essentially diphasic in figure 2. This is a particularly important point for discussion. Somewhere within the first 600 msec occurs the subject's willed intervention. The timing of this event will inevitably vary from trial to trial and this presumably accounts for the smoothing out of the record between 250 msec and 450 msec. Averaging results in the loss of significant details for mechanisms in individual trials if the timing of sequential events in each trial is variable. Echoing an earlier comment, we should also not forget the role of the sequential temporal pattern of afferent input following muscle stretch in determining the form of the reflex response. This point may be particularly relevant when labelling more or less poorly defined bursts of EMG as short latency (SL) and medium latency (ML) without knowledge of the timing of bursts of afferent input accompanying muscle stretches of this sort.

In his discussion Allum returns to a point made in his introduction that the concept of the servo-regulation of load disturbances is built on the premise that the forces generated by medium latency EMG responses should be powerful enough to return the limb to its original position. This is not the case. The words 'servo-regulation' of load disturbance do not imply perfect compensation, rather they imply that the limb moves less than it otherwise would in the presence of a disturbing force. Without errors servos do not work. It is true that, in the experiments described, the reflex responses observed were not of sufficient power to correct for the load disturbance, but that is not to deny the presence of a load-compensating servo system. I would argue that 18% at short latency and another 18% at medium latency represents a very effective automatic compensating response to be built upon by a final judged response of the subject.

A final point to be raised in relation to Allum's paper is the size of the external disturbance used in his study to elicit load-compensation reflexes. These produced only about 2° changes in ankle angle over a period of about 70 msec. This should be compared with an ankle angle change of some 25° over the same time period recorded by Dietz during running on the spot. The rates and extents of ankle angle change produced in Allum's study may not have been sufficient to elicit the load-compensation reflexes that were sought. The responses observed were perhaps more in the nature of responses to small jolts lasting a few milliseconds rather than those that would be produced by changes in load likely to be experienced in everyday life.

In view of the preceding critique by Rack, I have only one comment to make about the papers of Houk, Crago and Rymer, and Hoffer and Andreassen on the question of the regulation of muscle stiffness. Using simple engineering parlance, the controlled variable in a negative feedback system is determined by the nature of the feedback signal. The lower motoneuron is subject to an enormous variety of different sorts of feedback signals derived from activity in an enormous number of different afferents from different sites and of different modality. On this basis, alone, the search for *the* controlled variable in the face of the known enormous diversity of synaptic input to motoneurons is naïve and places the experimenter in a conceptual strait-jacket. Most of the feedback signals reach the lower motoneuron by way of more or less complicated, interacting and segmental interneuronal systems. My impression from the literature would be that these interneuronal systems and their descending control are perfectly designed to be able to match the content of the feedback signals reaching the lower motoneuron to be appropriate for the motor task to be carried out. On some occasions we can imagine that muscle length is an important controlled variable. On other occasions it might be muscle force, in which case the efficacy of force feedback might be increased and length feedback reduced. Sometimes stiffness may be controlled, in which case force and length feedback will be important. The possible combinations are infinite. No two decerebrate cats are the same. Certain interneuronal paths will be more or less active in some than others. The opportunity for different results and the finding of different controlled variables in different laboratories is enormous and the scope for generation of false controversies endless.

Our subject will be advanced not by those seeking to support or deny unlikely rigid hypotheses, but by those who rejoice in the complexity of the system, and who seek within it, function expressed in a useful and flexible way matched to circumstances.

REPLY TO STEPHENS'S CRITIQUE BY HOUK, CRAGO AND RYMER

Dr Stephens raises the possibility that motor servo function may involve a variety of 'control variables', rather than a single one—stiffness. While this question is off the topic of our present paper, the experimental evidence regarding this issue has been discussed in two recent reviews (Houk, 1978b, 1979). It is concluded that motor servo function may be much less 'flexible' than was formerly believed. While we do not deny the potential merits of flexible control, supported by Stephens on strictly teleological grounds, the absence of clear evidence for these ideas warns us to be cautious about adopting them. It is also important to point out that the theoretical basis for stiffness regulation (cf. Houk, 1978a) is not at all rigid and can easily accommodate most of the 'flexibility' wishes of Stephens.

Received on November 13th, 1980

References

Houk, J. C. (1978a). *Cybernetics 1977* (ed. G. Hauske and E. Butenandt), Munich, Oldenbourg, 35–46

Houk, J. C. (1978b). *Progress in Clinical Neurophysiology*, Vol. 4 (ed. J. E. Desmedt), Basel, Karger, 193–215

Houk, J. C. (1979). *Ann. Rev. Physiol.*, **41**, 99–114

REPLY TO STEPHENS'S CRITIQUE BY ALLUM

Unfortunately, Dr Stephens, in his critique, has succeeded in confusing a number of issues which could be resolved experimentally rather than by talented penmanship. Perhaps the false controversy that he raises is that a medium latency (ML) stretch response at 120 msec which may have some functional significance must necessarily be effective in correcting a disturbance. Clearly, there are two issues here: (1) functional significance and (2) effective compensation. To term the 18% force compensation of the ML response 'very effective' is just juggling with words.

According to Dr Stephens our understanding of human motor reflexes is best advanced by examining the different components of the afferent 'cocktail'. This truism is not followed by an explanation of the technical difficulties of microneurography which would dominate a study of reflex responses to load disturbances, nor are the types of load disturbances which would excite only one type of muscle afferent described. Furthermore, he does not concern himself with the classical approach to studying afferent feedback gains, namely open- and closed-loop measurements, and the difficult question of proper controls that this approach entails (Nichols and Houk, 1976; Rymer and Hasan, 1980).

When one seeks to extend our understanding by considering the temporal relationship between force and EMG, Stephens argues that this can only be done in isolation from the same muscle. The facts are that plantar flexion forces are produced by the triceps surae (TS) muscle group, each member of which has the same EMG latencies in response to stretch (Gottlieb and Agarwal, 1980). If Stephens is suggesting that force transducers should be mounted on human tendons (cf. Walmsley *et al.*, 1978) then he should know that this is not possible.

Stephens feels that I attach 'great significance to the fact that he (I) finds no increase in "muscle force" consistent with the occurrence of short latency (SL) soleus EMG activity'. This is not what I wrote. My words are 'no *observable* change in muscle force follows SL activity'. In the discussion I point out that this could be because SL activity compensates for yielding once the short-range stiffness amplitude is exceeded. Thus a smooth transition occurs and the muscle force then resembles the strength trajectory just as in the cat experiments of Nichols and Houk (1976). Is Stephens's suggestion that no observable force could be generated by other muscles, a logical explanation? Following the same thread he claims that a mystery is generated because SL EMG activity in soleus has a larger amplitude than the ML activity despite the observable force increase associated with the latter. No mystery is involved, as figure 11 of Gottlieb and Agarwal (1980) shows that this amplitude difference is reversed in the lateral gastrocnemius. My intention was to compare the onset and amplitude of force events with the onset of EMG events in TS using soleus as a representative muscle, not to compare the EMG amplitude of one muscle in the TS group with the total plantar flexion force.

Finally, exception is taken to the size (2°) and velocity (25°/sec) of disturbances used. These exceeded by a factor of 1.5–2 the estimated short-range stiffness of soleus and had rates of stretch comparable to those used to study muscle afferents in anaesthetised cats. Whether these represent the type of jolt received when our toes land unexpectedly on a fallen branch during walking can be left to the reader to decide. After Stephens's critique, should we now consider muscle belly taps to intrinsic hand muscles (Buller *et al.*, 1980) as something other than 'contrivances of the physiological laboratory'?

Received on November 4th, 1980.

References

Buller, N. P., Garnett, R. and Stephens, J. A. (1980). *J. Physiol., Lond.*, **303**, 337–349
Gottlieb, G. L. and Agarwal, G. C. (1980). *J. Neurophysiol.*, **43**, 86–101
Nichols, T. R. and Houk, J. C. (1976). *J. Neurophysiol.*, **39**, 119–141
Rymer, W. Z. and Hasan, Z. (1980). *Brain Res.*, **184**, 203–209
Walmsley, B., Hodgson, J. A. and Burke, R. E. (1978). *J. Neurophysiol.*, **41**, 1203–1216

Muscle stiffness and locomotion

U. PROSKE* AND B. WALMSLEY*

SUMMARY

Recent measurements of tension in the soleus and medial gastrocnemius muscles of the walking cat show that soleus undergoes very rapid tension changes and at most speeds of walking and running develops more tension than gastrocnemius. Since in locomotion both muscles are stretched during their periods of activity, we have tried to account for the observed changes by measuring the stiffness of muscle fibres and of the tendon during stretch of contracting muscle. We found the tendon stiffness of both muscles to be about the same, while, for a similar level of isometric tension, muscle fibres of medial gastrocnemius were twice as stiff as in soleus. This could be accounted for by soleus fibres having twice as many sarcomeres. The peak level of force during the rapid stretch, using amplitudes typical for locomotion, was found to occur when the stretch was applied during the rising phase of a tetanic contraction, rather than during the subsequent plateau. It was concluded that soleus was a better choice than gastrocnemius for most forms of locomotion because it could maintain a relatively high stiffness over large distances of stretch and because it could operate over a wide range of joint angles, developing tension at relatively low energy cost. Medial gastrocnemius really only becomes important when very large, rapidly rising forces are required, such as occurs during jumping.

In recent years the study of locomotion has been greatly helped by the techniques of telemetry and high-speed movie photography. It has become possible to record in freely moving animals the electromyographic (EMG) activity of muscle groups (Engberg and Lundberg, 1969), and more recently, muscle length changes together with activity patterns of muscle receptors

* Department of Physiology, Monash University, Clayton, Victoria

(Prochazka *et al.*, 1976). However, estimation of muscle forces can only be inferred indirectly from the EMG. This gap has now been filled by the experiments of Walmsley *et al.* (1978), who were able to record tension using strain gauges attached to the tendons of the medial gastrocnemius and soleus muscles in freely moving cats.

Despite the large amount of information available on the structure, motor unit composition and reflex connections of these two ankle extensors, the measurements of tension during locomotion provided several surprising results. Examples of the recorded length and tension changes in the two muscles are shown in figure 1. We would like to draw particular attention to the surprisingly brief duration of the tension change in soleus. When the twitch tension in soleus is measured under isometric conditions (dashed line)

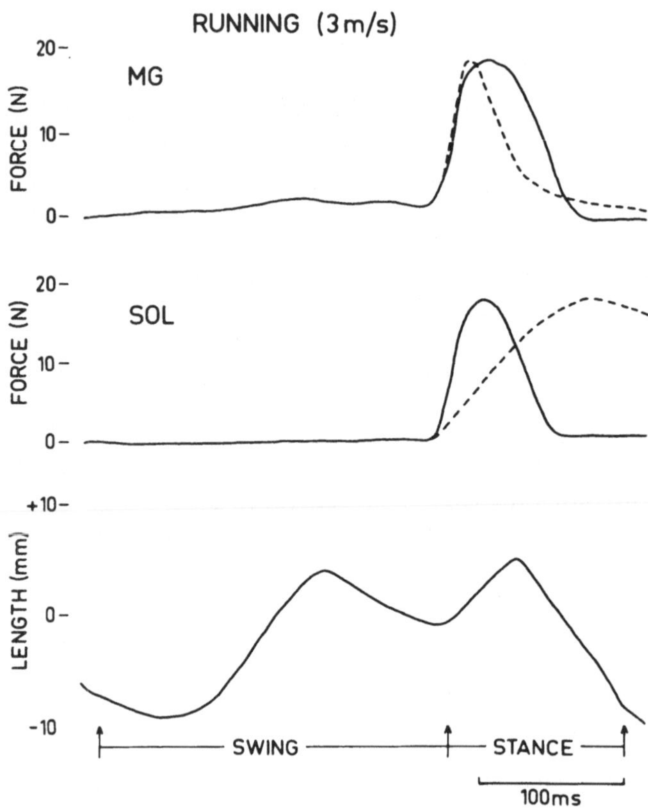

Figure 1. Records of length and tension changes in the medial gastrocnemius (MG) and soleus (SOL) muscles of the cat during running on a treadmill at 3 m/s. Tension changes were recorded using strain gauges implanted separately into each of the two tendons of insertion. Length was estimated from videotaped recordings synchronised with the taped tension data. The dashed lines indicate the time-courses of the respective isometric twitches. (After Walmsley *et al.*, 1978.)

the time-course is several times longer than during locomotion. In medial gastrocnemius the reverse is the case, the isometric twitch being somewhat briefer. While it is relatively easy to explain slow tension changes during locomotion, it is not immediately apparent how the rapid changes seen in soleus are achieved.

The explanation, of course, lies in the fact that in locomotion soleus is being rapidly stretched during its period of activity. In order to be able to predict the observed tension profile it is therefore necessary to study the behaviour of contracting muscle during sudden changes in length. This leads to the second observation made by Walmsley *et al.* (1978) and which is illustrated in figure 2. When forces in the two muscles are compared at different speeds of locomotion on a treadmill, soleus is seen to contribute more force at all speeds up to 3 m/s. Medial gastrocnemius only really begins to dominate during jumping. Slow-twitch muscles such as soleus have traditionally been thought to have a purely postural role (see for example, Buller, 1976). It is clear from figure 2 that soleus is doing a great deal more. It occurred to us that since soleus undergoes rapid length changes during locomotion, a clue to its role might be provided by the behaviour during stretch and shortening.

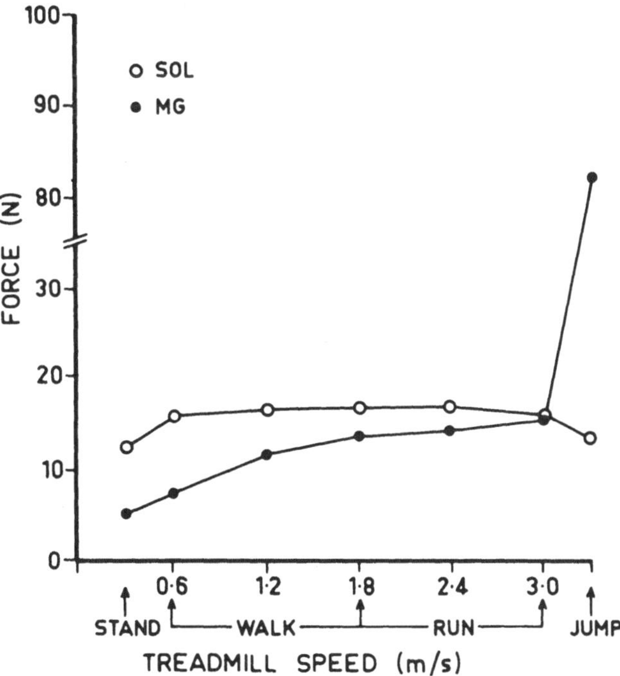

Figure 2. The mean force output of MG and SOL over a range of treadmill speeds. (After Walmsley *et al.*, 1978.)

We therefore carried out a series of experiments on soleus and medial gastrocnemius with the aim of explaining the tension changes seen during locomotion. We chose to measure muscle stiffness during rapid stretches since during most of their period of activity the two muscles are lengthening (*see* figure 1). The problem in measuring muscle stiffness is that when a muscle is stretched the total observed stiffness does not simply consist of the stiffness of contracting muscle fibres, but includes the stiffness of the tendon. An ingenious method which separates tendon and muscle stiffness is that of Morgan (1977). It relies on the fact that the stiffness of muscle fibres is directly proportional to the level of isometric tension prior to stretch. Tendon stiffness, on the other hand, remains the same at all levels of tension. The reciprocal of stiffness—compliance—is a more convenient form of expression since compliances in series can be simply added together. By plotting the product of compliance and tension, against tension, for different levels of tension, a straight line relationship is obtained whose slope represents the tendon compliance and the intercept on the ordinate relates to the compliance of muscle fibres.

There is one further complication. During large stretches muscle does not behave as a simple spring and this shows up in the tension record. If stretch is applied at a time when the muscle has reached a constant level of isometric tension, the initial steep rise in tension, called the 'short-range stiffness' (Rack and Westbury, 1974), gives way, after some distance of movement, to a more gradual tension change. In other words, during large stretches the muscle initially resists the movement with high stiffness, but beyond the limit of the short-range stiffness, it becomes more compliant. The Morgan method restricts measurements to the short-range stiffness and this must be taken into account when interpreting stiffness values for the large movements typical of locomotion.

Measurements of stiffness of soleus and medial gastrocnemius muscles in four cats of similar weight gave tendon compliance values which were similar for both muscles (0.06–0.08 mm/N), while muscle fibres in gastrocnemius were twice as stiff as in soleus (gastrocnemius α_0=0.34 mm, soleus α_0= 0.57 mm—*see* Walmsley and Proske, 1980).

Muscle stiffness, within the short-range stiffness, is thought to arise from stress of the cross-links between actin and myosin filaments (*see*, for example, Flitney and Hirst, 1978). Since at any one level of tension the number of cross-bridges which are in the attached position is thought to be the same for all mammalian skeletal muscles (i.e. same number of attachment sites per sarcomere), the difference in stiffness values must represent different numbers of sarcomeres in series in fibres of the two muscles—the more sarcomeres in series, the more compliant the fibre. The only proviso is that the stiffness measurements have been made using stretches which are sufficiently rapid to prevent significant breakdown of cross-links during the stretch (Gregory *et al.*, 1978). We have made sarcomere counts on single fibres selected from different portions of the two muscles. The values obtained were 7000–8000 for gastrocnemius and 12 000–15 000 for soleus, exactly the kind of difference predicted by the stiffness measurements.

As mentioned earlier, during large stretches, contracting muscle does not maintain a constant level of stiffness. Since the short-range stiffness in soleus extends, at the most, over only about 2 mm of movement, and since during locomotion movements of more than twice that amplitude are more typical, in order to predict tension during locomotion it is necessary to examine the changes in the muscle after the short-range stiffness has been exceeded.

In figure 3 the tension changes in soleus during stretch applied during the rising phase and plateau of an isometric contraction are shown. The stretch (1 mm), remains largely within the limits of the short-range stiffness. As predicted from theory, the amplitude of the tension change during stretch

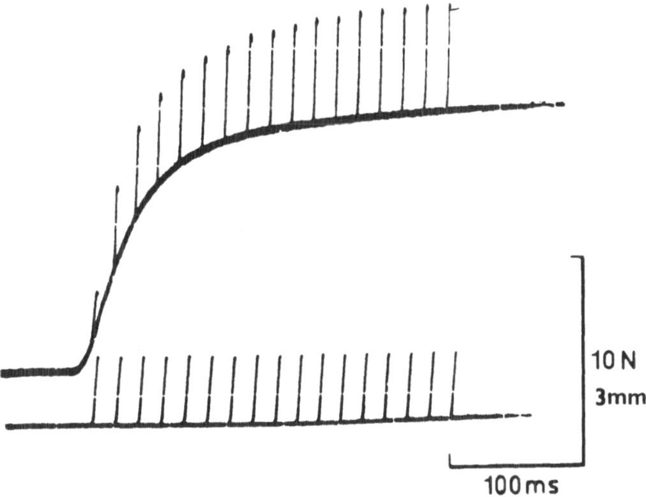

Figure 3. Tension changes during rapid stretches (1 mm at 80 mm/s) applied during the rising phase and plateau of a tetanically contracting soleus muscle. Tension is shown uppermost with the applied length change below. The waveform applied to the stretcher was actually triangular, but for simplicity the release phase has been omitted from the records.

increases with the level of tension and reaches its highest value during the plateau of the tetanus. If stretch amplitude is now increased to 4 mm, as in figure 4, the tension during stretch reaches its peak not at the top of the plateau as expected but during the rising phase of the contraction. The conclusion drawn from these observations is that when contracting muscle is stretched through movements of sufficiently large amplitude, the peak tension reached during stretch depends on whether the level of tension is con-

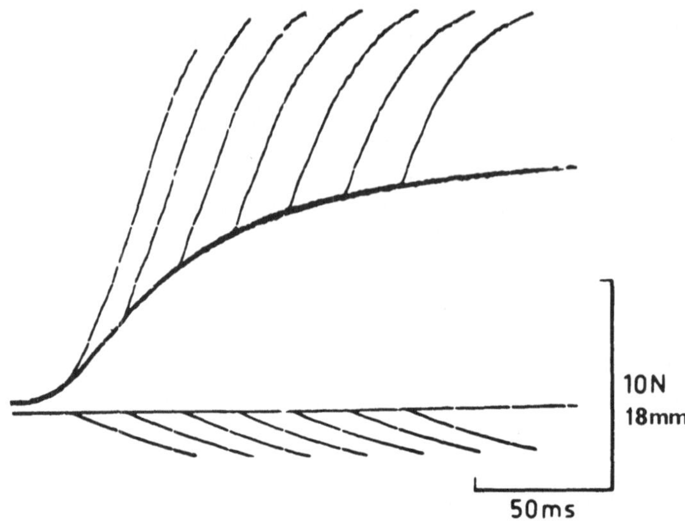

Figure 4. Tension changes during rapid stretches (4 mm at 80 mm/s) applied during the rising phase and plateau of a tetanically contracting soleus muscle. Tension is shown uppermost and the applied length change (inverted) below. Changes during the release following stretch have been omitted.

stant or whether it is changing at the time of the stretch. It can be seen in figure 4, that for stretches during the rising phase of the contraction, there is no sign of a discontinuity marking the limit of the short-range stiffness. Our interpretation of this observation is that during the rising phase the rate of cross-bridge formation is sufficiently high to prevent any drop in tension resulting from mechanical rupture of bridges by stretch (which marks the limit of the short-range stiffness).

What are the conclusions that we can draw from our observations? The rapid rise in tension in soleus during locomotion is the result of the muscle undergoing stretch at a time which coincides with its period of peak activity. Subsequently tension begins to fall as the rate of stretch slows down and activation declines.

Why is soleus used in preference to medial gastrocnemius at most speeds of locomotion? Soleus muscle fibres are nearly twice as long as in gastrocnemius. The disadvantage of this is that for two muscles of the same size, the muscle with the longer fibres will normally develop less tension, since the packing density of fibres will be less. On the other hand, the longer-fibred muscle will have a broader length–tension curve, enabling it to develop tension over a wider range of joint angles. Furthermore, it will have its short-range stiffness persisting over a greater distance of movement than the short-fibred muscle. This means that during quiet standing a muscle such as soleus is more likely to be able to oppose a disturbing force since it will be able to maintain a greater overall stiffness. Furthermore, during locomotion, muscle and tendon together behave more nearly like a simple spring providing opportunity for the storage of elastic energy (Morgan *et al.*, 1978).

Finally, it must be remembered that soleus and medial gastrocnemius have quite different motor unit compositions (Burke *et al.*, 1973; Burke *et al.*, 1974; Proske and Waite, 1974). Soleus is composed of fatigue-resistant motor units with relatively uniform histochemical characteristics and mechanical properties. Medial gastrocnemius, on the other hand, contains many motor units which would only be able to sustain tension during brief bursts of activity. For two muscle fibres of identical composition the one with twice as many sarcomeres would be expected to expend twice as much energy. However, since soleus fibres use relatively energy-efficient oxidative metabolic processes, the cost incurred in doubling fibre length is less than for fibres in gastrocnemius which depend largely on glycolytic metabolism.

In conclusion, soleus is the better choice for most forms of locomotion because of the inherent stiffness of its muscle fibres, the wide range of joint angles over which it can operate and the low cost in terms of energy consumption. Medial gastrocnemius really only comes into its own when it is required rapidly to develop very large levels of tension such as occur during jumping.

Received on July 7th, 1980.

REFERENCES

Buller, A. J. (1976). The mechanisms of postural control in the limbs and trunk. In *Mastication* (eds. D. J. Anderson and B. Matthews), paper No. 9, 66–71

Burke, R. E., Levine, D. N., Salcman, M. and Tsairis, P. (1974). Motor units in cat soleus muscle: Physiological, histochemical and morphological characteristics, *J. Physiol., Lond.*, **238**, 503–514

Burke, R. E., Levine, D. N., Tsairis, P. and Zajac, F. E. III (1973). Physiological types and histochemical profiles in motor units of the cat gastrocnemius, *J. Physiol., Lond.*, **234**, 723–748

Engberg, I. and Lundberg, A. (1969). An electromyographic analysis of muscular activity in the hindlimb of the cat during unrestrained locomotion, *Acta physiol. scand.*, **75**, 614–630

Flitney, F. W. and Hirst, D. G. (1978). Cross-bridge detachment and sarcomere 'give' during stretch of active frog's muscle, *J. Physiol., Lond.*, **276**, 449–465

Gregory, J. E., Luff, A., Morgan, D. L. and Proske, U. (1978). The stiffness of amphibian slow and twitch muscle during high speed stretches, *Pflügers Arch.*, **375**, 207–211

Morgan, D. L. (1977). Separation of active and passive components of short-range stiffness of muscle, *Am. J. Physiol.*, **232**, C45–C49

Morgan, D. L., Proske, U. and Warren, D. (1978). Measurements of muscle stiffness and the mechanism of elastic storage of energy in hopping kangaroos, *J. Physiol., Lond.*, **282**, 253–261

Prochazka, A., Westerman, R. A. and Ziccone, S. P. (1976). Discharge of single hindlimb afferents in the freely moving cat, *J. Neurophysiol.*, **39**, 1090–1104

Proske, U. and Waite, P. M. E. (1974). Properties of types of motor units in the medial gastrocnemius muscle of the cat, *Brain Res.*, **67**, 89–101

Rack, P. M. H. and Westbury, D. R. (1974). The short range stiffness of active mammalian muscle and its effect on mechanical properties, *J. Physiol., Lond.*, **240**, 331–350

Walmsley, B., Hodgson, J. A. and Burke, R. E. (1978). Forces produced by medial gastrocnemius and soleus muscles during locomotion in freely moving cats, *J. Neurophysiol.*, **41**, No. 5, 1203–1215

Walmsley, B. and Proske, U. (1980). Comparison of the stiffness of the soleus and medial gastrocnemius muscles in cats, *J. Neurophysiol.*, to be published

Examination of stretch reflexes in biceps and triceps muscles of the human arm using pseudo-random stimulation

S. W. JOHNSON*, P. A. LYNN*, S. MILLER† AND G. A. L. REED*

SUMMARY

The reflex electromyographic responses to stretch of biceps and triceps muscles in the human arm have been examined using small-amplitude random displacements of the forearm, which produce flexion and extension at the elbow joint. The method of input–output cross-correlation allows the reflex system to be characterised in terms of its impulse response. Only those components of the output signals (biceps or triceps electromyogram) which are systematically related to the random muscle stretch contribute to the response, so that by using a task of sufficient duration a small response can be recovered from a substantial level of background noise. The approach offers several advantages: the use of small displacements for which the response appears linear, separation of automatic reflex components from voluntary components, and a potentially simple method for examining reflex responses in neurological patients.

Studies of movement control in man have revealed that there is a complex interaction between the voluntary control of limb movement and stretch reflexes (Hammond, 1956, 1960; Hagbarth, 1967; Evarts and Granit, 1976). It is now well established that disturbances of the limb evoke reflex responses in the appropriate muscles at various latencies, the responses at longer latencies possibly involving supraspinal and even cortical circuits (Marsden *et al.*, 1973). The reflexes are further complicated by the non-linearities inherent in the muscle spindles (Matthews and Stein, 1969) and by fusimotor bias (e.g. Vallbo, 1971). Experimental studies designed to clarify the features of stretch reflexes in man have generally followed the classical approach to system identification by using deterministic forcing functions in the form of

*Department of Electrical and Electronic Engineering, University of Bristol
†Department of Anatomy, University of Newcastle upon Tyne

controlled mechanical disturbances: step functions, impulses or ramps (Hammond, 1960; Hammond *et al.*, 1956; Marsden *et al.*, 1976, 1977; Evarts and Granit, 1976); or sinusoidal inputs (e.g. Berthoz and Metral, 1970; Neilson, 1972; Joyce *et al.*, 1974).

An alternative approach reviewed in this paper is based upon continuous random displacements of a limb segment which are then cross-correlated with the evoked reflex responses. This technique seems to have two main advantages. Firstly, use of random displacements of the limb segment, too fast to be tracked by the subject, should clarify those elements of the movement control system which are automatic. Secondly, clear responses can be obtained even when the responses are small and the signal-to-noise ratio is low. Thus it is possible to use small-amplitude inputs and thereby avoid driving the system into serious non-linearity. Random excitation has been used in one other recent study of the stretch reflex behaviour (Dufresne *et al.*, 1978). However, in this case a controlled torque input was used, producing a limb displacement of relatively large amplitude and predominantly low frequency, and the results appear generally complementary to those reported here. A preliminary note and a description of the present method, together with a discussion of the results in some normal subjects, have been published (Johnson *et al.*, 1979, 1980). These results together with some additional observations on normal subjects and two hemiplegic stroke patients are reviewed here.

The full details of the experimental methods are given in Johnson *et al.* (1980). In brief, the subject sits with the forearm of one side supported on a hinged rest allowing rotation at the elbow joint. The support is driven by an electromechanical actuator capable of producing small-amplitude elbow rotation corresponding to vertical displacements of up to a few millimetres at the wrist. The experiment is controlled with a digital computer which outputs to the actuator a pseudo-random binary sequence (PRBS, e.g. Hoffman de Visme, 1971). Surface electromyogram signals (EMG) are obtained from biceps and triceps brachii muscles using miniature skin-mounted preamplifiers (Johnson *et al.*, 1977). After initial bandlimiting to 40–400 Hz and full-wave rectification the EMG signal is processed in two separate ways. Firstly, it is low-pass filtered with a cut-off frequency of 1 Hz and displayed to the subject, who is asked to maintain a constant EMG level during each trial by reference to a display on an oscilloscope. The rectified EMG signal is also filtered by a fourth-order Bessel filter with a cut-off of 3 dB at 20 Hz. This signal is regarded as the response of the reflex system to the applied displacement and is sampled by the computer at 100 Hz.

Responses have been recorded from 12 normal subjects and two hemiplegic stroke patients. In figure 1 the responses of 20 separate trials from the biceps muscle of a normal subject are superimposed. These responses were recorded consecutively under identical conditions during a 1 h period. The records in figure 1(b) show corresponding measurements of the rig displacement recorded simultaneously in the manner indicated, and include the effect of the

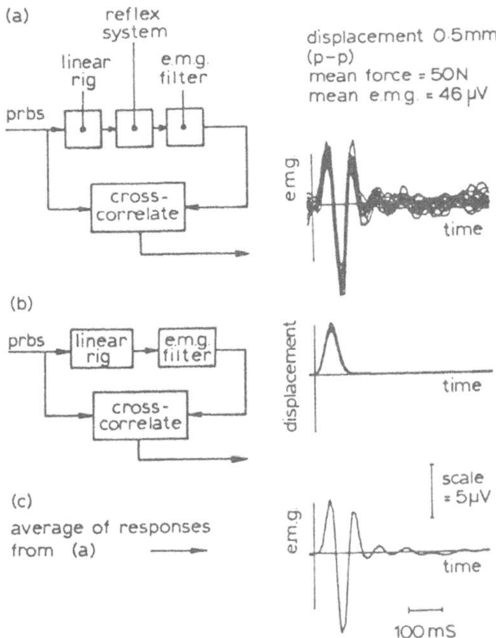

Figure 1. Responses from normal subject. (a) Superimposition of 20 consecutively recorded responses from single subject. (b) Corresponding recordings of the equivalent applied displacement obtained in the manner illustrated. (c) Average responses from (a). (This figure is reproduced from Johnson *et al.* (1980) by kind permission.)

filters used for EMG processing. This waveform can be regarded as an equivalent applied displacement which gives rise to the evoked responses. Figure 1(c) shows the average of the 20 trials. Being highly consistent, the initial components of the response are preserved in the average. Johnson *et al.* (1980) have examined the dependence of the amplitude of the EMG responses upon the amplitude of displacement. For peak-to-peak displacements at the wrist of up to 0.5 mm the response amplitude is scaled roughly in proportion to the input; above 0.5 mm the response tends to saturate.

Figure 2 illustrates EMG samples from biceps and the lateral head of triceps under conditions where the subject maintains various levels of contraction in either of the muscles. Displacement-related responses occur predominantly in the muscle which is active, the amplitude corresponding to the mean level of EMG recorded. No reflex is observed when the muscles are relaxed. When triceps is active the reflex in triceps is reversed in phase; this is expected because the phase of the applied stretch is reversed in this muscle.

The present technique has been used to investigate stretch reflexes in two hemiplegic stroke patients, of which one example is shown in figure 3. Two main features can be observed in the patient's responses in figure 3. Persistent

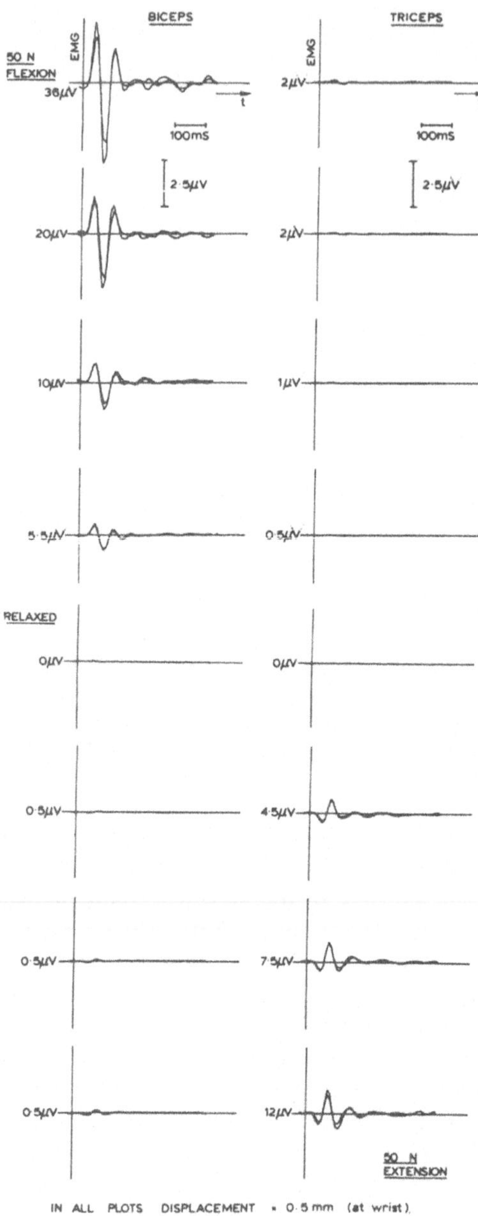

Figure 2. Dependence of the responses from biceps and triceps muscles of a normal subject on different levels of mean EMG (indicated in μV). Displacement amplitude at the wrist is 0.5 mm in all cases. In each trace two consecutive records are shown superimposed.

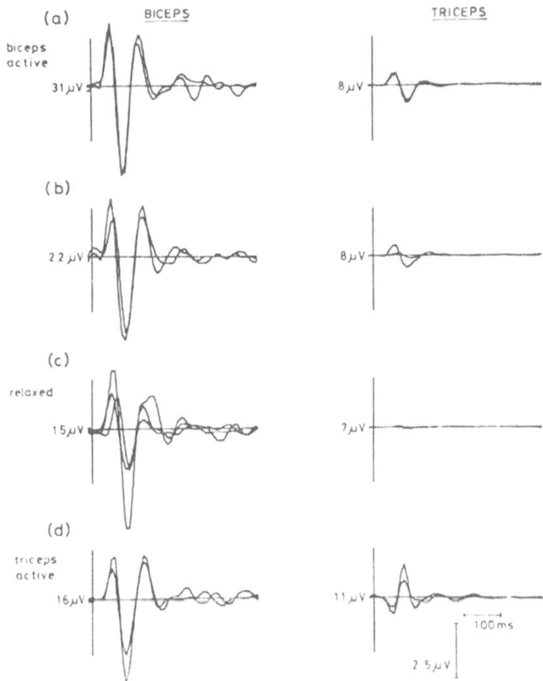

Figure 3. Responses from biceps and triceps muscles of the affected arm of a hemiplegic stroke patient (female, 66 years, 5 months after stroke, right sided hemiplegia, right handed). Different levels of contraction are attempted by the patient: (a) and (b), elbow flexion; (c), relaxed about elbow joint; (d), elbow extension. The applied displacement amplitude remains constant at 0.5 mm at the wrist. In each case two consecutive records (three for (c)) are shown superimposed.

background EMG activity and displacement-related responses occur in biceps whatever action the patient intends, e.g. figures 3(c) and 3(d). The response in triceps occurring in phase with that of biceps (Figures 3(a) and 3(b)) is also larger than that of the normal subject of figure 2.

The onset of the response occurs in normal subjects at 15–35 ms with an average of about 24 ms. These figures suggest involvement of the monosynaptic stretch reflex. Timing of the second peak is consistent with that observed in experiments involving sudden stretch of biceps (Hammond, 1954, 1956, 1960; Hagbarth, 1967; Evarts and Granit, 1976) for which polysynaptic or long loop pathways have been suggested. In some subjects the response, however, is markedly oscillatory being reminiscent of an underdamped control system and may reflect oscillations in the spinal reflex at the frequency of normal physiological tremor (Lippold, 1970; Joyce and Rack, 1974). The small fluctuations following the second peak in figures 1(a) and 1(c) also suggest such behaviour. In order to assist interpretation of these different features, Johnson *et al.* (1980) have developed a model of the stretch reflex. The responses from normal subjects could be fitted by a model in which a short latency monosynaptic reflex pathway is highly active and possibly acting in parallel with longer latency pathways of lower gain.

ACKNOWLEDGEMENTS

We wish to thank Dr Langton Hewer, Director, Avon Neurological Stroke Rehabilitation Unit, for his help and encouragement and to acknowledge support from the UK Medical and Science Research Councils and the Wellcome Foundation.

Received on August 1st, 1980.

REFERENCES

Berthoz, A. and Metral, S. (1970). Behaviour of a muscle group subjected to a sinusoidal and trapezoidal variation in force, *J. appl. Physiol.*, **29**, 378–384

Dufresne, J. R., Soechting, J. F. and Terzuolo, C. A. (1978). Electromyographic response to pseudorandom torque disturbances of forearm position, *Neurosci.*, **3**, 1213–1226

Evarts, E. V. and Granit, R. (1976). Relations of reflexes and intended movements, *Prog. Brain Res.*, **44**, 1–14

Hagbarth, K. E. (1967). E.M.G. studies of stretch reflexes in man. In *Recent Advances in Clinical Neurophysiology* (ed. L. Widen), *Electroenceph. clin. Neurophysiol.*, Suppl. 25, Amsterdam, Elsevier, 74–79

Hammond, P. H. (1954). Involuntary activity in biceps following the sudden application of velocity to the abducted forearm, *J. Physiol., Lond.*, **127**, 23–25

Hammond, P. H. (1956). The influence of prior instruction to the subject on apparently neuromuscular response, *J. Physiol., Lond.*, **132**, 17–18

Hammond, P. H. (1960). An experimental study of servo action in human muscular control, *Proc. 3rd Int. Conf. Med. Electron.*, London, I.E.E.

Hammond, P. H., Merton, P. A. and Sutton, G. G. (1956). Nervous gradation of muscular contraction, *Br. med. Bull.*, **12**, 214–218

Hoffman de Visme, G. (1971). *Binary Sequences*, London, English Universities Press

Johnson, S. W., Lynn, P. A., Miller, S. and Reed, G. A. L. (1977). Miniature skin-mounted preamplifier for measurement of surface electromyographic potentials, *Med. biol. Eng. Comput.*, **15**, 710–711

Johnson, S. W., Lynn, P. A., Miller, S. and Reed, G. A. L. (1979). Reflex electromyographic response to random limb displacement, *J. Physiol., Lond.*, **292**, 2–3P

Johnson, S. W., Lynn, P. A., Miller, S. and Reed, G. A. L. (1980). Identification of the stretch reflex using pseudo-random excitation: Electromyographic response to displacement of the human forearm, *Med. biol. Eng. Comput.*, to be published

Joyce, G. C. and Rack, P. M. H. (1974). The effects of load and force on tremor at the normal human elbow joint, *J. Physiol., Lond.*, **240**, 375-396

Joyce, G. C., Rack, P. M. H. and Ross, H. F. (1974). The forces generated at the normal human elbow joint in response to imposed sinusoidal movements of the forearm, *J. Physiol., Lond.*, **240**, 351–374

Lippold, O. C. J. (1970). Oscillation in the stretch reflex arc and the origin of the rhythmical 8–12 Hz component of physiological tremor, *J. Physiol., Lond.*, **206**, 359–382

Marsden, C. D., Merton, P. A. and Morton, H. B. (1973). Is the human stretch reflex cortical rather than spinal? *The Lancet*, **i**, 759–761

Marsden, C. D., Merton, P. A. and Morton, H. B. (1976). Servo action in the human thumb, *J. Physiol., Lond.*, **257**, 1–44

Marsden, C. D., Merton, P. A. and Morton, H. B. (1977). The sensory mechanism of servo action in human muscle, *J. Physiol., Lond.*, **265**, 521–535

Matthews, P. B. C. and Stein, R. B. (1969). The sensitivity of muscle spindle afferents to small sinusoidal changes in length, *J. Physiol., Lond.*, **200**, 723–743

Neilson, P. D. (1972). Frequency response characteristics of the tonic stretch reflexes of biceps brachii muscles in intact man, *Med. biol. Eng.*, **10**, 460–472

Vallbo, A. B. (1971). Muscle spindle response at the onset of isometric voluntary contractions in man. Time difference between fusimotor and skeletomotor effects, *J. Physiol., Lond.*, **318**, 405–431

Area display of H, M and T Waves

S. HOMMA*, Y. NAKAJIMA*, K. HAYASHI*, M. SHITO†
and K. SATO†

SUMMARY

H and M waves (elicited by electrical stimulation of the tibial nerve) and the T wave (induced by a tap on the Achilles tendon) were recorded topographically from the triceps surae muscles in man. One stimulus point which evoked the H wave with the lowest stimulus intensity, and another which elicited the M wave only with relatively lower stimulus intensity, were chosen. In the latter case, when triceps surae was contracted voluntarily, the stimulus elicited a new wave which appeared around 30 msec after the stimulus. This latency suggests that the new wave was also an H wave. The normal H wave and the new H wave are called H_T and H_P, respectively.

H_T waves started in the soleus and in the middle parts of the medial and lateral gastrocnemius muscles, and moved downwards. H_P waves appeared first above the H_T wave and in some cases moved upwards. H_T and H_P waves seem to be elicited in different parts of the muscles. Therefore we propose that the H_T waves are potentials of tonic muscle fibres innervated by small alpha-motoneurons of the tonic stretch reflex and that the H_P waves are those of phasic muscle fibres innervated by large alpha-motoneurons of the phasic stretch reflex. T waves were almost the same in the origin and the direction of the movement as H_P waves; therefore we also concluded that the tendon reflex may belong to reflex activity of the phasic stretch reflex.

INTRODUCTION

About a quarter of a century ago, Goldman *et al.* (1957) simultaneously recorded electrocardiograms with 16 electrodes attached on the chest wall of a human subject and topographically displayed them on a cathode-ray oscilloscope after conversion of the amplitude into luminous intensity. By cinema-

* Department of Physiology, Chiba University
† San-Ei Instrument Co Ltd, Tokyo

tographically recording the intensity-transformed amplitude distribution on the oscilloscope at high speed, he was able to learn how the isoelectric potentials moved. Since then, we have tried to record topographically evoked electromyograms (EMGs) using the same analytical methods. Initially, we used a number of preamplifiers, all of which were of equally high efficiency, but our trial was not successful because of the difficulty of eliminating stimulus artifacts due to electrical stimulation (Homma and Watanabe, 1957).

Owing to continuing progress in electrical technology, highly efficient amplifiers and stimulators with very low artifacts have become available. Furthermore, since it has become possible to digitise biological potentials within a very short interval, i.e. within a few microseconds, we can easily compare potentials recorded from several channels at almost the same time. This technical progress has made it possible to analyse topographically evoked EMGs and to detect the direction of movement of the isoelectric potentials.

In this study, we have topographically recorded H and M waves (elicited by electrical stimulation of the tibial nerve) and the T wave (induced by a tap on the Achilles tendon) from the triceps surae muscle of man.

METHODS

The subjects were five healthy medical students. The right tibial nerve was stimulated with a rectangular pulse of 0.5 msec duration every second. The stimulator and the isolator were Type SC-6 and Type IS/V (Medelec, UK), respectively. Twelve surface electrodes were attached over the triceps surae muscle, and the electromyographic signals were amplified by 12 AA6M preamplifiers (Medelec, UK), with a band-pass frequency ranging from 8 Hz to 16 kHz. The amplified EMGs were further analysed by a 7T07 signal processor (San-Ei Instrument Co, Japan). The analysed data were displayed on a colour cathode-ray tube. The signal processor classified the amplitudes of the potentials into 11 levels, positive divisions being displayed with a reddish tint (red, pink, pink dots, brown dots, yellow and white) and negative ones with a bluish tint (deep blue, blue, blue dots, light blue and green). Potential gradients between electrodes were calculated according to the sampling theorem proposed by Miyagawa (1959). Positive divisions of evoked potential amplitude were used as an index. The colour distribution on the cathode-ray tube was recorded cinematographically, and then the movement was recorded on videotape at a slow-motion ratio of 1111 to 1. This videotape was shown at this symposium. In this paper, movement of the peak-positive isoelectric contours corresponding to the red tint is considered.

On stimulating the tibial nerve in the popliteal fossa, one stimulus point was chosen which evoked the H wave with the lowest stimulus intensity and another was chosen which elicited only the M wave with relatively lower

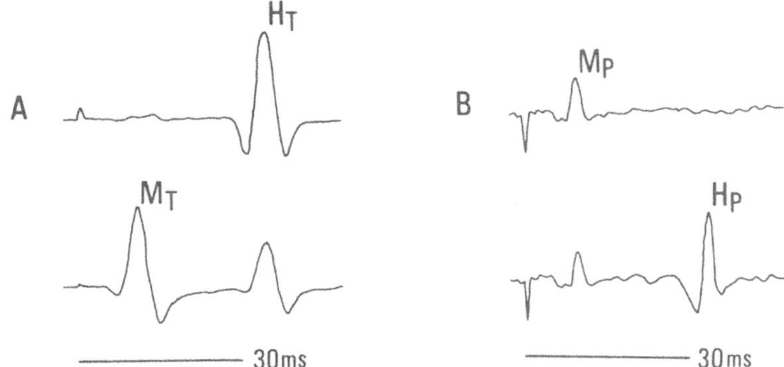

Figure 1. H and M waves. A (upper): one stimulus point which evoked the H wave with the lowest stimulus intensity. A (lower): a slightly higher stimulus intensity elicits the M wave and conversely decreases the amplitude of the H wave. B (upper): another stimulus point which elicits the M wave only with relatively lower stimulus intensity. B (lower): during voluntary contraction of the triceps surae muscle, the stimulus elicited a new wave which appeared around 30 msec after stimulus. The H and M waves are designated H_T, M_T, H_P and M_P as shown in this figure.

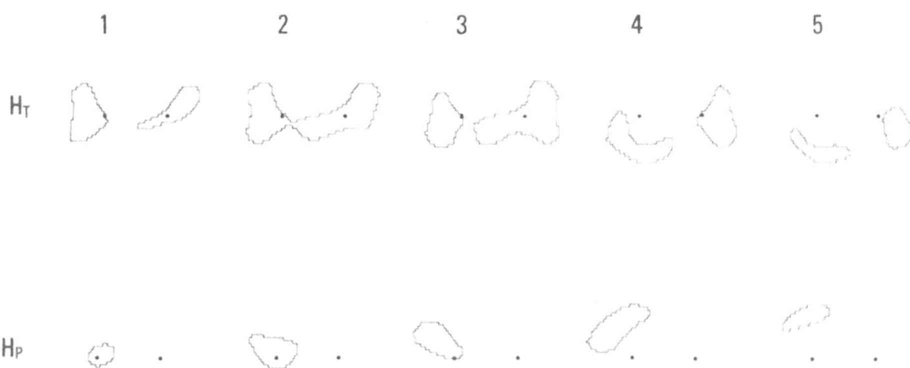

Figure 2. Area display of H_T and H_P waves. H_T shows isoelectric contours of the peak-positive potential of the H_T wave at various latencies from the stimulation of the nerve. Values after the stimulus (msec): H_T–1, 33.4; H_T–2, 33.8; H_T–3, 34.2; H_T–4, 34.6; and H_T–5, 34.8. The contours begin on S and MG middle and LG middle, and then move downwards. Values after the stimulus (msec) for the area display of the H_P wave: H_P–1, 30.4; H_P–2, 32.0; H_P–3, 33.0; H_P–4, 33.4; and H_P–5, 34.0. The contour begins on MG middle and then moves upwards.

stimulus intensity. In the former case, a slightly higher stimulus intensity elicited the M wave and conversely decreased the amplitude of the H wave (figure 1A). In this paper, these are called H_T and M_T waves, respectively.

In the latter case, if voluntary contraction was performed in the triceps surae muscle, the stimulus elicited a new wave which appeared around 30 msec after the stimulus (figure 1B). This latency suggests that the new wave is an H wave. The M wave and the new H wave in this case are called M_P and H_P waves, respectively. The T wave was also elicited around 30 msec after the tendon tap. These four kinds of evoked EMGs and the T wave were topographically analysed and discussed in relation to the spinal stretch reflex.

RESULTS

Area display of the H_T wave

The upper half of figure 2 shows isoelectric contours of the peak-positive potential of the H_T wave at various latencies from the stimulation of the nerve. In this case, the amplitude of the peak-positive contour was set at more than 0.48 mV. In figure 2, H_T–1 shows that 33.4 msec after the stimulus the potentials began on the upper part of the soleus, and on the middle parts of the medial and lateral gastrocnemius muscle areas. The contours became wider 0.4 msec and 0.8 msec after the onset of the H_T wave with a latency of 33.8 msec and 34.2 msec, respectively (figure 2, H_T–2 and H_T–3), and during the next 1.2 msec and 1.4 msec, the contour on the medial gastrocnemius area moved down (figure 2, H_T–4 and H_T–5).

Area display of the H_P wave

A stimulus point was chosen which only elicited the M wave with a relatively lower stimulus intensity. If voluntary contraction of the triceps surae muscle was performed, the stimulus elicited a new wave which was called the H_P wave. The lower half of figure 2 shows isoelectric contours of the peak-positive potentials of the H_P wave at various latencies from the stimulation of the nerve.

Figure 2, H_P–1 shows that 30.4 msec after the stimulation, the potentials began in the middle part of the medial gastrocnemius muscle area. The contour spread 2.4 msec after the onset of the H_P wave with a latency of 32.8 msec (figure 2, H_P–2) and moved sideways and upwards (figure 2, H_P–3, H_P–4 and H_P–5).

Locations of initiation and termination of evoked EMGs

In figure 3 the isoelectric contours at the onset and just prior to the disappearance of the peak-positive potentials are shown, and directions of the movements of the contours are indicated by arrows. Two examples (A and B)

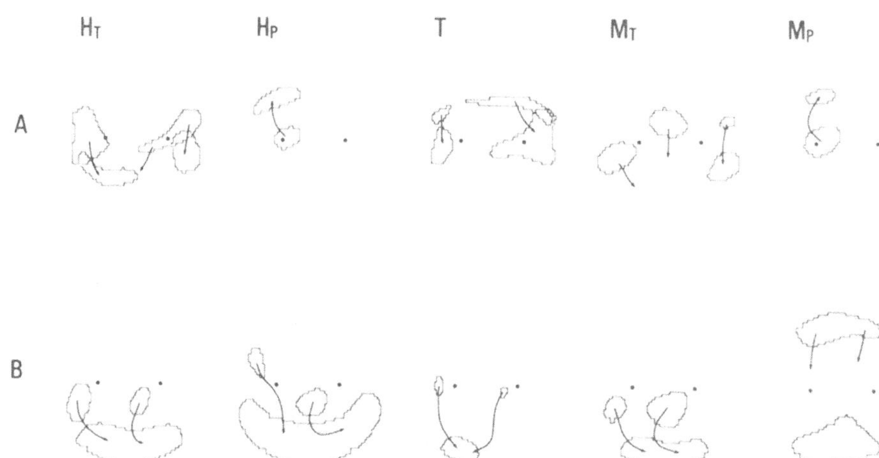

Figure 3. Isoelectric contours of the onset, and just prior to, the disappearance of the peak-positive potentials. Directions of the movements of the contours are shown by arrows. A and B show two examples from two subjects. See text for details.

obtained from two subjects are shown. Evoked potentials which began to appear on the soleus muscle are designated as S; those on the upper part of the medial and lateral gastrocnemius muscles as MG upper and LG upper; those on the middle part as MG middle and LG middle; and those on the lower part as MG lower and LG lower, respectively.

Figure 3A shows that the H_T wave was initially elicited on S, MG middle and LG middle, that all contours moved downwards, and that the H_P wave began on MG middle and moved upwards. In figure 3B, the H_T wave was initially elicited on S and MG middle and both contours moved downwards. On the other hand, the H_P wave began on S and MG upper and both contours moved downwards. The initial location of the H_P wave on the triceps surae muscle appeared to be higher than that of the H_T wave and some of the H_P waves moved upwards.

T waves induced by a tap on the Achilles tendon appeared first on MG and LG upper parts in figure 3A, and on MG and LG middle parts in figure 3B. All T waves moved downwards. Judging by the area of initiation, the T wave appeared to be recorded from the same muscle area as the H_P waves. The M_T waves appeared on S, MG and LG middle parts in figure 3A and on S and mid-MG in figure 3B. Since all M_T waves moved down, they are thought to be similar in origin to the H_T waves.

The M_P wave appeared on MG middle in figure 3A and on MG and LG upper parts in figure 3B. The M_P waves were similar in origin to the H_P waves with respect to the initial location and the direction of movement. However, M_P waves appeared above M_T waves.

DISCUSSION

These results show that the locations where the H_T wave appears are on the soleus and on the middle part of the medial and lateral gastrocnemius muscles, and that the location of the H_P wave which is evoked under voluntary contraction of the muscle is different from that of the H_T wave. Homma pointed out that the H wave elicited by a lower stimulus intensity is a result of action potentials of tonic muscle fibres (Homma and Kano, 1962). This was confirmed by Buchthal and Schmalbruch (1970). This means that such an H wave is a result of activity of a stretch reflex consisting of group Ia afferent axons with lower excitatory thresholds and alpha-motor axons with higher excitatory thresholds. Since alpha-motor axons with higher excitatory thresholds are thought to belong to small motoneurons (Granit *et al.*, 1956; Eccles *et al.*, 1958), the H wave has been thought to reflect tonic stretch reflex activity. On the other hand, tibial nerve stimulation at the point which can easily evoke the M wave, elicited the M_P wave on the upper part of the gastrocnemius muscle, the location of which is quite different from that where the M_T wave appeared. In this case, voluntary contraction of the triceps surae muscle induced an H wave. Since this H wave was elicited on a higher part of the muscle than the M_T wave, we concluded that the H_P wave should not be attributed to activity of small alpha-motoneurons. The H_P wave induced during voluntary contraction has been thought of as follows. Antidromic volleys elicited in efferent axons by electrical stimulation of the tibial nerve collide with, and eliminate, efferent activity produced by voluntary action. As a result orthodromic volleys elicited in group Ia afferent axons can pass through the motoneurons and arrive at the muscle (Magladery, 1955). Therefore the reflex arc in this case is thought to consist of the same size of group Ia afferent axons and alpha-motor axons, which correspond to large alpha-motoneurons. Therefore the H_P wave is considered to reflect activity of the phasic stretch reflex.

The T wave initially appeared in the middle and upper parts of MG and LG. Therefore tendon reflexes elicited by a tap on the Achilles tendon may belong to reflex activity of the phasic stretch reflex analogous to the H_P wave.

Received on July 8th, 1980.

REFERENCES

Buchthal, F. and Schmalbruch, H. (1970). Contraction times of twitches evoked by H-reflexes, *Acta physiol. scand.*, **80,** 378–382

Eccles, J. C., Eccles, M. and Lundberg, A. (1958). The action potentials of the alpha motoneurones supplying fast and slow muscles, *J. Physiol., Lond.*, **142,** 274–291

Goldman, S., Lidovitch, S. and Davison, R. (1957). Some examples of electronic mapping of the electrical activity of the heart. On 16 mm film, New York, Syracuse University

Granit, R., Henatsch, H. D. and Steg, G. (1956). Tonic and phasic ventral horn cells differentiated by post-tetanic potentiation in cat extensors, *Acta physiol. scand.*, **37,** 114–126

Homma, S. and Kano, M. (1962). Electrical properties of the tonic reflex arc in the human proprioceptive reflex. In *Symposium on Muscle Receptors* (ed. D. Barker), Hong Kong, Hong Kong University Press, 167–174

Homma, S. and Watanabe, S. (1957). On the scanning coordinate recorder, *Low freq. Pulse Med.*, **1,** 1–4

Magladery, J. W. (1955). Some observations on spinal reflexes in man, *Pflügers Arch. ges. Physiol.*, **261,** 302–321

Miyagawa, H. (1959). Sampling theorem of stationary stochastic variables in multidimensional space. *J. Inst. electron. commun. Engng Japan*, **42,** 421–427

State-dependent responses during locomotion

S. ROSSIGNOL*, C. JULIEN*, L. GAUTHIER* and J. P. LUND*

SUMMARY

Reflex responses to cutaneous stimulation vary with the phase of locomotion. In high decerebrate cats walking on a treadmill, strong cutaneous stimulation to one hindlimb induces a crossed extension during the contralateral stance and a crossed flexion during the contralateral swing. During 'fictive' locomotion in acute spinal cats pretreated with Nialamide and DOPA, essentially the same reversal of crossed reflexes can be observed. However, some responses appear to be 'wrong' for the phase of locomotion and in some animals the crossed responses cannot be reversed at all. This was not seen in true locomotion. It is suggested that during actual walking, reflex patterns are largely selected by a central generator for locomotion and that the selection is reinforced by peripheral inputs originating from the moving limbs.

INTRODUCTION

'The reflex reactions vary with the state of the centres' (Graham Brown, 1911). This simple statement, written 70 years ago, summarises well the conclusion reached today by a number of investigators. Reflex responses to an identical stimulus have been shown to vary markedly with postural attitude (Sherrington, 1900; Graham Brown, 1911, 1914a; Graham Brown and Sherrington, 1912; Magnus, 1909a, 1909b, 1910, 1924; Ranson and Hinsey, 1930; Grillner, 1973; Grillner and Rossignol, 1978; Rossignol and Gauthier, 1980), as well as with the phase of various rhythmic movements such as locomotion (Forssberg *et al.*, 1975, 1976, 1977; Forssberg, 1979; Grillner *et al.*, 1977; Duysens and Pearson, 1976; Duysens, 1977a, 1977b; Duysens and

*Centre de Recherche en Sciences Neurologiques, University of Montreal.

Stein, 1978; Duysens and Loeb, 1980; Duysens *et al.*, 1980; Miller *et al.*, 1977; Schomburg *et al.*, 1977; Schomburg and Behrends, 1978; Prochazka *et al.*, 1978; Wand *et al.*, 1980; Rossignol and Gauthier, 1978; Gauthier and Rossignol, 1980), mastication (Lund and Rossignol, 1980a, 1980b; Lund *et al.*, 1980) and respiration (*see* von Euler, 1980 for review). Not only has it been found that the amplitude of a given reflex response may be phase-dependent but also that the nature or the direction of the response itself may also be phase-dependent. Indeed, the same stimulation can alternately influence the agonist and then the antagonist muscles activated in opposite phases of the rhythmic movements. The neural mechanisms by which different responses are selected are still not known. However, after reviewing the results of various experiments some possible mechanisms that may participate in the control of these state-dependent responses will be discussed.

METHODS

Cats were first anaesthetised by an i.v. injection of a short-acting barbiturate, Methohexital Sodium, 7 mg/kg. After tracheotomy, anaesthesia was continued with ether or halothane. One carotid artery was cannulated for blood pressure monitoring and the other was ligated. An external jugular vein was also cannulated for injections of fluids and drugs. After various surgical procedures the animals were decerebrated at a precollicular level at an angle of 30° (postmammillary preparations) or 50° (premammillary preparations). The nervous tissue anterior to the section was partially or totally removed. In preparations which did not walk spontaneously on the treadmill a concentric bipolar electrode was lowered under visual guidance through the inferior colliculus (4 mm lateral, 5–7 mm deep and 1–1.5 mm posterior to the intercollicular sulcus (Shik *et al.*, 1966)).Trains of constant-current square-wave pulses (1 msec duration, 30 Hz) were given at intensities varying from 00 to 150 μA for periods not exceeding 2 min.

Cuff electrodes were placed on one or both sides around the superficial peroneal nerve dissected at the ankle. Trains of 10 pulses, 1 msec in duration, 100 Hz in frequency, were applied to one nerve at intensities sufficient to evoke well-developed responses in both hindlimbs (40 μA–5 mA depending on the amount of shunting in the cuff electrode). During walking, the integrated version of one electromyographic recording was used to trigger the stimulator at preset points in the cycle. Electromyograms (EMGs) were recorded differentially with pairs of copper wires, insulated except for 1 mm at the tips, directly inserted into the muscles.

In 'fictive' locomotion experiments the animals were similarly prepared, but cuff electrodes were placed around various nerves of both hindlimbs to record electroneurograms (ENGs). The limbs were kept in a normal standing position over the treadmill belt. After paralysis with Flaxedil (5 mg/kg), the

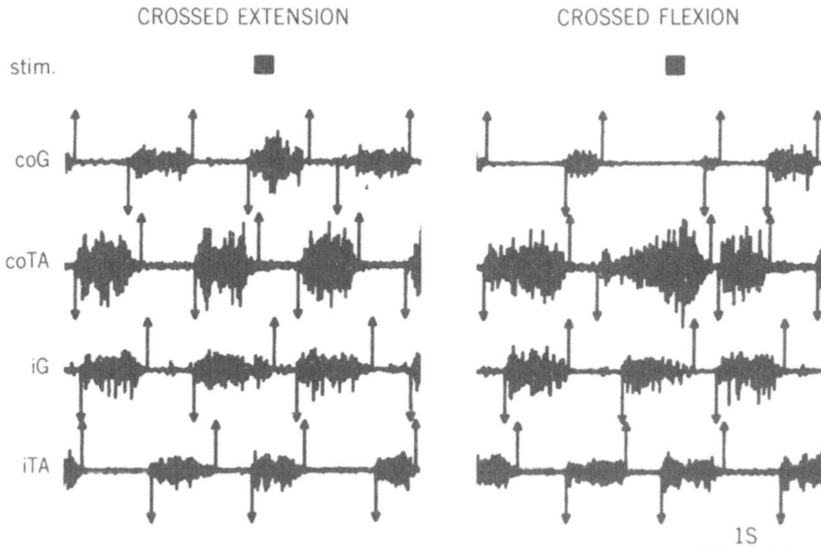

Figure 1. Two patterns of crossed hindlimb reflexes during locomotion in a high decerebrate cat. The EMG recordings are taken from an ankle extensor (gastrocnemius lateral (G)) and an ankle flexor (tibialis anterior (TA)), ipsilateral (i) and contralateral (co) to the stimulated superficial peroneal nerve (stim.). The stimulation is a 100 msec train of 1 msec pulses, 100 Hz, at an intensity of 1 mA. The speed of the treadmill was set at 0.85 m/sec. All EMG channels were digitised at 1 kHz from tape, high-pass filtered with a cut-off at 7 Hz, and displayed on a graphics terminal. Down-going and up-going arrows, respectively, indicate the beginning and end of each EMG burst as detected by a computer algorithm. On the left, the stimulation occurs during the ipsilateral swing and the contralateral stance. The coG burst during which the stimulation occurs is increased in amplitude. On the right, the stimulation occurs during the ipsilateral stance and the contralateral swing. The coTA burst is increased in amplitude and duration.

spinal cord was cut at the last thoracic segment. Nialamide (50 mg/kg) and L-DOPA (50–70 mg/kg) were injected i.v. (Grillner and Zangger, 1979).

EMGs and ENGs were recorded on magnetic tape and filmed from an oscilloscope. Some analyses were performed on a PDP-11/34 computer (figure 1). The EMGs were digitised at 1 kHz and were displayed on a Tektronix 4010 terminal together with arrows indicating the beginning and end of each EMG burst as detected by the computer (Blanchette, Rossignol and Smith, unpublished report).

RESULTS

Contralateral responses during walking

The stimulation used in the experiments was always strong enough to evoke an ipsilateral flexion response in all parts of the step cycle. If given during the ipsilateral swing, these responses would increase the ongoing ipsilateral flexor activity in amplitude and/or duration. When given during stance, the stimuli would block the ipsilateral extensor activity and trigger a flexor burst, thus resetting a new ipsilateral cycle (figure 1, right, *see* iTA).

In the contralateral limb, the patterns of responses observed varied according to the moment of stimulation in the step cycle. With stimuli during ipsilateral swing (figure 1, left) there was an increase in amplitude of the ongoing contralateral extensor burst with little change in its overall duration, so that the changes in total cycle length were usually negligible. Such extensor responses could be seen with stimuli delivered before and during most of the extensor burst.

When the stimuli arrived during the ipsilateral stance, thus resetting a new ipsilateral swing, there was a marked increase in the amplitude and/or duration of the contralateral flexor burst (figure 1, right). The subsequent extensor burst was usually smaller in both amplitude and duration, so that the overall contralateral cycle was slightly shorter than normal, whereas the ipsilateral cycle was markedly shortened because of resetting. This mismatch between the two hindlimbs was compensated for in the next cycle, where the contralateral limb performed a step shorter in duration than that of the ipsilateral limb. Crossed flexion responses were seen to occur with stimuli arriving before the onset and during most of the contralateral flexor burst.

Thus, in general, two patterns of response were seen: crossed extension during contralateral stance and crossed flexion during contralateral swing. One or the other type of response could be reliably evoked when the stimuli occurred before the onset and during about the first half of the respective locomotor burst. Towards the end of each burst and in the interval between the two bursts, one or the other response could be elicited for a stimulation given at the same point in the cycle. Thus with stimuli towards the end of the contralateral flexor burst, either a crossed flexion or a crossed extension, could be evoked.

Since the responses are buried within the locomotor bursts, a computer was used to measure the latencies of 25 crossed extension and 25 crossed flexion responses by detecting the point of amplitude increase of the EMG after the stimulation, taking 25 normal bursts as a template for comparison. These studies revealed that the latencies of ipsilateral flexion responses measured during swing (145 msec ± 15) or stance (154 msec ± 11) were similar to those of crossed flexor responses (146 msec ± 64) and crossed extensor responses (122 msec ± 45). This clearly indicates that the pattern of crossed responses is not secondary to ipsilateral responses but is part of a simultaneous response involving both hindlimbs. These rather long latencies are reminiscent of the 'late discharges' reported by others (Andén *et al*., 1967; Grillner and Shik, 1973).

The duration of responses could also be evaluated by measuring the period during which the EMG was of higher amplitude than the template. On average the duration of crossed responses was about 200–250 msec. The duration of these responses, however, varied depending on where they occurred in the cycle.

Reflex reversal during 'fictive' locomotion

The same protocol was followed using paralysed spinal cats pretreated with Nialamide and L-DOPA. The rhythm generated in such conditions is generally slow (Grillner and Zangger, 1979) but there was usually a clear alternation between the activity recorded in the flexor (St) and extensor (G) nerves on both sides as during locomotion. Having shown before that extensive denervation of the leg or complete de-afferentation (Rossignol and Gauthier, 1980) can markedly change the pattern of crossed responses, nerves were left intact except those being recorded from. These were ligated distal to the cuff electrodes. Since the initial position of the limbs may also dictate the resonse patterns, the limbs were left in a normal standing position.

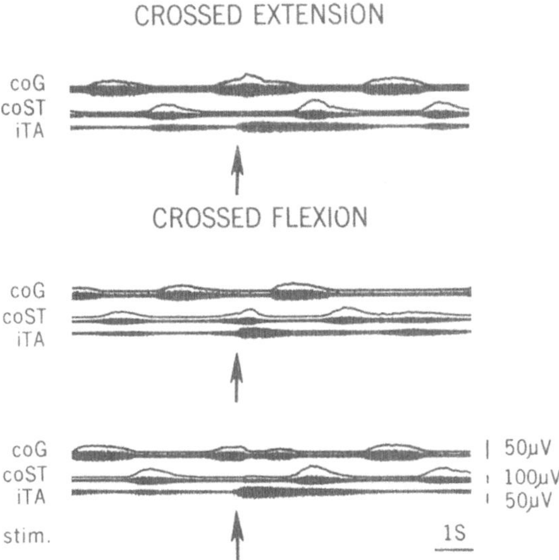

Figure 2. Patterns of crossed hindlimb reflexes during 'fictive' locomotion. The ENG recordings are taken from a contralateral ankle extensor (coG), a contralateral knee flexor and hip extensor semitendinosus (coST), and an ipsilateral ankle flexor (iTA). The smooth traces above coG and coST are the rectified versions of the respective ENGs. The arrows indicate the stimulation (200 msec train, 1 msec pulses at 100 Hz, 1 mA). The top record shows a crossed extension response and the middle record a crossed flexion response. The record at the bottom shows a response which appears 'wrong' for the phase: inhibition of the on-going coG, a small burst in coST and a return of the activity in coG.

The position of the limb as well as the intact afferent signals from the limb may have tonically influenced the central rhythm in the present situation. This might account for the very slow rhythms obtained at times as well as occasional unco-ordination between the two sides.

In cats which have developed a stable rhythmicity, stimuli given during the contralateral extensor burst could increase the amplitude of the burst as well as its duration (figure 2, top). Similarly, during the contralateral flexor burst the same stimulus could increase the amplitude and/or duration of the burst (figure 2, middle). The distribution of the types of responses within the cycle markedly resembled that obtained during walking. Indeed, stimuli occurring before and during the greater part of the activity recorded in a nerve could increase the activity in that nerve. Effects on the duration of the contralateral cycle were not identical to those in the walking animal. Whereas stimuli during the initial part of the contralateral flexor burst tended to shorten the cycle duration as in the normally walking cat, the stimuli given in the second half of the extensor activity increased the cycle duration. This was contrary to what had been observed in the walking animal. In some preparations these responses could last several seconds, thus effectively blocking the rhythmicity.

Although the response patterns could in general be well predicted by the time relationship between the stimulus and the activity in one or the other nerve, it often occurred that the stimuli would generate a response which appeared to be 'wrong' for the phase of the rhythm. Indeed, it could be observed quite often that a stimulus given during the flexor activity would block this activity and induce a short 150–250 msec burst in the extensor. The flexor activity would then resume. The same type of responses could also be seen with stimuli given in the period of extensor activity. Again the contralateral extensor burst would stop, a flexor burst would appear and then the extensor activity would resume (figure 2, bottom). In both cases, the overall duration of the interrupted burst would be almost the same as the preceding unstimulated burst.

Finally, in other preparations the crossed reflex response was always the same, independent of the phase of the rhythm. In these cases only crossed extension with inhibition of crossed flexor activity was obtained, or else the reverse, i.e. always crossed flexion with inhibition of the extensor. This had not been seen in the normally walking animal (Rossignol and Gauthier, 1978; Gauthier and Rossignol, 1980). Although the pattern of crossed responses can reverse when the animal is not actually walking, thus suggesting that the central rhythm alone can switch between the two types of responses, one is struck by the fact that quite often these responses appear to be 'wrong' for the phase of the rhythm and also that at times the responses cannot be reversed at all. Such is not the case in decerebrate cats walking on the treadmill.

In two experiments with fictive locomotion, clear examples of reflex reversal were seen on the ipsilateral side using rather weak stimuli of the SPN (30–40 μA). As can be seen in figure 3, the stimulation can induce a clear

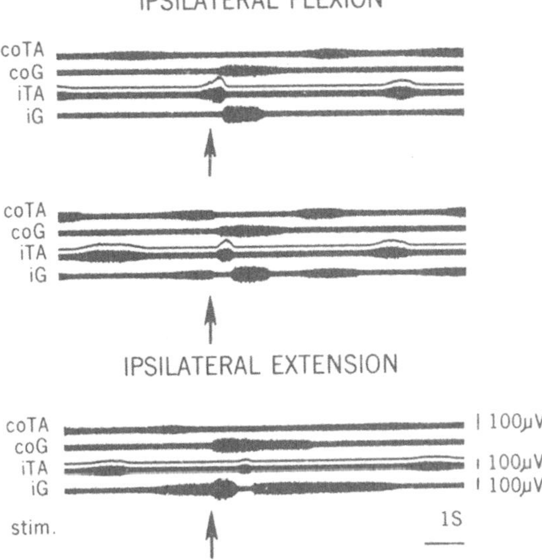

Figure 3. Reversal of the ipsilateral responses during 'fictive locomotion'. In the top record, the stimulation (arrow) of the ipsilateral SPN (40 μA) produces an ipsilateral flexion (iTA) and a contralateral extension (coG). In the middle record, the stimulation blocks the ipsilateral extensor activity (iG), evokes a flexion burst (iTA) and a crossed extension burst (coG). In the bottom record, the stimulation enhances the on-going ipsilateral extension activity (iG) and generates a crossed extensor response (coG).

increase of either the flexor burst (figure 3, top) or the extensor burst (figure 3, bottom). It is remarkable that, contralaterally, the response was then always extension. At other times, ipsilateral flexion was generated while the ipsilateral extensor burst was inhibited (figure 3, middle).

Entrainment of contralateral rhythm by cutaneous stimulation

In some preparations, the central rhythm with DOPA was disorganised and irregular (figure 4, top). It was found, as shown in figure 4 (bottom), that the contralateral rhythm could be regularised by giving a train of pulses to the ipsilateral peroneal nerve at regular intervals of 2–3 sec. It is noticeable here that the contralateral cycle starts with the flexor burst. These, as well as the extensor bursts following flexion, were of durations similar to those of the spontaneous bursts.

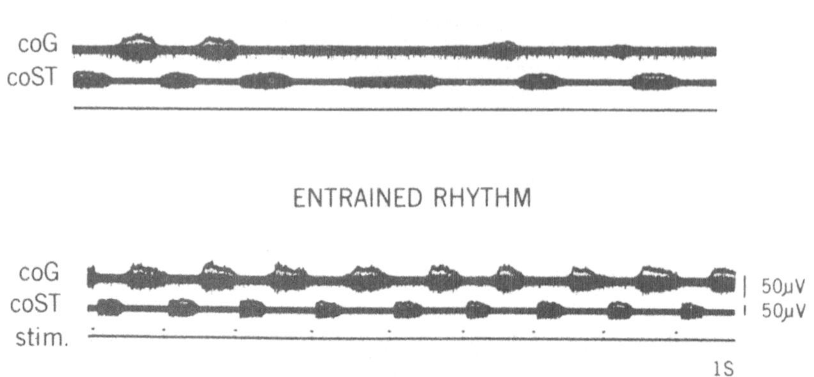

Figure 4. Entrainment of the central rhythm by cutaneous stimulation. The spontaneous rhythm of the top record was irregular, as seen from missing bursts and bursts of unequal duration. In the bottom record, a 100 msec train (1 msec pulses at 100 Hz, 5 mA) was applied at 2 per second to the contralateral superficial peroneal nerve. Note that the contralateral cycle regularly follows the stimulation and starts with the flexor burst.

DISCUSSION

This paper shows that 'reflex reactions vary with the state of the centres' (Graham Brown, 1911). Although it is not known by which mechanisms these state-dependent reflex responses are selected, it is perhaps worthwhile at least to discuss the evidence for or against a number of possible mechanisms. One should, however, be cautious before discarding any hypothesis since little is known of the actual properties of neuronal circuits during behaviour, and also because a number of traditional concepts derived from acute studies cannot be so easily extrapolated to behaving animals. For example, the contralateral flexion responses or the ipsilateral extension responses to cutaneous stimuli have often been labelled as 'abnormal' responses (Graham Brown and Sherrington, 1912; Graham Brown, 1914b), whereas in the present situation these are as normal as the better-described contralateral extension or ipsilateral flexion responses (Sherrington, 1910). Since these so-called 'abnormal' responses are evoked only in certain conditions, probably in dynamic conditions where they are needed, less attention has been paid to the neural circuits subserving them. Thus, whereas there is good evidence that during a crossed extension response the Ia inhibitory interneurons may participate in the concomitant flexor inhibition (Bruggencate and Lundberg, 1974), it is not known which interneurons may be responsible for exciting the flexor motoneurons during a crossed flexion response and whether Ia inhibitory interneurons are involved in the inhibition of crossed extensor motoneurons.

Nevertheless there is abundant experimental proof that cutaneous pathways may reach both ipsilateral flexor (Sherrington, 1910) and extensor motoneurons (Wilson, 1963) as well as contralateral flexor and extensor motoneurons (Evdokimov and Safyants, 1971). Thus the block diagram of figure 5 represents, in an oversimplified manner, a skin input reaching two different interneurons (IN) synapsing, respectively, on flexor and extensor motoneurons.

With the type of nerve stimulation used here there is obviously an activation of groups of axons which may have different central connections. One could imagine, for instance, that the stimulation of a whole nerve with a large receptive field may activate axons leading to excitation of flexors or extensors. This is unlikely since a more localised stimulation using a pair of wires

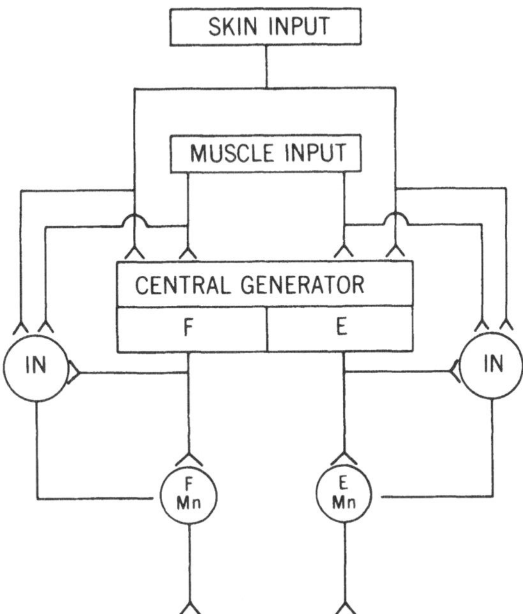

Figure 5. Block diagram suggesting some control mechanisms for reflex reversal. Skin inputs reach flexor (F) and extensor (E) motoneurons (Mn) through chains of interneurons (IN) and through a central generator for locomotion. Muscle inputs also reach these interneurons as well as the interneurons interposed in the cutaneous pathways. The central generator, besides organising the motor outputs of the motoneurons, controls the interneurons. Thus a cutaneous stimulation would reach one or the other group of motoneurons, depending on the control of transmission in the interneurons exerted by both central and peripheral signals.

inserted under the skin around the ankle can lead to the two types of crossed responses described here. Also, as far as ipsilateral reflex reversal is concerned, a natural stimulation such as a touch (Forssberg *et al.*, 1975) or air puffs (Forssberg, 1979) on the dorsum of the foot, which presumably activate one kind of receptor, are equally effective. In such a case, one can presume that one type of input may reach both types of motoneurons through different pathways as illustrated in figure 5. If this is the case, then how is one or the other route selected?

Perhaps the simplest explanation would be that the state of excitability of the antagonist groups of motoneurons would determine whether the cutaneous stimulation will produce a flexion or an extension. If one raises the excitability of one group of motoneurons, for example by stretching the relevant muscle (von Uexküll, 1904; Magnus, 1909a, 1909b, 1910, 1924), or by inducing the animal to maintain an active posture (Graham Brown, 1911), the same stimulation will excite whichever motoneurons are the most excitable at the time.

Although the excitability level of motoneurons may play a definite role in some characteristics of the responses, such as their latencies or their amplitude, there are several pieces of evidence indicating that this cannot be a predominant factor in the selection of the type of response.

In acute spinal cats pretreated with Clonidine, the crossed responses depend on the initial position of the contralateral limb at the time of stimulation (Rossignol and Gauthier, 1980). If the limb is extended manually at all joints, crossed flexion may appear. If it is flexed, then crossed extension can occur. After de-afferenting the contralateral limb or performing an extensive denervation, only crossed extension responses remain. Unless denervation or de-afferentation produces an overwhelming bias in favour of extensor motoneurons, a fact which has never been well documented, it would seem that in these conditions the crossed responses are not symmetrically controlled by raising the excitability of motoneurons through stretches of the muscles.

Experiments by Andersson *et al.* (1978) also showed that during fictive locomotion, the synaptic potentials evoked by cutaneous inputs are not principally regulated by the cyclical changes in the level of the motoneurons' membrane potentials. The synaptic potency appears to be gated at a premotoneuronal level (*see* Schomberg and Behrends, 1978; Forssberg, 1980).

In this work it is also clear that sometimes during fictive locomotion the responses occurring in one phase of the cycle are 'wrong' for that phase. If the reflex responses were driven mainly by motoneuronal excitability one would not expect such a mismatch between the responses and the ongoing activity of motoneurons. The responses should always enhance whatever activity there is at the moment of stimulation.

Finally, another piece of evidence can be drawn from the work on the jaw-opening reflex during natural mastication in the rabbit lightly anaesthetised with urethane (Lund and Rossignol, 1980a, 1980b; Lund *et al.*,

1980). This work shows that the amplitude of the jaw-opening reflex recorded in the digastric muscle is modulated throughout mastication, being maximal during jaw closure and minimal during jaw opening. Were the reflex driven by motoneuronal excitability, it should be expected to be maximal during jaw opening.

Thus, a simple model in which one cutaneous input would reach two antagonist groups of motoneurons and exert its effect on whichever is most active at the time of stimulation, is not tenable. The evidence suggests that the selection of the responses of motoneurons is made at a premotoneuronal level.

One central question is whether the responses dealt with here are responses produced by the neurons which are themselves responsible for generating the rhythmicity (collectively grouped as the central generator in figure 5), or whether they are produced through independent neuronal pathways which are heavily controlled by those neurons generating the rhythmicity. This question cannot be entirely resolved at this time.

On the other hand, as shown in figure 4, the peripheral cutaneous stimulation, when properly timed, can entrain the central rhythm (*see also* Grillner, 1979; Andersson *et al.*, 1980). The characteristics of the bursts would suggest that the skin input has access to the central generator as shown in figure 5 (Jankowska *et al.*, 1967). In this instance the stimulation would initiate the step, starting with a crossed flexion burst. As Graham Brown (1914a) put it, 'It might be supposed that the peripheral afferent stimulus, as it were, captures one side of the pair of linked half centres'. In this context the responses seen during real and fictive locomotion would be 'an exaggeration of one or the other phase of the fundamental rhythmic activity' (Graham Brown, 1914a). On the other hand, it is difficult to reconcile the fact that the responses may at times overrule the established rhythmic pattern and momentarily impose a response which is contrary to the ongoing activity. It seems reasonable then to conclude that although the cutaneous stimulation may have access to the central generator to the point of entraining it, it can also, through other pathways, generate responses in the absence of a central rhythm (Grillner and Rossignol, 1978; Rossignol and Gauthier, 1980) and, indeed, responses that may interfere with the central rhythm. Thus in figure 5 the interneurons are set apart from the central generator, but are controlled by it.

This central control of the interneurons does not appear to be absolute, as was demonstrated in fictive locomotion by responses which appeared inappropriate for the phase of the rhythm. Such responses are probably avoided by the integration of afferent inputs signalling limb position in walking animals. In acute spinal cats pretreated with Clonidine (Rossignol and Gauthier, 1980), it appears that the stretch of contralateral flexor muscles can determine the appearance of flexor responses thus preventing the crossed extension which would otherwise occur by default (*see* de-afferentation experiments, above). Since muscle afferents may interact with cutaneous pathways, especially Ib afferents (Eccles *et al.*, 1963, 1964) one may presume that during locomotion the activated muscle afferents (Prochazka *et al.*, 1977; Loeb, 1980) may contribute to reinforce the central selection of the appropriate responses.

In conclusion, the reflex responses do vary with the state of the centres and at this time it seems that this control is exerted centrally and reinforced by peripheral muscular inputs which ensure that the responses are appropriate for the state of the limbs at the time of stimulation.

ACKNOWLEDGEMENTS

This work was supported by a Group Grant from the Medical Research Council of Canada. The excellent assistance of M. Leduc, G. Blanchette, R. Bouchoux, C. Gagner and E. Rupnik is gratefully acknowledged.

Received on August 5th, 1980

REFERENCES

Andén, N. E., Jukes, M. G. M., Lundberg, A. and Vyklicky, L. (1967). The effect of DOPA on the spinal cord. I: Influence on transmission from primary afferents, *Acta physiol. scand.*, **67**, 373–386

Andersson, O., Forssberg, H., Grillner, S. and Lindquist, M. (1978). Phasic gain control of the transmission in cutaneous reflex pathways to motoneurons during 'fictive' locomotion, *Brain Res.*, **149**, 503–507

Andersson, O., Forssberg, H., Grillner, S. and Wallen, P. (1980). Peripheral feedback mechanisms acting on the central pattern generators for locomotion in fish and cat, *Can. J. Physiol., Pharmacol.*, to be published

Bruggencate, G. Ten and Lundberg, A. (1974). Facilitatory interaction in transmission to motoneurones from vestibulospinal fibres and contralateral primary afferents, *Expl Brain Res.*, **19**, 248–270

Duysens, J. (1977a). Reflex control of locomotion as revealed by stimulation of cutaneous afferents in spontaneously walking premammillary cats, *J. Neurophysiol.*, **40**, 737–751

Duysens, J. (1977b). Fluctuations in sensitivity to rhythm resetting effects during the cat's step cycle, *Brain Res.*, **133**, 190–195

Duysens, J. and Loeb, G. E. (1980). Modulation of ipsi- and contralateral reflex responses in unrestrained walking cats, *J. Neurophysiol.*, to be published

Duysens, J., Loeb, G. E. and Weston, B. J. (1980). Crossed reflex reversal in the unrestrained walking cat, *Brain Res.*, to be published

Duysens, J. and Pearson, K. G. (1976). The role of cutaneous afferents from the distal hindlimb in the regulation of the step cycle of thalamic cats, *Expl Brain Res.*, **24**, 245–255

Duysens, J. and Stein, R. B. (1978). Reflexes induced by nerve stimulation in walking cats with implanted cuff electrodes, *Expl Brain Res.*, **32**, 213–224

Eccles, R. M., Holmqvist, B. and Voorhoeve, P.E. (1964). Presynaptic depolarization of cutaneous afferents by volleys in contralateral muscle afferents, *Acta physiol. scand.*, **62,** 474–484

Eccles, J. C., Schmidt, R. F. and Willis, W. D. (1963). Depolarization of the central terminals of cutaneous afferent fibres, *J. Neurophysiol.*, **26,** 646–661

von Euler, C. (1980). The contribution of sensory inputs to the pattern generation of breathing, *Can. J. Physiol., Pharmacol.*, to be published

Evdokimov, S. A. and Safyants, V. I. (1971). Synaptic effects of contralateral somatic afferents on spinal motoneurons, *Neurophysiol.*, **3,** 316–321

Forssberg, H. (1979). Stumbling corrective reaction: A phase dependent compensatory reaction during locomotion, *J. Neurophysiol.*, **42,** 936–953

Forssberg, H. (1980). Phasic gating of cutaneous reflexes during locomotion. This publication

Forssberg, H., Grillner, S. and Rossignol, S. (1975). Phase-dependent reflex reversal during walking in chronic spinal cats, *Brain Res.*, **85,** 103–107

Forssberg, H., Grillner, S. and Rossignol, S. (1977). Phasic gain control of reflexes from the dorsum of the paw during spinal locomotion, *Brain Res.*, **132,** 121–139

Forssberg, H., Grillner, S., Rossignol, S. and Wallen, P. (1976). Phasic control of reflexes during locomotion in vertebrates. In *Neural Control of Locomotion* (eds R. M. Herman, S. Grillner, P. S. G. Stein and D. G. Stuart), New York, Plenum Press, 647–674

Gauthier, L. and Rossignol, S. (1980). Contralateral hindlimb responses to cutaneous stimulation during locomotion in high decerebrate cats, *Brain Res.*, to be published

Graham Brown, T. (1911). Studies in the physiology of the nervous system. VIII: Neural balance and reflex reversal with a note on progression in the decerebrate guinea pig, *Q. Jl expl Physiol.*, **4,** 273–288

Graham Brown, T. (1914a). On the nature of the fundamental activity of the nervous centres; together with an analysis of the conditioning of rhythmic activity in progression, and a theory of the evolution of function in the nervous system, *J. Physiol.*, **58,** 18–46

Graham Brown, T. (1914b). Studies in the physiology of the nervous system. XVIII: The successive effects of the compounding of reflexes where the 'pure' reactions are abnormal (ipsilateral extension or contralateral flexion) in decerebrate preparations, *Q. Jl expl Physiol.*, **7,** 383–405

Graham Brown, T. and Sherrington, C. S. (1912). The rule of reflex response in the limb reflexes of the mammal and its exceptions, *J. Physiol., Lond.*, **44,** 125–130

Grillner, S. (1973). Locomotion in the spinal cat. In *Control of Posture and Locomotion* (eds R. B. Stein, K. G. Pearson, R. S. Smith and J. B. Redford), New York, Plenum Press, 515–533

Grillner, S. (1979). Interaction between central and peripheral mechanisms in the control of locomotion, *Prog. Brain Res.*, **50,** 227–235

Grillner, S. and Rossignol, S. (1978). Contralateral reflex reversal controlled by limb position in the acute spinal cat injected with clonidine i.v., *Brain Res.*, **144,** 411–414

Grillner, S., Rossignol, S. and Wallen, P. (1977). The adaptation of a reflex response to the ongoing phase of locomotion in fish, *Expl Brain Res*, **30,** 1–11

Grillner, S. and Shik, M. L. (1973). On the descending control of the lumbosacral spinal cord from the 'mesencephalic locomotor region', *Acta physiol. scand.*, **87,** 320–333

Grillner, S. and Zangger, P. (1979). On the central generation of locomotion in the low spinal cat, *Expl Brain Res.*, **34,** 241–261

Jankowska, E., Jukes, M. G. M., Lund, S. and Lundberg, A. (1967). The effect of DOPA on the spinal cord. VI: Half-centre organization of interneurons transmitting effects from the flexor reflex afferents. *Acta physiol. scand.*, **70**, 389–402

Loeb, G. E. (1980). Somatosensory unit input to the spinal cord during normal walking, *Can. J. Physiol., Pharmacol.*, to be published

Lund, J. and Rossignol, S. (1980a). Dissociation of the amplitude of the jaw opening reflex and the masticatory cycle, *Proc. XXVIII Int. Physiol. Congr.*, to be published

Lund, J. and Rossignol, S. (1980b). Modulation of the amplitude of the digastric jaw opening reflex during the masticatory cycle, *Neurosci.*, to be published

Lund, J., Rossignol, S. and Murakami, T. (1980). Interactions between the jaw opening reflex and mastication, *Can. J. Physiol., Pharmacol.*, to be published

Magnus, R. (1909a). Zur Regelung der Bewegungen durch das Zentralnervensystem. Mitteilung I, *Pflügers Arch. ges. Physiol.*, **130,** 81–86

Magnus, R. (1909b). Zur Regelung der Bewegungen durch das Zentralnervensystem. Mitteilung II, *Pflügers Arch. ges. Physiol.*, **132,** 253–269

Magnus, R. (1910). Zur Regelung der Bewegungen durch das Zentralnervensystem. Mitteilung III, *Pflügers Arch. ges. Physiol.*, **134,** 545–583

Magnus, R. (1924). *Körperstellung*, Berlin, Springer, 24–49

Miller, S., Ruit, J. B. and van der Meche, F. G. A. (1977). Reversal of sign of long spinal reflexes dependent on the phase of the step cycle in the high decerebrate cat, *Brain Res.*, **128,** 447–459

Prochazka, A., Sontag, K. H. and Wand, P. (1978). Motor reactions to perturbations of gait: proprioceptive and somesthetic involvement, *Neurosci. Lett.*, **7,** 35–39

Prochazka, A., Westerman, R. A. and Ziccone, S. P. (1977). Ia afferent activity during a variety of voluntary movements in the cat, *J. Physiol., Lond.*, **268,** 423–448

Ranson, S. W. and Hinsey, J. C. (1930). Reflexes in the hindlimbs of cats after transection of the spinal cord at various levels, *Amer. J. Physiol.*, **94,** 471–495

Rossignol, S. and Gauthier, L. (1978). Patterns of contralateral limb responses to nociceptive stimuli during locomotion, *Soc. Neurosci. Abstr.*, **4,** 304

Rossignol, S. and Gauthier, L. (1980). An analysis of mechanisms controlling the reversal of crossed spinal reflexes, *Brain Res.*, **182,** 31–45

Schomburg, E. D. and Behrends, H. B. (1978). Phasic control of the transmission in the excitatory and inhibitory reflex pathway from cutaneous afferents to motoneurones during fictive locomotion in cats, *Neurosci. Lett.*, **8,** 277–282

Schomburg, E. D., Roesler, J. and Meinck, H. M. (1977). Phase dependent transmission in the excitatory propriospinal reflex pathways from forelimb afferents to lumbar motoneurones during fictive locomotion, *Neurosci. Lett.*, **4,** 249–252

Sherrington, C. S. (1900). On the innervation of antagonistic muscles. Sixth notes, *Proc. R. Soc. B.*, **67,** 66–67

Sherrington, C. S. (1910). Flexion-reflex of the limb, crossed extension reflex, and reflex stepping and standing. *J. Physiol., Lond.*, **40,** 28–121

Shik, M. L., Severin, F. V. and Orlovsky, G. N. (1966). Control of walking and running by means of electrical stimulation of the mid-brain, *Biofizika*, **11,** 659–666

von Uexküll, J. (1904). Die ersten Ursachen des Rhythmus in der Tierreihe, *Ergebn. Physiol.*, **3,** 1–11

Wand, P., Prochazka, A. and Sontag, K. H. (1980). Neuromuscular responses to gait perturbations in freely moving cats, *Expl Brain Res.*, **38,** 109–114

Wilson, V. J. (1963). Ipsilateral excitation of extensor mononeurones, *Nature, Lond.*, **198,** 290–291

Phasic gating of cutaneous reflexes during locomotion

HANS FORSSBERG*

INTRODUCTION

In the course of a movement, mechanical conditions vary, and a reflex response that is effective in one phase of the movement might be ineffective or even disruptive in another phase of the movement. Reflex responses studied in active states, during ongoing movements, have been found to be influenced by the course of the movement. The phasic modulation of skin reflexes during locomotion is perhaps the best studied example. These reflexes have been elicited from the dorsum of the paw and recorded in both the ipsilateral and contralateral hindlimbs (Forssberg *et al.*, 1975, 1977; Prochazka *et al.*, 1978; Forssberg, 1979a,b; Wand *et al.*, 1980; Duysens and Loeb, 1980) or in the forelimbs (Miller *et al.*, 1977). Stimulation of other parts of the limb has also elicited phase-dependent reflexes (Duysens and Pearson, 1976; Duysens and Stein, 1978). The phasic modulation of reflexes has also been found in other motor behaviours such as swimming (Grillner *et al.*, 1977) and respiration (Lipski *et al.*, 1977; Berger and Mitchell, 1976).

In this paper the focus is on one phase-dependent reflex elicited from the dorsum of the hindpaw and recorded in the same limb in the walking intact cat, and the neural origin of the phasic gating will also be discussed.

STUMBLE CORRECTIVE REACTION

When the foot dorsum of an intact cat is touched during locomotion a reaction is elicited. The quality of the reaction depends entirely on which phase of the step cycle is involved. During the swing phase, when the obstacle impedes the forward movement of the limb, the foot is lifted above the obstacle by a

* Department of Physiology III, Karolinska Institutet, Stockholm

flexion in all joints. The same stimulus applied during the stance phase does not, however, evoke flexion, but an enhanced flexion may instead be induced in the following swing phase (Forssberg *et al.*, 1975, 1977; Prochazka *et al.*, 1978; Forssberg, 1979a,b). From a more proximal region on the frontal side of the limb, a similar reaction is evoked and even abdominal hairs may give a reaction in which the foot is lifted over the obstacle without touching it. Skin receptors are the main source for the reaction as it is depressed after cutaneous anaesthesia (Forssberg *et al.*, 1975; Prochazka *et al.*, 1978) and even a weak puff of air may elicit the reaction (Forssberg, 1979a). Proprioceptors, however, induce activity which may have a stabilising effect (Prochazka *et al.*, 1978).

Reflex responses

Five different types of reflex response contribute to the reaction of the stimulated limb. They have been identified by applying a weak electrical pulse to the dorsum of the paw (Forssberg *et al.*, 1975, 1977; Forssberg, 1979a; Duysens and Loeb, 1980). Such a stimulus evoked the same phase-dependent reaction as touching the skin. In flexor muscles, one early reflex burst occurred with a latency of about 10 msec after the stimulus, and a second burst occurred at a latency of about 25 msec (figure 1). Two corresponding reflex bursts could be found in the extensor muscles—the first after 10 msec and the second with a more varying latency (25–50 msec) (figure 1). Also an 'inhibitory' reflex effect was induced in the extensors, with an onset at 10 msec after stimulus. The early flexor burst and the extensor inhibition may correspond to the same effects in the flexor reflex (Sherrington, 1910). Sherrington noted further that the 'normal' flexor reflex effects could be changed in certain conditions to a flexor inhibition and an extensor excitation (Sherrington and Sowton, 1911a,b). The latter effect could then correspond to the early extensor burst. Intracellular recordings have also demonstrated combined EPSP–IPSPs in extensor motoneurons after cutaneous stimulation (Burke *et al.*, 1970).

The later reflex burst could be consistent with a transmission to supraspinal centres and back. Of special interest are the pathways utilised in the tactile placing reaction, as the reaction is very similar to the flexor form of the stumbling corrective reaction and as the latencies are of the same magnitude (25–30 msec) (Amassian *et al.*, 1972; Norrsell and Lundberg in Lundberg, 1973). It is, however, notable that there are cutaneous reflexes in the passive spinal cat with latencies corresponding both to the early and the late reflexes of the stumbling corrective reaction (Hagbarth, 1952), and that tactile placing reactions as well as the stumbling corrective reaction can be seen in chronic spinal animals (Forssberg *et al.*, 1974, 1975, 1977).

The discussion of the pathways involved in the stumbling corrective reaction is mainly of interest because it shows how reflex pathways long

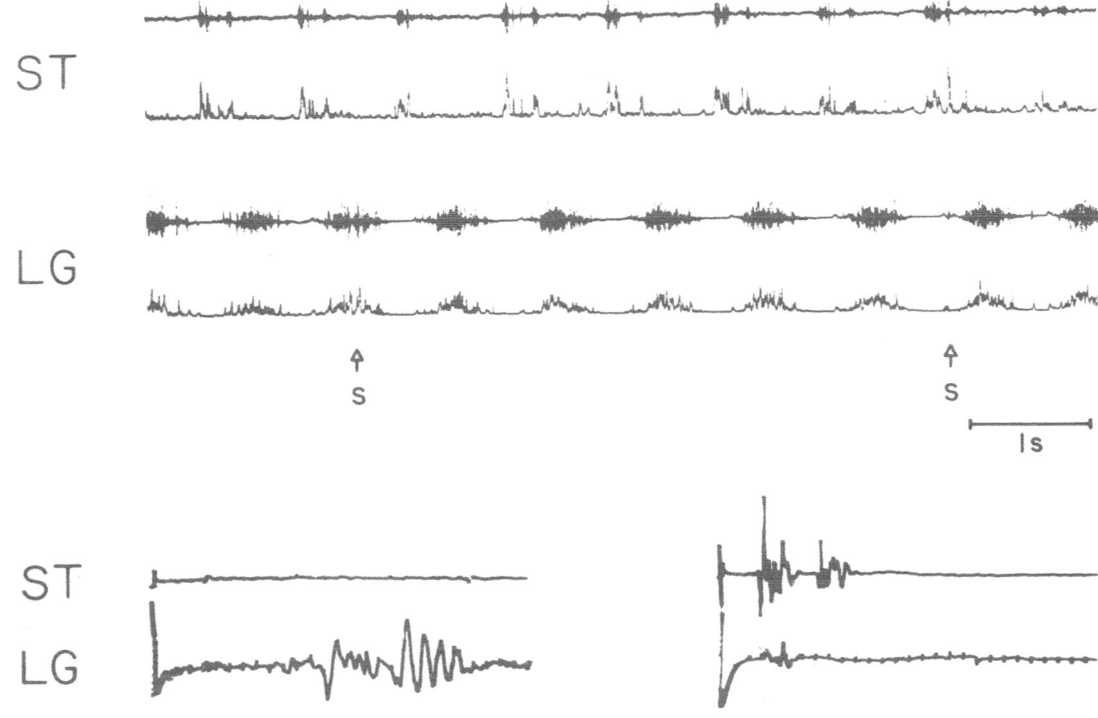

Figure 1. Reflex responses in the semitendinosus (ST) and the lateral gastrocnemius (LG) muscles after electrical stimulation (1 pulse, 1 mA, 1 msec) of the dorsum of the paw in walking intact cats. *Top:* EMG recordings from several consecutive step cycles are shown in the original form, and in a rectified and filtered form. The paw is stimulated twice (at the arrows), first during the stance and some cycles later during the swing phase. *Bottom*: reflex responses shown on an expanded time base.

known in different passive preparations assume a meaningful function when they are activated in an animal performing movements.

Supraspinal influence

The stumbling corrective reaction was found in chronic spinal cats (Forssberg *et al.*, 1975, 1977) before it was identified in intact cats (Prochazka *et al.*, 1978; Forssberg, 1979a). With the technique of electrical stimulation, all five types of reflex responses could be found in the spinal, as well as in the mesencephalic, preparation (Forssberg, 1979a). The amplitude of the reflex responses did, however, vary substantially in the different preparations. In the spinal cat the late extensor burst was large, while the extensor inhibition as well as the late flexor burst was small. Supraspinal centres thus enhance the extensor inhibition and the late flexor burst, and depress the late extensor reflex. This could be due to tonic or phasic changes of the transmission in relevant spinal pathways, but one cannot exclude the existence of complementary supraspinal loops.

Phasic modulation of the reflex bursts

The size of the four reflex bursts varied, depending on the phase of the step cycle during which the foot was stimulated. During the swing phase, the two flexor bursts and the early extensor burst were elicited (figure 1). During stance these responses were depressed, and instead, extensor inhibition occurred and was followed by the late extensor burst (figure 1). The phasic modulation of the four excitatory responses is shown in figure 2, in which the amplitude of the rectified and filtered EMG responses is plotted in the different phases of the step cycle. The flexor responses are largest during the beginning of swing, and fall towards the end of swing when the two extensor responses begin to rise. The early extensor burst is maximal during this period and falls when the later response increases to reach a maximum during mid-stance.

Neural mechanism of the phasic gating

The simplest explanation for the modulation of the reflex bursts would entail a gating effect at the motoneuronal level due to the periodic oscillation of the membrane potential during the step cycle (Edgerton *et al.*, 1976). Such a mechanism could not explain the appearance of the early extensor burst during the swing or the absence of the reflex bursts in ST during the second locomotor burst (figure 2). Another explanation, involving a more selective mechanism, would be that the individual reflex pathways are controlled at a pre-motoneuronal level. There could, in addition, be an interaction with other peripheral afferents, phasically activated by the movements of the limb,

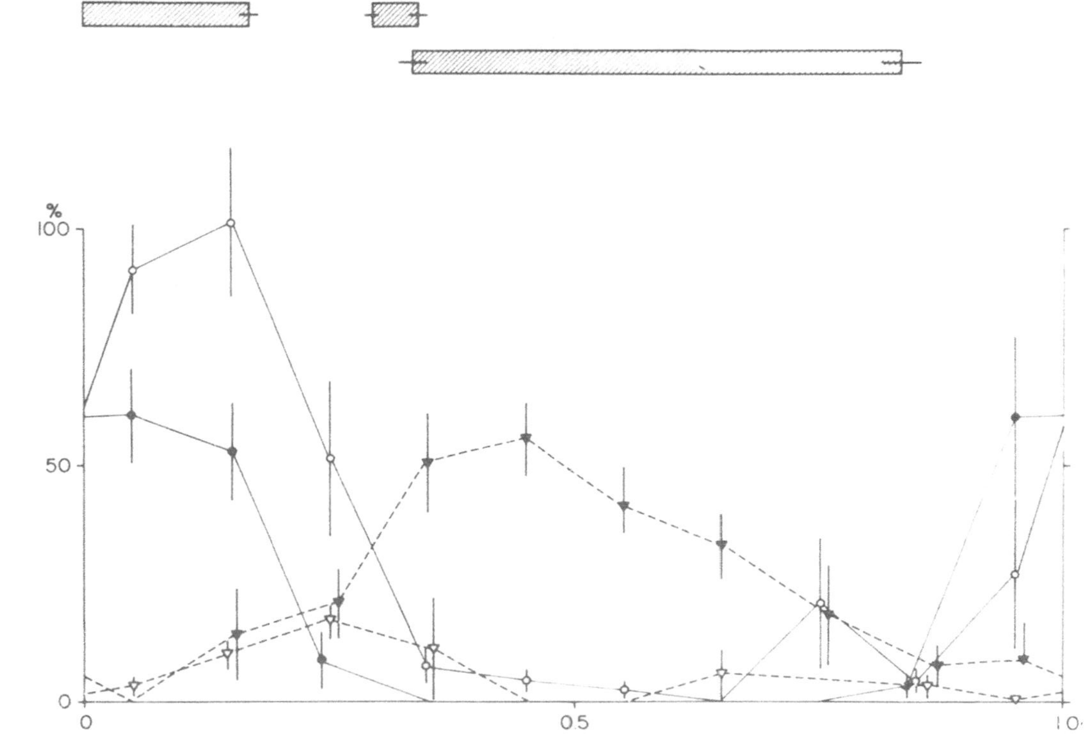

Figure 2. Phase dependence of the reflex responses in semitendinosus (ST) and lateral gastrocnemius (LG) muscles during treadmill locomotion in an intact cat. The paw is electrically stimulated (2 mA, 1 msec). The mean amplitude of rectified and filtered EMG responses (± SE) is calculated for each tenth of the step cycle, and expressed as a percentage of the mean maximal activity of 10 normal locomotor bursts, and plotted versus the phase of stimulation in the step cycle. The step cycles are normalised to the onset of the main activity in ST. The schematic EMGs represent 10 normal locomotor bursts (± SD). The symbols marked ST and LG are used in the graph, in which open symbols indicate the early reflex responses (*ca.* 10 msec) and the filled the later responses (25–50 msec) (from Forssberg, 1979a).

or, the pathways could be influenced by the spinal locomotor generator, which could control the reflex pathways as well as the efferent activity to the motoneurons.

To investigate the neuronal mechanism of this phasic gating, intracellular recordings were made in cats performing 'fictive locomotion' (Grillner and Zangger, 1974). In this preparation, activity in the locomotor generator can be induced in immobilised spinal animals and the efferent activity recorded in nerve filaments. By recording from flexor and extensor motoneurons during stimulation of the paw, it is possible to see whether the transmission is gated at a pre-motoneuronal level, or whether it is simply dependent on changes in the excitability level of the motoneurons. In acute spinal cats performing 'fictive locomotion', the paw stimulus usually evoked an early EPSP (7–8 msec) in the flexor motoneurons, and a combined EPSP–IPSP in the extensor motoneurons (*see* figure 3C, Burke *et al.*, 1970). Late extensor EPSPs were only occasionally evoked and the late flexor EPSPs in a few motoneurons (n=3) during early flexor activity. The amplitude of the 'early' EPSPs of both flexor and extensor motoneurons was regularly largest during the flexor period, and significantly larger in nine out of 22 flexor motoneurons and in 15 of 37 extensor motoneurons (figure 4, Edgerton *et al.*, 1976; Andersson *et al.*, 1977, 1978; *see also* Schomburg and Behrends, 1978a,b). The IPSPs of the extensor motoneurons were often largest during the extensor period, but they varied more.

In chronic spinal cats performing 'fictive locomotion' a similar modulation of the early EPSP was seen in both flexor and extensor motoneurons. However, a second large EPSP, with a latency of 20–40 msec, was also regularly seen. This EPSP was dramatically larger during the extensor period of the locomotor activity (figure 3D, Forssberg *et al.*, 1977; Andersson *et al.*, 1978).

The phasic modulation of the EPSPs in acute, as well as chronic, spinal cats performing 'fictive locomotion' are thus the same as the modulations of the reflex bursts in walking intact and spinal cats (figures 3A–3D). Note especially the early extensor burst/EPSP, which in all preparations are largest in antiphase, i.e. when the antagonists are active in the step cycle. The functional meaning of this burst is obscure as is the occurrence of a late extensor EPSP in the chronic spinal preparations, but not in the acute ones. The late flexor pathway apparently needs supraspinal facilitation to be effective (see above).

The modulation of the EPSPs means that the reflex must be influenced at a pre-motoneuronal level. However, the amplitude was seldom changed by more than 40%, which means that the excitability level of the motoneuron also has a modulating effect. Since no phasic afference was present, the gating of the reflex prior to the motoneuron must be exerted by the locomotor generator itself. This presumably occurs at an interneuronal level, but a presynaptic inhibition of alternative reflex pathways or effects on distal branches of the dendritic tree of the motoneuron cannot be excluded.

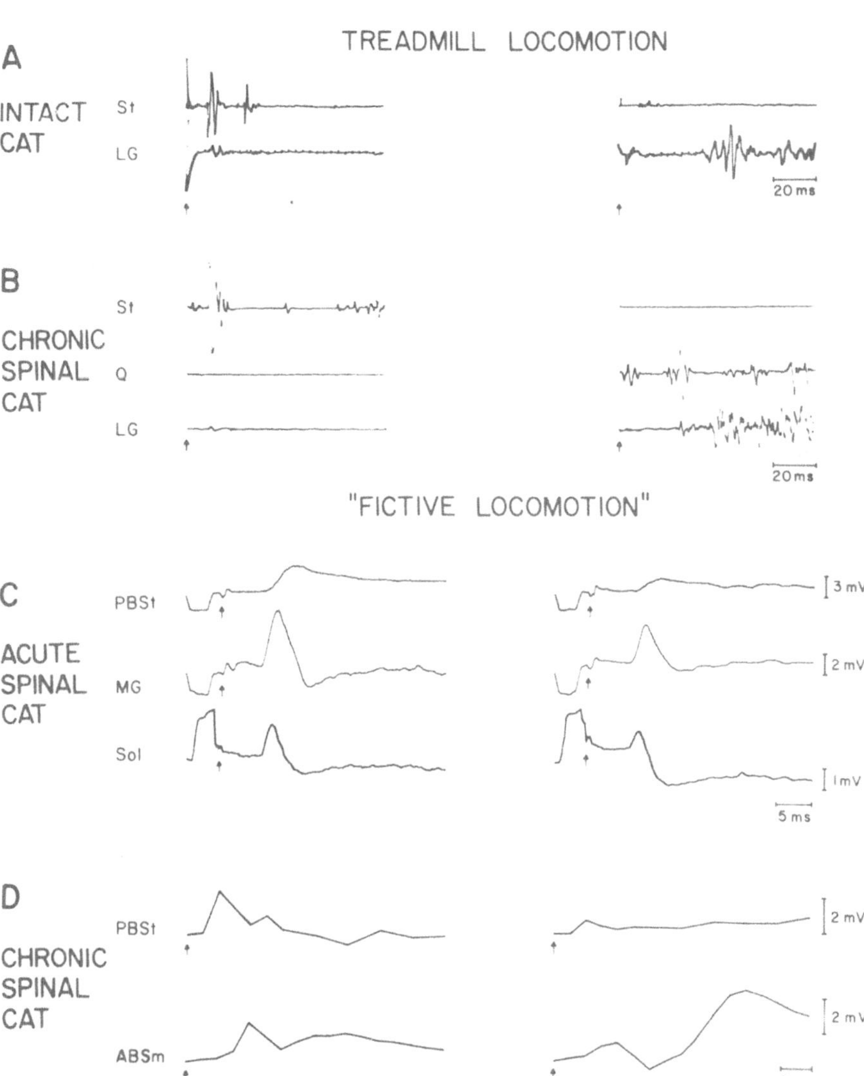

Figure 3. Reflex responses evoked by an electrical pulse (1 mA, 0.5 msec–5 mA, 5 msec) during different periods of locomotor activity in different preparations. In A and B single reflex responses are recorded in the muscles by EMG. In C and D the responses are intracellularly recorded from the motoneurons. In C the responses are averaged on an HP 21 MX computer (n = 12–28) and in D calculated from oscillograph recordings (n = 18–37). The stimulus is indicated by an arrow. In C a conductance pulse (constant current) is injected through the microelectrode prior to the stimulus (from Forssberg, 1979b).

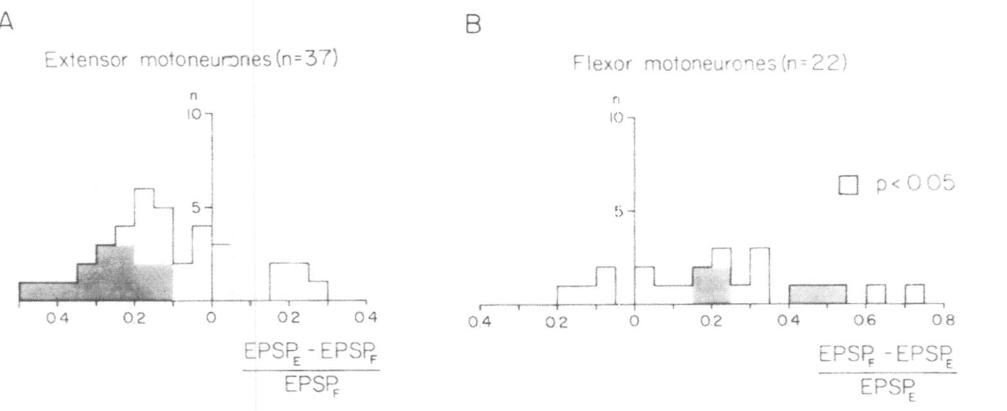

Figure 4. The relative modulation of the early EPSPs in acutely spinalised cats, performing 'fictive locomotion'. The differences of the averaged amplitudes (*see* figure 3C) between the flexor and extensor periods are normalised and plotted as histograms for extensor and flexor motoneurons. (*See* formulae in the figure.) Shadowed units indicate motoneurons in which the responses are significantly ($P<0.05$) different in the two periods (Student's *t*-test) (from Forssberg, 1979b).

So far, phasic gating has been documented for cutaneous reflexes during locomotion, but there are results suggesting that Ia pathways are also centrally gated (Schomburg and Behrends, 1978a) as are reflexes from chemoreceptors during respiration (Lipski *et al.*, 1977). The gating at the motoneuronal level will, of course, be present in all reflexes and in all types of phasic activity, but it is reasonable to assume that the central gating is also a general principle, utilised to control reflexes of importance for a well-adapted performance in different motor behaviours.

ACKNOWLEDGEMENT

This work was supported by the Swedish Medical Research Council (14X–3026).

Received on June 10th, 1980.

REFERENCES

Amassian, V. E., Weiner, H. and Rosenblum, M. (1972). Neural systems subserving the tactile placing reaction: A model for the study of higher level control of movement, *Brain Res.*, **40**, 171–178

Andersson, O., Forssberg, H., Grillner, S. and Lindquist, M. (1978). Phasic gain control of the transmission in cutaneous reflex pathways to motoneurones during 'fictive' locomotion, *Brain Res.*, **149**, 503–507

Andersson, O., Forssberg, H. and Lindquist, M. (1977). The neural mechanism of the phase dependent reflex reversal. Abstract of Satellite Symposium on Neurophysiological Mechanisms of Locomotion. *XXVII Int. Congr. Physiol. Sciences,* Paris, p. 11

Berger, A. J. and Mitchell, R. A. (1976). Lateralized phrenic nerve responses to stimulating respiratory afferents in the cat, *Amer. J. Physiol.*, **230**, 1314–1320

Burke, R. E., Jankowska, E. and Ten Bruggencate, G. (1970). A comparison of peripheral and rubrospinal synaptic input to slow and fast twitch motor units, *J. Physiol., Lond.*, **207**, 709–732

Duysens, J. and Loeb, G. E. (1980). Modulation of ipsi- and contra-lateral reflex responses in unrestrained walking cats, *J. Neurophysiol,* to be published

Duysens, J. and Pearson, K. G. (1976). The role of cutaneous afferents from the distal hindlimb in the regulation of the stepcycle in thalamic cats, *Expl Brain Res.*, **24**, 245–255

Duysens, J. and Stein, R. B. (1978). Reflexes induced by nerve stimulation in walking cats with implanted cuff electrodes, *Expl Brain Res.*, **32**, 213–224

Edgerton, V. R., Grillner, S., Sjöström, A. and Zangger, P. (1976). Central generation of locomotion in vertebrates. In *Neural Control of Locomotion* (eds. R. Herman, S. Grillner, P. Stein and D. G. Stuart), New York, Plenum Press, 439–464

Forssberg, H. (1979a). Stumbling corrective reaction: A phase dependent compensatory reaction during locomotion, *J. Neurophysiol.*, **42**, 936–953

Forssberg, H. (1979b). On integrative motor functions in the cat's spinal cord, *Acta physiol. scand.*, supplement 474

Forssberg, H., Grillner, S. and Rossignol, S. (1975). Phase dependent reflex reversal during walking in chronic spinal cats, *Brain Res.*, **85**, 103–107

Forssberg, H., Grillner, S. and Rossignol, S. (1977). Phasic gain control of reflexes from the dorsum of the paw during spinal locomotion, *Brain Res.*, **132**, 121–139

Forssberg, H., Grillner, S. and Sjöström, A. (1974). Tactile placing reactions in chronic spinal kittens, *Acta physiol. scand.*, **92**, 114–120

Grillner, S., Rossignol, S. and Wallén, P. (1977). The adaptation of a reflex to the ongoing phase of locomotion in fish, *Expl Brain Res.*, **30**, 1–11

Grillner, S. and Zangger, P. (1974). Locomotor movements generated by the deafferented spinal cord, *Acta physiol. scand.*, **91**, 38A–39A

Hagbarth, K.-E. (1952). Excitatory and inhibitory skin areas for flexor and extensor motoneurones, *Acta physiol. scand.*, **26**, supplement 94

Lipski, J., McAllen, R. M. and Spyer, K. M. (1977). The carotid chemoreceptor input to the respiratory neurones of the nucleus of tractus solitarius, *J. Physiol.*, **269**, 797–810

Lundberg, A. (1973). The significance of segmental spinal mechanisms in motor control, *Proc. Symp. Papers 4th Int. Biophysics Congr.*, Moscow

Miller, S., Ruit, J. B. and Van der Meché, F. G. A. (1977). Reversal of sign of long spinal reflexes dependent on the phase of the step cycle in the high decerebrate cat, *Brain Res.*, **128**, 447–459

Prochazka, A., Sontag, K. H. and Wand, P. (1978). Motor reactions to perturbations of gait: proprioceptive and somesthetic involvement, *Neurosci. Lett.*, **7**, 35–39

Schomburg, E. D. and Behrends, H. B. (1978a). The possibility of phase-dependent monosynaptic and polysynaptic Ia excitations to homonymous motoneurones during fictive locomotion, *Brain Res.*, **143**, 533–537

Schomburg, E. D. and Behrends, H. B. (1978b). Phasic control of the transmission in the excitatory and inhibitory reflex pathways from cutaneous afferents to α-motoneurones during fictive locomotion in cats, *Neurosci. Lett.*, **8**, 277–282

Sherrington, C. S. (1910). Flexion-reflex of the limb, crossed extension-reflex and reflex stepping and standing, *J. Physiol., Lond.*, **40**, 28–121

Sherrington, C. S. and Sowton, S. C. M. (1911a). Chloroform and reversal of reflex effect, *J. Physiol., Lond.*, **42**, 383–388

Sherrington, C. S. and Sowton, S. C. M. (1911b). Reversal of the reflex effects of an afferent nerve by altering the character of the electrical stimulus applied, *Proc. R. Soc. B*, **83**, 435–446

Wand, P., Prochazka, A. and Sontag, K. H. (1980). Neuromuscular responses to gait perturbations in freely moving cats, *Expl. Brain Res.*, **38**, 109–114

Changes in segmental and propriospinal reflex pathways during spinal locomotion

E. D. SCHOMBURG*, H.-B. BEHRENDS* and H. STEFFENS*

SUMMARY

In high spinal immobilised cats the transmission in segmental reflex pathways from cutaneous afferents and in descending propriospinal pathways from forelimb afferents to lumbar motoneurons was investigated during fictive locomotion. Synaptic responses evoked via these pathways showed basically the same dependence on the phase of the step cycle—generally EPSPs were favoured during the phase of activity of the corresponding motoneurons, while IPSPs (except for the early propriospinally mediated IPSP in FDL motoneurons) tended to increase or to occur during the phase of inactivity of the motoneurons. By this facilitation of excitatory influences and depression of inhibitory influences to motoneuron pools, the 'locomotor generator' may functionally support those muscles which are in their active phase, independently of the source of the stimulated afferents.

INTRODUCTION

Segmental reflex effects from cutaneous afferents and other flexor afferents upon lumbar α-motoneurons are not uniform. Excitatory and inhibitory pathways have been shown to project in parallel to the different motoneuron pools (Eccles and Lundberg, 1959; Holmqvist, 1961; Holmqvist and Lundberg, 1961), whereby the transmission in these pathways was controlled by supraspinal structures (Lundberg and Voorhoeve, 1962; Hongo *et al.*, 1969). The observation that reflexes evoked from the skin were phase-dependently reversed during walking in spinal cats (Forssberg *et al.*, 1975) raised the question of whether the excitatory and inhibitory pathways could

*Physiologisches Institut der Universität, Göttingen

be opened phasically in an alternating manner by the spinal locomotor structures.

The descending spinal reflex effects from cutaneous afferents or medium- and high-threshold muscle afferents of the forelimb to lumbar α-motoneurons also included excitatory and inhibitory components. However, here, with the exception of sartorius and flexor digitorum longus motoneurons, excitation clearly dominated, and an interpretation of the functional meaning was complicated by the fact that the effects occurred in a quite uniform non-reciprocal manner in flexor and extensor motoneurons regardless of the side of stimulation (Schomburg *et al.*, 1978). Observations in walking animals gave the first hints that the movement of the limbs could modulate the propriospinal reflexes (Miller *et al.*, 1977).

In order to investigate the influence of spinal locomotion on the transmission in the segmental and propriospinal reflex pathways with intracellular recording, we used the technique of 'fictive locomotion', i.e. a spinal locomotor-like rhythm induced pharmacologically in spinal curarised cats (Jankowska *et al.*, 1967a,b; Viala and Buser, 1971; Grillner and Zangger, 1974, 1979). Preliminary reports have been published (Schomburg *et al.*, 1977; Schomburg and Behrends, 1978b).

METHODS

The experiments were performed on anaemically decapitate, high-spinal, immobilised (Pancuronium, Organon) cats (technique cf. Schomburg *et al.*, 1978).

The ventral roots L_6, L_7 and S_1 were cut, and L_7 ventral root α-efferents were functionally isolated for recording. The efferents were identified by means of their short-latency activation by stimulation of the corresponding muscle nerve at low group I strength. Intracellular recording from lumbar α-motoneurons was performed with 3M K-citrate microelectrodes (2–4 MΩ). The recorded motoneurons were identified by antidromic stimulation of the ventral roots and stimulation of group Ia afferents of the corresponding muscle nerve.

The following hindlimb nerves and ipsilateral and contralateral forelimb nerves were cut peripherally and mounted for stimulation (abbreviations used are given in parentheses)—quadriceps (Q), sartorious (Sart.), posterior biceps and semitendinosus (PBSt), anterior biceps and semimembranosus (ABSm), gastrocnemius-soleus (GS), plantaris (Pl), flexor digitorum and hallucis longus (FDL), tibialis anterior and extensor digitorum longus (DP), peroneus longus, brevis and tertius (SPM), plantar section of the tibial nerve (Tib.), saphenus (Saph.), peroneus superficialis cutaneous branch (SPC), sural (Sur.), ulnar (U), median (M), superficial radial (SR) and deep radial (DR). Stimulation was performed with rectangular pulses (duration

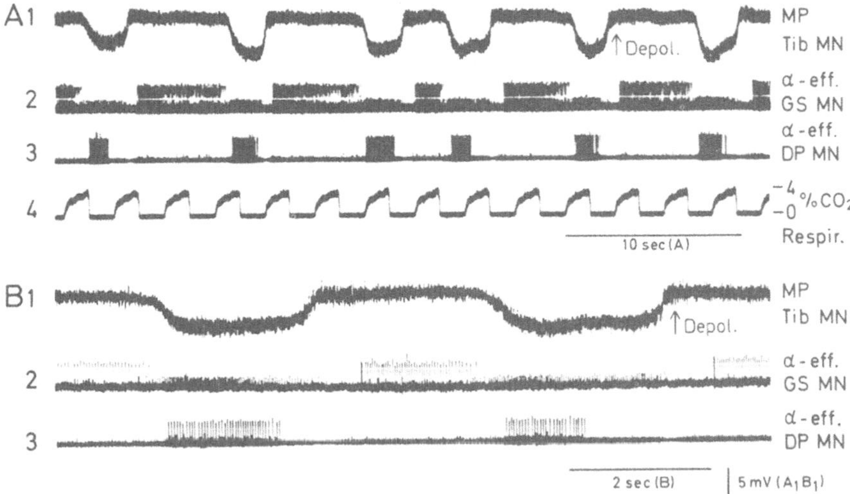

Figure 1. Cyclic fluctuations of the membrane potential of a Tib motoneuron (MP, A_1B_1) correlated with alternating extensor (GS, A_2B_2) and flexor (DP, A_3B_3) activity during fictive spinal locomotion. Simultaneous intracellular d.c. recording of a Tib motoneuron (A_1B_1, MP, Tib Mn) and filament recording of a GS (A_2B_2) and a DP (A_3B_3) α-efferent fibre. Respiration monitored as fluctuations of CO_2 concentration in the tracheal cannula (A_4), to exclude coupling between respiratory rate and locomotor rhythm. Note different time resolutions in A and B. High spinal cat pretreated with Nialamide and L-DOPA.

0.1 msec); the stimulus strength is indicated in multiples of the threshold of the nerve.

Fictive locomotion was induced by intravenous injection of 40–100 mg/kg Nialamide (Pfizer) and 30–60 mg/kg L-DOPA (Roche or Fluka). Rhythmic spinal motor activity normally occurred 15–60 min after injection of L-DOPA and was realised by the rhythmic activity in the ventral root filaments and corresponding fluctuations of the motoneuronal membrane potential (figure 1). In order to prove an alternating flexor–extensor activity, several efferent fibres were generally recorded simultaneously (cf. figure 1) or the monosynaptic reflexes of an extensor and a flexor were tested throughout the cycle (cf. Schomburg *et al.*, 1977). The phase during which a motoneuron is depolarised and synergistic efferents are active will be called the 'active phase', while the phase of inactivity will be termed the 'reciprocal phase'. Different cutaneous hindlimb nerves and forelimb nerves were stimulated at constant intervals (0.5–2 sec) and the responses recorded intracellularly from the motoneurons were analysed in relation to the phase of the step cycle. In the figures the times of occurrence of the stimuli which evoked the demonstrated responses are marked by arrows. The membrane potential of the motoneurons was recorded continuously in d.c. while the responses to the nerve stimulation were displayed on a faster time scale and with higher amplification in a.c. (bandwidth 0.1 Hz to 3 kHz or 10 kHz). The terms 'ipsilateral' and 'contralateral' are used with respect to the side of the recorded motoneurons.

RESULTS

Occurrence of fictive locomotion

In about 23 out of 42 experiments a stable, long-lasting fictive locomotion was observed, which was maintained without additional stimulation. Only results from these experiments were used. Figure 1 shows recordings from such an experiment. The efferent fibres of the extensor (GS, A_2, B_2) and the flexor (DP, A_3, B_3) motoneuron show a rhythmic alternating pattern, to which the membrane potential of the intracellularly recorded Tib. motoneuron (A_1, B_1) is clearly correlated. It can be seen that the cycle duration may vary even under stable locomotor conditions and that the mean flexion phase is shorter than the extension phase, a pattern which was often observed (cf. also figures 4 and 6). Respiration was also recorded (A_4), to prove the independence of respiratory rate and locomotor rhythm.

Phasic modulation of segmental responses to cutaneous nerve stimulation

The segmental responses in lumbar motoneurons to cutaneous nerve stimulation (Sur, SPC, Saph.) complied with the flexor reflex pattern, i.e. EPSPs were mainly observed in flexor motoneurons while IPSPs dominated in extensor motoneurons (Eccles and Lundberg, 1959). However, the principal feature of the phase-dependent modulation of PSPs was similar regardless of whether they occurred in flexor or extensor motoneurons—EPSPs and discharges were favoured during the active phase of the motoneuron while IPSPs often tended to occur or to be larger during the reciprocal phase.

The characteristic modulation of EPSPs during the step cycle is shown in figure 2. The phases of depolarisation of this flexor (PBSt) motoneuron revealed a constant alternation with the activity of an extensor efferent, which was activated by low threshold group I afferents from the GS and the Pl nerves (records A and C). The EPSPs evoked in this PBSt motoneuron by stimulation of the SPC (Ab and B) and Sur (Cb and D) were clearly decreased during the reciprocal phase (Bd and Dd) and at the transition between the active to the reciprocal phase (Be and Dc). The difference was particularly evident in the averaged records (Bi and Di), which were taken from samples during phases of maximal depolarisation (active phase) and maximal hyperpolarisation (reciprocal phase). EPSPs evoked by the Saph. nerve showed the same phase dependence. The phasic modulation was most evident if the EPSPs

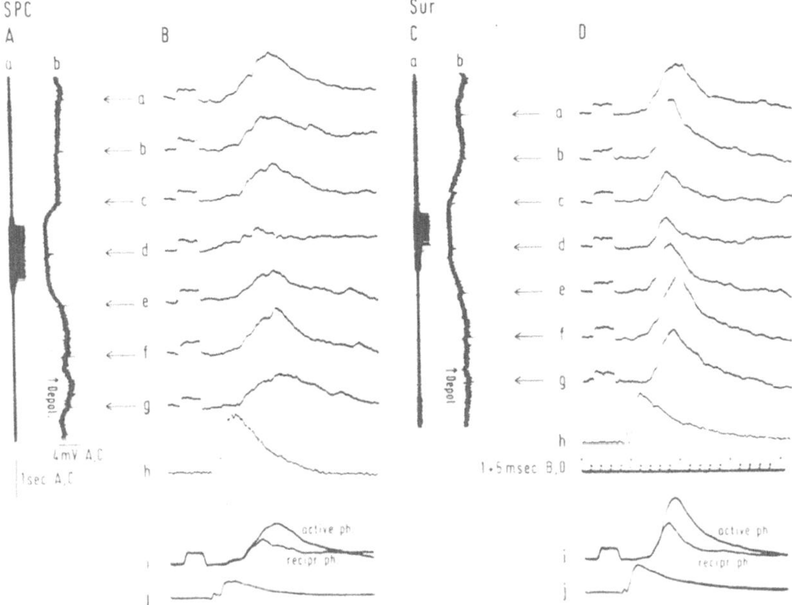

Figure 2. Phase-dependent segmental transmission from different cutaneous afferents (SPC, Sur) to a flexor motoneuron (PBSt) during fictive spinal locomotion. Intracellular recording of a PBSt motoneuron (Ab and Cb in d.c., B and D in a.c.) and simultaneous recording of a presumed extensor α-efferent fibre, which was activated by low threshold group I afferents of the GS and Pl nerve (Aa, Ca). Stimulation of the SPC (A and B) and Sur (C and D) nerve with constant intervals, stimulus strength 1.5T. The intracellular responses are displayed in B (a–g) and D (a–g) with higher time resolution and amplification (calibration pulse 1 mV). The position of the response within the step cycle is marked by arrows. h and j: incoming volleys recorded from the dorsal root entry zone. i and j: averaged superimposed recordings (16 samples each) taken from the phase of maximal depolarisation (active phase) or hyperpolarisation (reciprocal phase—recipr. ph.). For the incoming volleys no difference with respect to the phase can be seen.

were small (stimulus strength below 1.5–2T) but could also be observed with EPSPs evoked by stimuli of higher strength (5–20T). If the motoneurons were not spontaneously firing, discharges were almost exclusively triggered by EPSPs during the active phase.

IPSPs were modulated in two different ways during locomotor activity—they could increase or decrease during the active phase. Since the former pattern seems to be a passive reaction of the IPSP due to the change of the membrane potential, the latter reaction, which was mainly observed with small IPSPs, was of special interest. Figure 3 shows an IPSP evoked in a Pl motoneuron by low threshold (1.5T) stimulation of the Sur nerve, which was clearly larger during the reciprocal phase (g–j and p–s in A) than during the

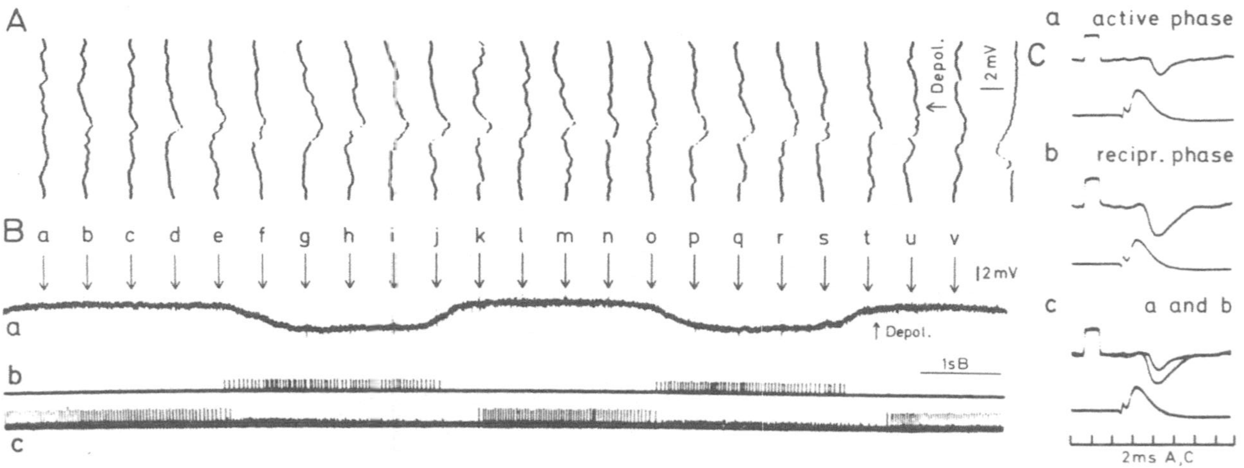

Figure 3. Phase-dependent modulation of an IPSP evoked by cutaneous afferents (Sur) in an extensor (Pl) motoneuron during fictive spinal locomotion. General recording technique as in figure 2. Intracellular recording of a Pl motoneuron in A and Ba. Simultaneous recording from DP (Bb) and GS (Bc) α-efferent fibres. Responses to stimulation of the Sur nerve with 1.5T in A(a–v). The calibration pulse is cut off in these recordings. An incoming volley is shown below Av. Averaged responses (33 samples each) in C; upper trace of each pair—intracellular response; lower trace—incoming volley.

active phase (a–d, k–o and t–v in A). The superimposed averaged records in Cc show that besides an increase in amplitude, a decrease in latency occurred during the reciprocal phase.

When cutaneous nerve stimulation evoked EPSPs followed by IPSPs, the EPSPs and IPSPs typically (eight out of 11 cases) showed a shifting of their share in the response. The EPSPs became larger during the active phase, while the IPSPs tended to be larger during the reciprocal phase.

Only five (ABSm, DP and SPM) out of more than 40 extensively tested motoneurons showed a complete reversal from almost pure EPSPs to IPSPs. In all these cells, during the active phase, EPSPs and discharges were observed. IPSPs occurred throughout (1 ABSm cell), or during particular parts, of the reciprocal phase. Figure 4 shows this pattern in a DP motoneuron. In the step cycle shown in G and H the position of responses to stimulation of the sural nerve (A–F), which were evoked during different cycles, is marked. The EPSPs became smaller when the stimulus occurred at a later part of the reciprocal phase, and the response turned into an IPSP at the transition to the active phase. The latency of the IPSP was slightly longer than that of the EPSP. The testing of this cell over 64 step cycles revealed that only in the period directly before and after D (12 stimuli) were IPSPs evoked, while throughout the other step cycle (51 stimuli), EPSPs occurred.

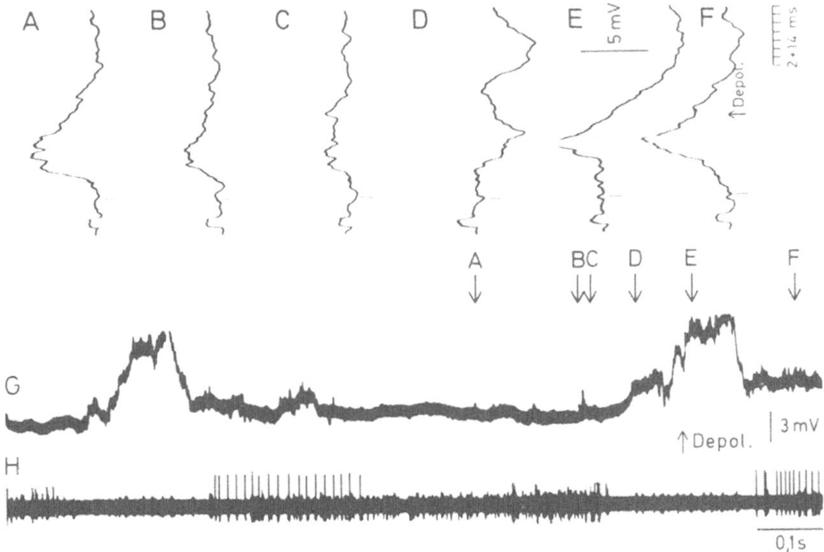

Figure 4. Reversal of a synaptic response to cutaneous nerve stimulation (Sur) during a particular phase of the step cycle during fictive spinal locomotion. Intracellular recording of a DP motoneuron (A–G; a.c. in A–F, d.c. in G). In H (large unit) recording of a Pl efferent fibre, simultaneously with G. The responses to Sur nerve stimulation (strength 8T, recurrence frequency 1 Hz) are gathered from different step cycles. Their position in relation to the phase of the cycles is marked in the cycle shown in G and H.

Phase-dependent transmission in propriospinal pathways from forelimb afferents to hindlimb motoneurons

Stimulation of medium- to high-threshold cutaneous afferents (group II and III) and muscle afferents (high group II range and group III) of the forelimb evoked EPSPs, and, less frequently, IPSPs in most hindlimb motoneurons with a central latency of 5.6 msec to about 16 msec (Schomburg *et al.*, 1978). These EPSPs and IPSPs were modulated during locomotor activity in a quite similar way to the segmental responses to cutaneous nerve stimulation—the EPSPs decreased during the reciprocal phase of a motoneuron while the IPSPs often became more pronounced during that phase.

In figure 5 the EPSPs, which were evoked in an ABSm motoneuron by stimulation of the contralateral superficial radial nerve (20T), were completely absent during the reciprocal phase of the motoneuron (E–I). As the stimulus recurrence frequency should not be faster than about 1 Hz (Schomburg *et al.*, 1978), in this case the extremely long duration of the step cycle (about 15 sec) allowed the demonstration of phase modulation within one cycle. Similar effects were observed with faster locomotor rhythms.

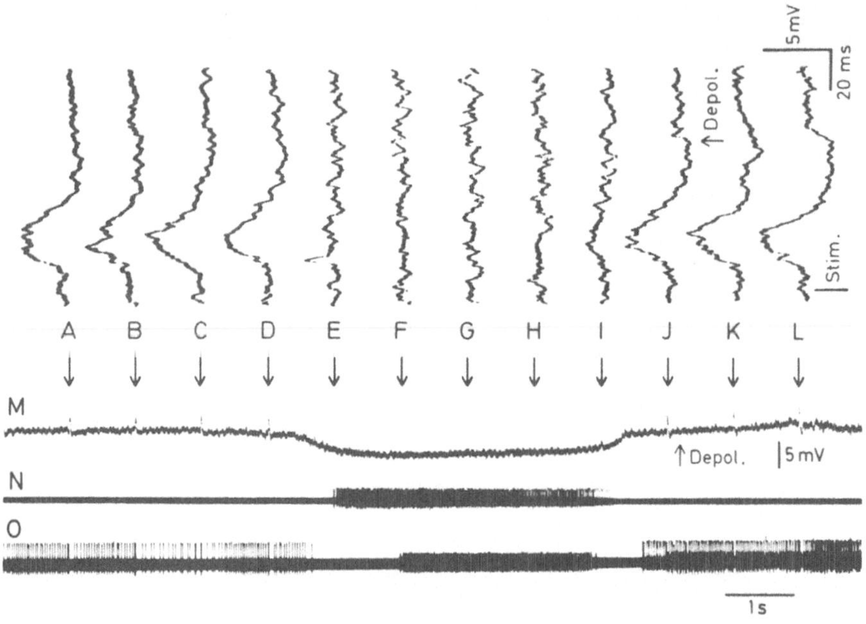

Figure 5. Phase-dependent transmission in the descending excitatory reflex pathway from forelimb afferents to a hindlimb motoneuron. General recording technique as in figure 2. Intracellular recording of an ABSm motoneuron (A–M; a.c. in A–L, d.c. in M). Simultaneous filament recording of a DP (N) and GS (large unit in O) α-efferent fibre. The responses to stimulation of the contralateral superficial radial nerve (strength 20T) are displayed separately in A–L; position marked in M. Stimulus marked in L.

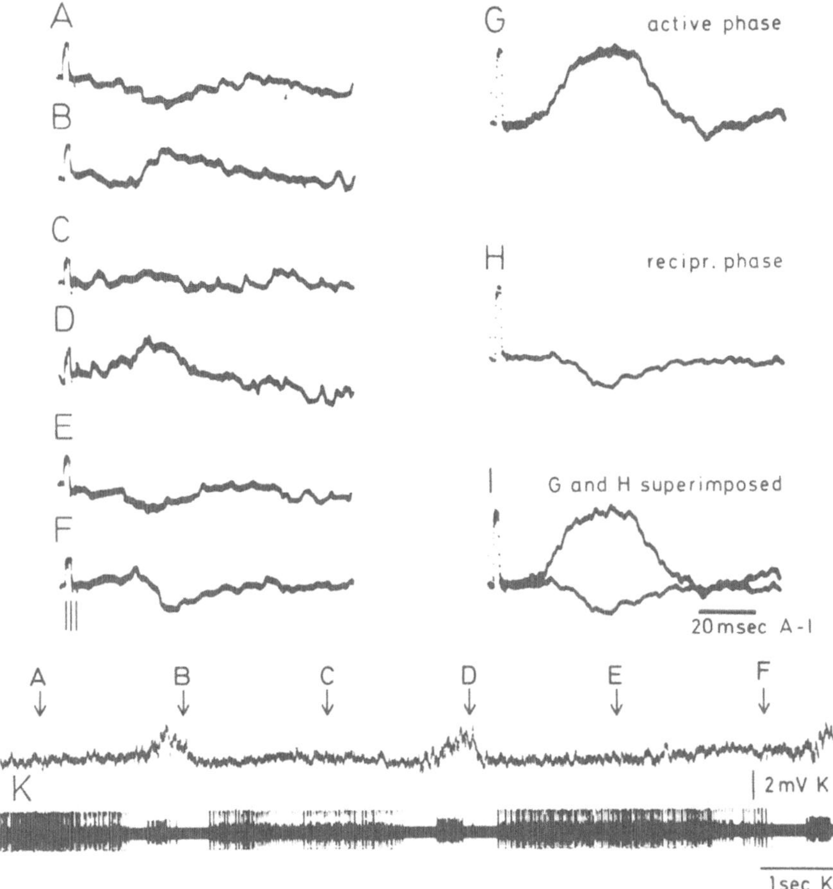

Figure 6. Phase-dependent reversal of a synaptic response to forelimb nerve stimulation in a hindlimb motoneuron. General recording technique as in figure 2. Intracellular recording of a DP motoneuron (a.c. in A–F, d.c. in K, upper trace). Simultaneous recording of a GS efferent fibre (large unit in K, lower trace). The responses to stimulation of the ipsilateral superficial radial nerve (strength 20T) are displayed in A–F; their position is marked in K. G–I averaged responses (20 samples each) taken during the active (G) and reciprocal phase (H). Calibration pulse in A–I 1 mV; stimulus (train of three pulses) is marked below F.

The clearest example of a different phase-dependent transmission for the propriospinally mediated EPSPs and IPSPs is shown in figure 6. This DP motoneuron showed quite short active phases. During these phases, stimulation of the ipsilateral superficial radial nerve evoked EPSPs (B and D), while during the reciprocal phases (A, C, E and F) IPSPs occurred after the same stimulus. From the superimposed averaged records (I) it appears that the latency of the IPSP was slightly longer than that of the EPSP, as was observed in previous experiments under static conditions (Schomburg *et al.*, 1978).

It is noteworthy that the responses to ipsilateral (figure 6) and contralateral (figure 5) nerve stimulation were modulated in the same way, i.e. the cyclic modulation of the EPSPs and/or IPSPs was only dependent on the phase of the step cycle of the hindlimb to which the motoneuron projected. Moreover, the phase dependence of the transmission of reflexes to flexor and extensor motoneurons had the same characteristics, except that EPSPs were often more pronounced in extensor motoneurons and IPSPs tended to be more frequent in flexor motoneurons (cf. Schomburg *et al.*, 1978).

The only stable exception from the described pattern was observed in FDL motoneurons. They received, often in addition to the EPSPs and IPSPs described above, early IPSPs (3.1 msec to about 8 msec), which were evoked by low-threshold cutaneous afferents (group I and II) and medium-threshold (mainly group II) muscle afferents only from the ipsilateral forelimb (Schomburg *et al.*, 1975; Schomburg *et al.*, 1978). These short latency IPSPs, which were specific for FDL motoneurons, showed a phase dependence which was the reverse of that of the late IPSPs, i.e. they were mainly increased during the active phase of the FDL motoneurons.

DISCUSSION

The central finding of our investigations is that synaptic responses, which were evoked in hindlimb motoneurons by stimulation of different cutaneous hindlimb nerves and ipsilateral and contralateral cutaneous and muscle nerves of the forelimb, showed basically the same dependence on the phase of the step cycle during spinal locomotion—EPSPs were favoured during the active phase of the corresponding motoneurons, while IPSPs (except the early propriospinally mediated IPSPs in FDL motoneurons) tended to increase or to occur during the reciprocal phase. Thus the main mechanism appears to lie in a shift of balance between the excitatory and inhibitory pathways, rather than in an alternate switching between these pathways. The latter may also occur but seems to be the exception.

The observed phase-dependent influence upon the reflex transmission is presumed to take place at an interneuronal level. Tests with constant-current pulses injected through the micro-electrode during different phases of the step cycle revealed only insignificant conductance changes of the motoneuronal membrane, which cannot explain the observed distinct changes, particularly in the direction of the synaptic responses. Moreover, the occasional changes of latency indicate an alteration of interneuronal transmission. Extracellular field potentials, which were recorded after the intracellular tests, never showed variations which could explain the intracellularly recorded findings.

The coincidence of the phase dependence of the reflex transmission from hindlimb and ipsilateral and contralateral forelimb afferents to hindlimb motoneurons indicates that both pathways were probably influenced by the same mechanism, the 'locomotor generator' responsible for the corresponding hindlimb. Theoretically this coincidence could be mediated by phasically modulated common interneurons intercalated in the pathways. However, despite intensive trials, a significant convergence from the corresponding pathways on to common segmental interneurons could not be demonstrated.

Since the phasic modulation of the responses to ipsilateral and contralateral forelimb nerve stimulation only depended on the phase of the ipsilateral hindlimb step cycle, an influence from cervical locomotor structures can be excluded, and it seems unlikely that the pathways mediating these descending propriospinal responses are identical with the pathways co-ordinating forelimb and hindlimb locomotor activity.

The phase-dependent reflex reactions to mechanical stimulation of the dorsum of the paw observed in intact or chronic spinal cats during locomotion have been explained as reactions for avoiding an obstacle. These 'stumbling corrective reactions' are thought to be controlled by the spinal locomotor generator (Forssberg *et al.*, 1975, 1977; Forssberg, 1979). The finding that synaptic responses to stimulation of cutaneous hindlimb afferents of different origin (Saph., Sur. and SPC) and of different forelimb afferents (cutaneous and muscle) were phasically modulated in the same manner during fictive spinal locomotion, indicates that the 'locomotor generator' possibly has a less specific function. By facilitating the excitatory pathways and depressing the inhibitory pathways to active motoneuron pools, and vice versa for antagonistic motoneurons, the 'locomotor generator' would functionally support those muscles which are in their active phase, independently of the source of the stimulated afferents. This behaviour does not imply a non-specific nociceptive mechanism, since it was also present for synaptic responses evoked by very low stimulus strengths (below 1.5T) from cutaneous hindlimb nerves and from forelimb afferents below group III threshold. The observed possible phase-dependent facilitation of polysynaptic group Ia excitation to homonymous motoneurons during the active phase (Schomburg and Behrends, 1978a) would support this functional concept. At the moment, it cannot be excluded that electrical stimulation of cutaneous nerves with its coactivation of afferents of different receptors may evoke reflex responses, which are modulated in a different way by the 'locomotor centre' than the responses evoked by mechanical stimulation. However, this does not seem to be very likely, since reflex reactions to weak electrical and mechanical skin stimulation were modulated in a quite similar way during spinal locomotion (Forssberg *et al.*, 1977; Andersson *et al.*, 1978).

The early IPSP in FDL motoneurons which could be evoked by weak mechanical or electrical stimulation of the forelimb seems to have a different functional meaning from that of the later IPSPs. The early IPSP was not only evoked by other afferents from the ipsilateral side, but its phase-dependent modulation also showed a reversed pattern compared with the later IPSPs, i.e. it was more pronounced during the active phase. This pathway is presumably a specific propriospinal, oligosynaptic inhibitory pathway to FDL motoneurons, which can probably be activated under normal locomotor conditions and which becomes efficient during the extension phase. This propriospinal inhibition, which is also present in the dog, would ensure the avoidance of plantar flexion of the end-phalanges of the digits (and additionally in cats, claw protrusion) during parts of the hindlimb extension phase (Schomburg *et al.*, 1975; Schomburg *et al.*, 1978).

ACKNOWLEDGEMENT

The work was supported by the Deutsche Forschungsgemeinschaft (SFB 33).

Received on June 2nd, 1980.

REFERENCES

Andersson, O., Forssberg, H., Grillner, S. and Lindquist, M. (1978). Phasic gain control of the transmission in cutaneous reflex pathways to motoneurones during 'fictive' locomotion, *Brain Res.*, **149,** 503–507

Eccles, R. M. and Lundberg, A. (1959). Synaptic actions in motoneurones by afferents which may evoke the flexion reflex, *Arch. ital. Biol.*, **97,** 199–221

Forssberg, H. (1979). Stumbling corrective reaction: A phase-dependent compensatory reaction during locomotion, *J. Neurophysiol.*, **42,** 936–953

Forssberg, H., Grillner, S. and Rossignol, S. (1975). Phase dependent reflex reversal during walking in chronic spinal cats, *Brain Res.*, **85,** 103–107

Forssberg, H., Grillner, S. and Rossignol, S. (1977). Phasic gain control of reflexes from the dorsum of the paw during spinal locomotion, *Brain Res.*, **132,** 121–139

Grillner, S. and Zangger, P. (1974). Locomotor movements generated by the deafferented spinal cord, *Acta physiol. scand.*, **91,** 384–394

Grillner, S. and Zangger, P. (1979). On the central generation of locomotion in the low spinal cat, *Expl Brain Res.*, **34,** 241–261

Holmqvist, B. (1961). Crossed spinal reflex actions evoked by volleys in somatic afferents, *Acta physiol. scand.*, **52,** Suppl. 181, 1–66

Holmqvist, B. and Lundberg, A. (1961). Differential supraspinal control of synaptic actions evoked by volleys in the flexion reflex afferents in alpha motoneurones, *Acta physiol. scand.*, **54,** Suppl. 186, 1–51

Hongo, T., Jankowska, E. and Lundberg, A. (1969). The rubrospinal tract. II: Facilitation of interneuronal transmission in reflex pathways to motoneurones, *Expl Brain Res.*, **7,** 365–391

Jankowska, E., Jukes, M. G. M., Lund, S. and Lundberg, A. (1967a). The effect of DOPA on the spinal cord. 5. Reciprocal organization of pathways transmitting excitatory action to alpha motoneurones of flexors and extensors, *Acta physiol. scand.*, **70,** 369–388

Jankowska, E., Jukes, M. G. M., Lund, S. and Lundberg, A. (1967b). The effect of DOPA on the spinal cord. 6. Halfcentre organization of interneurones transmitting effects from the flexor reflex afferents, *Acta physiol. scand.*, **70,** 389–402

Lundberg, A. and Voorhoeve, P. (1962). Effects from the pyramidal tract on spinal reflex arcs, *Acta physiol. scand.*, **56,** 201–219

Miller, S., Ruit, J. B. and van der Méche, F. G. A. (1977). Reversal of sign of long spinal reflexes dependent on the phase of the step cycle in the high decerebrate cat, *Brain Res.*, **128,** 447–459

Schomburg, E. D. and Behrends, H. B. (1978a). The possibility of phase dependent monosynaptic and polysynaptic Ia excitation to homonymous motoneurones during fictive locomotion, *Brain Res.*, **143,** 533–537

Schomburg, E. D. and Behrends, H. B. (1978b). Phasic control of the transmission in the excitatory and inhibitory reflex pathways from cutaneous afferents to alpha motoneurones during fictive locomotion in cats, *Neurosci. Lett.*, **8,** 277–282

Schomburg, E. D., Meinck, H.-M. and Haustein, J. (1975). A fast propriospinal inhibitory pathway from forelimb afferents to motoneurones of hindlimb flexor digitorum longus, *Neurosci. Lett.*, **1,** 311–314

Schomburg, E. D., Meinck, H.-M., Haustein, J. and Roesler, J. (1978). Functional organization of the spinal reflex pathways from forelimb afferents to hindlimb motoneurones in the cat, *Brain Res.*, **139,** 21–33

Schomburg, E. D., Roesler, J. and Kenins, P. (1978). On the function of the fast long spinal inhibitory pathway from forelimb afferents to flexor digitorum longus motoneurones in cats and dogs, *Neurosci. Lett.*, **7,** 55–59

Schomburg, E. D., Roesler, J. and Meinck, H.-M. (1977). Phase dependent transmission in the excitatory propriospinal reflex pathway from forelimb afferents to lumbar motoneurones during fictive locomotion, *Neurosci. Lett.*, **4,** 249–252

Viala, D. and Buser, P. (1971). Modalités d'obtention de rhythmes locomoteurs chez le lapin spinal par traitements pharmacologiques (DOPA, 5-HTP, D-Amphétamine), *Brain Res.*, **35,** 151–165

On the feedback control of the cat's hindlimb during locomotion

OLOF ANDERSSON* and STEN GRILLNER*

SUMMARY

A short account is given of the different feedback signals that control the central pattern generators for locomotion. Particular attention is given to the feedback signals that are elicited by static hip position and by dynamic hip movements.

During normal locomotion an animal must (1) produce the stereotypic propulsive movements characteristic of its species, (2) maintain its equilibrium during these movements and (3) adapt them not only to the environment but also to the animal's own goals. We will deal only with the neural generation of the stereotypic locomotor movements. In vertebrates, central pattern generators assume the main responsibility for the pattern of muscle activity during progression (cf. Grillner, 1975, 1980; Grillner and Zangger, 1975; Perret, 1976). These pattern generators are activated to a different degree with varying speeds of locomotion. Feedback from receptors in the moving limb is thus not responsible for the actual co-ordination of the locomotor bursts in the different muscles.

Feedback acting directly on the motoneurons, such as the monosynaptic effects from muscle spindle afferents, will inevitably add to or subtract from the general excitability of the motoneurons in each stepcycle. We will, however, focus on another type of feedback, which acts on the central pattern generators themselves. It will, among other things, affect the duration of the different phases of the stepcycle (Andersson *et al.*, 1980; Grillner, 1975, 1979, 1980).

Kittens made spinal at low thoracic level at an age of one or two weeks can be made to use their hindlimbs during locomotion. This becomes evident if a chronic spinal kitten is placed on a treadmill with the hindlimbs on the moving

*Karolinska Institutet, Stockholm

treadmill belt and with the forelimbs standing on a platform. Under such conditions the hindlimbs can support the weight of the hindpart despite a marked impairment of the control of equilibrium. When the treadmill is moved the hindlimbs can walk in an alternating fashion with a pattern of muscle activity similar to that of an intact cat. Not only can the hindlimbs perform walking movements but also they can adapt to the changes of the belt speed, often within a rather wide range (Grillner, 1973; Forssberg *et al.*, 1980a,b; cf. Forssberg and Grillner, 1973). In fact at high speed the co-ordination between the two hindlimbs may change from alternation as in a walk to more or less simultaneous movements as in a gallop. These findings show that peripheral signals are influencing the locomotor movements. In fact the stepcycle duration and some of its components are under a special feed-back control

Afferents influenced by the position in the hip joint are of importance in this context. If the backward movement of the limb is stopped during hip extension, the extensor activity will continue for a very long period even when the other limb(s) continue(s) to perform walking movements. On the other hand, if the limb is rapidly extended to its normal most caudal position a flexor activation will be induced (Grillner and Rossignol, 1978). A similar finding in the flexion phase was briefly reported by Orlovsky (1972). It is not known if receptors in the muscles around the hip joint and/or joint receptors are responsible (Grillner and Rossignol, 1978).

The interaction between the centrally generated pattern and the peripheral feedback can be analysed more clearly under a slightly different experimental situation. Decorticate or spinal animals (Perret, 1976; Grillner and Zangger, 1979; Vidal *et al.*, 1979) may be made to exhibit rhythmic activity in the efferents to different muscles after being paralysed by drugs that block the neuromuscular transmission. This activity corresponds to the efferent signals that would normally elicit locomotor movements. In such 'ficitive locomotion' the centrally generated activity has been dissociated from the movement. It is then possible to apply different movements to the limb to test their effects on the central network (Andersson *et al.*, 1978, 1980). If rhythmic movements mimicking the normal hip movements of the cat are imposed during fictive locomotion in the spinal preparation, the centrally generated efferent pattern is markedly modified. If there is a reasonable match (e.g. ±50%) between the frequency of the imposed movements and the resting frequency of the fictive locomotion, the latter activity becomes entrained by the movements. In fact flexor activity will occur during hip flexion and extensor activity during hip extension, i.e. the approximate phase relationships which occur during normal locomotion. In addition, efferents to different muscles at all joints of the limb, ranging from proximal hip muscles to distal foot muscles (such as extensor digitorum brevis), are co-ordinated in a way similar to that occurring during locomotion (Andersson *et al.*, 1978, 1980, and personal observations).

Thus it can be concluded that a phase relationship (between movement and efferent activity) approximately corresponding to that in locomotion can be maintained by feedback signals elicited from the moving limb. Extensive denervations show that only the small muscles around the hip (e.g. pectineus) and the joint afferents need to be intact to achieve entrainment, but it has not been possible to decide if muscle and/or joint afferents are required (Andersson and Grillner, 1980; Andersson *et al.*, 1978, 1980). In a related experimental situation muscle receptors have been found to be of importance (Rossignol and Gauthier, 1980).

To test the relevant feedback parameters further the spinal fictive locomotor preparation has been utilised in a slightly different way. During the continuous bursting pattern, well-defined ramp movements of the femur (with respect to the hip) were applied repeatedly in different parts of the fictive stepcycle (Andersson *et al.*, 1980; Andersson and Grillner, 1980). Such ramps were applied, for example, in every fifth or tenth locomotor cycle. The effects on the efferent pattern differ markedly with the direction of the ramp and with the phase in the cycle in which the movement is applied. A 'flexion ramp' applied in the second half of the flexor burst enhances this burst and prolongs it, but an extension ramp will, in contrast, cause a reduction and earlier cessation of the burst. Opposite effects, even though weaker, of the ramp movements could be encountered during the period of extensor activity. Prominent effects were also obtained on the 'cycle duration'. These effects were markedly dependent on when the ramp was applied. The strongest change in the phase response curve occurred with extension ramps applied in the later part of the ipsilateral flexor burst. This may be of particular importance for causing an effective entrainment.

The feedback acting on the central pattern generators that has been discussed above is summarised in figure 1. During the support phase (extension), afferent signals in the earlier part of the movement range will prolong the period of extensor activity, but in contrast, afferent activity in the later

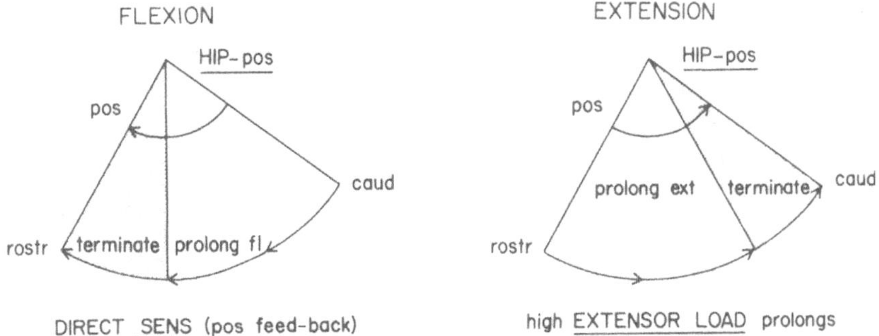

Figure 1. Schematic representation of the feedback effects on the central pattern generator controlling a single hindlimb in cat locomotion (from Grillner, 1979).

(caudal) part will instead act to induce flexion. The reverse is true for the flexion phase, although there is an asymmetry in that the effects that promote termination of flexion are much weaker. Figure 1 in addition indicates the directional sensitivity mentioned and also the finding that the load on the extensor muscles is an important signal. The higher the load on the ankle extensors the less likely it is that a flexion can be initiated (Pearson and Duysens, 1976; Duysens and Pearson, 1980).

The schematics of figure 1 not only indicate that the reflex effects may explain entrainment (with its control of the duration of the different phases of the stepcycle), but also that their main action may be to control the amplitude of the limb movements in each stepcycle (rostro-caudal extent of the movements in the hip). This may be useful not only when the limb is subjected to unexpected perturbations, but also in situations when the animal changes speed, moves in circles or walks up hill or down hill (Andersson *et al.*, 1980; Grillner, 1979, 1980).

ACKNOWLEDGEMENTS

This work was supported by the Swedish Medical Research Council (Project No. 3026), Magnus Bervalls Stiftelse and Karolinska Institutets fonder. The help of Mrs I. Klingebrant is gratefully acknowledged.

Received on October 13th, 1980.

REFERENCES

Andersson, O., Forssberg, H., Grillner, S. and Wallén, P. (1980). Peripheral feedback mechanisms acting on the central pattern generators for locomotion in fish and cat, *Can. J. Physiol., Pharmacol.*, to be published

Andersson, O. and Grillner, S. (1980). Phase dependent effects exerted on the spinal locomotor generator by hip movements, *Neurosci. Lett.*, Suppl. No. 5, S222

Andersson, O., Grillner, S., Lindquist, M. and Zomlefer, M. (1978). Peripheral control of the spinal pattern generators for locomotion in cat, *Brain Res.*, **150,** 625–630

Duysens, J. and Pearson, K. G. (1980). Inhibition of flexor burst generation by loading ankle extensor muscles in walking cats, *Brain Res.*, **187,** 321–332

Forssberg, H., and Grillner, S. (1973). The locomotion of the acute spinal cat injected with Clonidine i.v., *Brain Res.*, **50,** 184–186

Forssberg, H., Grillner, S. and Halbertsma, J. (1980a). The locomotion of the low spinal cat. 1: Coordination within a hindlimb, *Acta physiol. scand.*, **108,** 269–281

Forssberg, H., Grillner, S., Halbertsma, J. and Rossignol, S. (1980b). The locomotion of the low spinal cat. 2: Interlimb coordination, *Acta physiol. scand.*, **108,** 283–295

Grillner, S. (1973). Locomotion in the spinal cat. In *Control of Posture and Locomotion* (eds R. B. Stein, K. G. Pearson, R. S. Smith and J. B. Redford), New York, Plenum Press, 515–535

Grillner, S. (1975). Locomotion in vertebrates: Central mechanisms and reflex interaction, *Physiol. Rev.*, **55**, 247–304

Grillner, S. (1979). Interaction between central and peripheral mechanisms in the control of locomotion. In *Reflex Control of Posture and Movement* (*Progress in Brain Research*, Vol. 50) (eds R. Granit and O. Pompeiano), Elsevier/North-Holland Biomedical Press, 227–235

Grillner, S. (1980). Control of locomotion in bipeds, tetrapods and fish. In *Handbook of Physiology*, Vol. III: *Motor Control* (ed. V. Brooks), to be published

Grillner, S. and Rossignol. S. (1978). On the initiation of the swing phase of and locomotion in chronic spinal cats. *Brain Res.*, **146**, 269–278

Grillner, S. and Zangger, P. (1975). How detailed is the central pattern generator for locomotion?, *Brain Res.*, **88**, 367–371

Grillner, S. and Zangger, P. (1979). On the central generation of locomotion in the low spinal cat, *Expl Brain Res.*, **34**, 241–262

Orlovsky, G. N. (1972). Activity of rubrospinal neurons during locomotion, *Brain Res.*, **46**, 99–112

Pearson, K. G. and Duysens, J. (1976). Function of segmental reflexes in the control of stepping in cockroaches and cats. In *Neural Control of Locomotion*, Vol. 18 (eds R. Herman, S. Grillner, P. S. G. Stein and D. Stuart), New York, Plenum Press, 519–538

Perret, C. (1976). Neural control of locomotion in the decorticate cat. In *Neural Control of Locomotion*, Vol. 18 (eds R. Herman, S. Grillner, P. Stein and D. Stuart), New York, Plenum Press, 587–615

Rossignol, S. and Gauthier, L. (1980). An analysis of mechanisms controlling the reversal of crossed spinal reflexes, *Brain Res.*, **182**, 31–45

Vidal, C., Viala, D. and Buser, P. (1979). Central locomotor programming in the rabbit, *Brain Res.*, **168**, 57–73

A critique of the papers by Rossignol, Julien, Gauthier and Lund; Forssberg; Schomburg, Behrends and Steffens; and Andersson and Grillner

J. DUYSENS*

In this session we learned a great deal about the reflex control of locomotion as well as about the locomotory control of reflexes. From Andersson and Grillner, and Rossignol *et al.* we learned how afferent signals from the hip or from distal cutaneous sources can modulate the working of the central pattern generator for locomotion. On the other hand we learned from Fossberg and Schomburg *et al.* that this spinal generator can in turn modulate the effects which afferent input may have on motoneurons.

Nevertheless the wealth of data presented may seem somewhat bewildering or even confusing because the results of the various authors are not always easily integrated and because of difficulties with terminology. While the question of terminology will be addressed in the second part of this critique I will first try to integrate some—seemingly unrelated—observations and put them in perspective.

Schomburg *et al.* showed that stimulation of the sural nerve during fictive locomotion leads to EPSPs in flexor motoneurons except for a short period of about 100 msec at the beginning of the flexor activity period. Next, Forssberg demonstrated that stimulation of the paw during fictive locomotion elicits late flexor EPSPs only for stimuli given during early flexor activity. Why are reflex effects so different in the early and late parts of the period of flexor activity during locomotion? Is there a fundamental functional difference between these two parts of the period of flexor activity?

A close examination of the events occurring at the transition from stance to swing does indeed suggest that there may be such a difference. During the step cycle of the normal walking cat the first flexor EMG appears some 70 msec before the foot is lifted off the ground. No flexion movement occurs during this 70 msec period because of the inertia of the extension movement and the delay involved in the production of effective force. Consequently, the

*Laboratorium voor Neuro- en Pyschofysiologie, Leuven

function of sensory feedback during this period should be more related to an adjustment of the forthcoming flexor 'programme' than to a correction of an ongoing flexion movement. This is in contrast with the period of flexor activity following foot lift-off during which an actual movement of the limbs accompanies the flexor EMG activity, and short latency reflex effects would provide fast corrections to any type of perturbation. In a sense the situation following foot lift-off could be called 'closed loop' while the period before foot lift-off can be thought of as 'open loop'. If the period of flexor activity during locomotion consists of two distinctive parts one would expect this to be reflected in the organisation of the central pattern generator as well.

Previous data support such a notion since it was found in decerebrate cats that the last part of the period of flexor activity during walking can easily be suppressed by inhibitory cutaneous (Duysens, 1977) or proprioceptive (Duysens and Pearson, 1980) input, while the early flexor period cannot be further reduced. Instead, the first part of the period of flexor activity is either fully present or else completely absent, with no intermediary stages. These observations, along with Bayev's (1978) finding that the activity period in some flexors appears only at the initial period of the 'flexor' phase when there is a reduction in the intensity of fictive locomotion, suggest that the early part of the flexor burst is a period of 'minimal expression' of the flexor pattern generator. Once triggered, such a generator presumably goes through a first 'irreversible' phase during which some flexor excitatory pathways are suppressed (explaining the absence of early EPSPs in Schomburg *et al.*'s data) and others facilitated (explaining the appearance of late EPSPs in the results of Forssberg).

This 'irreversible' phase might actually start even before the onset of the period of flexor activity, since Andersson and Grillner's data show that the triggering of the flexor pattern generator is likely to be linked to full hip extension, which is reached some 70–80 msec before the onset of flexor EMG. Hence the transition from stance to swing has four phases. In the first phase, near the end of stance, there is presumably a gradual rise in the excitability of the flexor pattern generator which reaches threshold once the limb is sufficiently unloaded and the hip is fully extended (*see* Andersson and Grillner). The flexor pattern generator is then triggered and we enter the second period during which there is not yet any flexor EMG. The onset of flexor EMG marks the start of the third period which lasts till about foot lift-off. Finally flexion occurs and the swing phase begins in the fourth period. It is during this last period that one may expect that positive feedback signals of the type described by Andersson and Grillner and Rossignol *et al.* may have their full impact.

The second issue to be discussed in this critique concerns terminology. The term 'reflexes' in the context of this session on 'dependence of reflexes on phases of movement' has been interpreted by the various authors to mean

EPSPs, fixed latency EMG peaks, total EMG bursts or complex behavioural reactions. With such a variety of meanings it is becoming increasingly difficult to evaluate a notion such as 'reflex reversal', which really formed the starting point of this whole line of research. Should the term 'reflex reversal' apply when a given stimulus prolongs flexor bursts when given during the flexion phase and prolongs extensor bursts when given during the extension phase (*see* Rossignol *et al.*)? Or should the term be restricted to the case where a flexor EMG response at a given latency is replaced during part of the step cycle by an extensor EMG response at the same latency? In the normal cat one may observe a 'reversal' of the latter type despite a shortening of the duration of the corresponding EMG bursts (Duysens and Loeb, unpublished observation) so that in this case the two types of interpretation are not compatible. Perhaps it will soon be necessary to replace the words 'reflex reversal' with terms such as 'burst duration reversal', 'EPSP reversal', 'early response reversal', etc.

At a behavioural level the question of the existence of a reversal, in the sense of a totally different and opposite movement elicited by the same stimulus, is still unresolved. Although Forssberg does not use the term 'reversal' explicitly, he states that, 'the quality of the (stumbling) reaction differs completely depending on the phase of the step cycle', since flexion is evoked during swing while the same stimulus applied during stance does not evoke flexion. This description seems to me exact only if 'flexion' is used in the very restricted sense of a foot-lift. However, if 'flexion' is used in a general kinematic sense, as in 'flexion movement' then there is no basic difference between the behavioural reactions seen during swing and stance. Indeed, during stance the main response is one of extensor inhibition followed by rebound and the behavioural translation of this is an excessive yield of the stimulated limb. Such yield is no less a flexion movement than the reflexly induced flexion during swing.

It is also questionable as to whether one should define a set of reflex responses in terms of the behavioural context, as is the case with the phrase 'stumbling corrective response'. Just as there is no reason to develop terms such as 'walking', 'jumping' or 'standing' stretch reflexes, there may be little need to call a set of polysynaptic reflex responses 'stumbling' because they happen to be described in the context of obstructed locomotion. The danger would then exist that one would start thinking about such responses as specific for locomotion, while in fact—as Forssberg points out himself—passive animals show the same set of responses, both with respect to stimulus site and with respect to participation of various limb muscles.

Received on August 11th, 1980.

REFERENCES

Bayev, K. V. (1978). Central locomotor program for the cat's hindlimb, *Neurosci.*, **3**, 1081–1092

Duysens, J. (1977). Reflex control of locomotion as revealed by stimulation of cutaneous afferents in spontaneously walking premammillary cats, *J. Neurophysiol.*, **40**, 737–751

Duysens, J. and Pearson, K. G. (1980). Inhibition of flexor burst generation by loading ankle extensor muscles in walking cats, *Brain Res.*, **187**, 321–332

REPLY TO DUYSENS'S CRITIQUE BY ROSSIGNOL, JULIEN, GAUTHIER AND LUND

Broad expressions such as 'reflex reversal' always fail to describe the details. However, it seems better, at this stage of our knowledge, to keep a broad expression which is in continuity with the historical development of the subject, and which defines in a nutshell the main characteristics of the observations. Perhaps the (possibly sterile) debate on terminology could await more definitive studies on the underlying mechanisms resposible for these observations.

Received on October 24th, 1980.

REPLY TO DUYSENS'S CRITIQUE BY SCHOMBURG, BEHRENDS AND STEFFENS

We agree that the events occurring during locomotion at the transition from extension to flexion phase are of special interest. However, our data on phase dependent reflex transmission do not generally show notable changes at these transition phases. The observation of EPSPs in flexor motoneurons, which were replaced by IPSPs at the transition to the flexion phase, has to be regarded as an exception. A phase-dependent modulation of the size of EPSPs or IPSPs was the more general feature.

Received in October, 1980

A critique of the papers by Rossignol, Julien, Gauthier and Lund; Forssberg; Schomburg, Behrends and Steffens; and Andersson and Grillner

DOUGLAS G. STUART*

SUMMARY

This critique focuses on the need for detailed kinesiological studies of locomotion in general, and phase-dependent reflexes in particular. It is further emphasised that agreement is needed on the definition of reflex reactions during stepping and that the selective activation issue should be taken into account when ascribing reflex effects to the input of a single afferent species.

INTRODUCTION

The present reports on phase-dependent reflexes provide a valuable review of an area of research that has far broader implications for neurobiology than the still elusive problem of the precise role played by segmental proprioceptive reflexes in the control of locomotion. For example, there is recent description of both phase-dependent respiratory reflexes (von Euler, 1980) and state-dependent cardiovascular reflexes (Smith *et al.*, 1981). It is likely that the very definition of most somatic and visceral reflexes will eventually include both state dependencies and phase dependencies, once techniques are developed to study them in natural and near-natural circumstances (Stuart *et al.*, 1979). This viewpoint has long been accepted in the segmental motor control literature (for review, *see* Forssberg *et al.*, 1977) which, at least in the twentieth century, has had many similarities to developments in the respiratory motor control literature (e.g. Graham Brown, 1916; Sears, 1971; Grillner, 1975,

*University of Arizona Health Sciences Centre, Tucson

1979). However, the significance of state- and phase-dependent reflexes has been far less appreciated in other areas (e.g. cardiovascular responses to exercise) for reasons that probably relate more to the way textbooks are written and to the way scientists communicate with each other than to differences in the neuronal machinery that controls the reflex responses themselves.

The purpose of this critique is not to address specific points made in the preceding papers; rather, from the perspective of an interested and particularly supportive 'outsider' who frequently edits work in this field, I would like to voice some general concerns that I feel should be dispelled if the wider readership in neurobiology is to appreciate fully the recent work on mammalian locomotion.

KINESIOLOGICAL ISSUES

Locomotor capability of test preparations

The recent mammalian locomotion literature features use of six basic types of preparation, five of which are studied during treadmill or overground locomotion and one of which is paralysed for the analysis of 'fictive' locomotion (treated separately below).

Overground locomotion is studied in a relatively unrestrained state (e.g. Prochazka *et al.*, 1976) except that behavioural (training) techniques are used to bring out the basic gait patterns of interest. In contrast, treadmill locomotion is 'controlled' by belt speed and, in addition, either by training techniques (conscious preparations, e.g. Loeb *et al.*, 1977), by electrical stimulation of central structures (high decerebrate and thalamic preparations—reviewed in Lundberg, 1969; Grillner, 1975; Shik and Orlovsky, 1976; and Wetzel and Stuart, 1976), or, in the case of spinal preparations, by pharmacological activation of the isolated neuronal machinery combined usually with augmented sensory input such as pinching the tail (Grillner, 1975).

Similarities in the locomotor performance of these various preparations have been emphasised here and in the recent literature, but, in reality, these preparations differ far more substantially in the quality, vigour, rhythmicity and adaptability of motor performance than might be readily apparent to the wider readership in neurobiology. This point deserves emphasis because it would be unfortunate if, as in the de-afferentation literature, so much emphasis was placed on what locomotor capabilities the 'reduced' preparations have retained that information was not forthcoming on the significance of what subtle capabilities have been lost. By a quantitative comparison of the locomotor deficits of these various reduced preparations, it might be possible to bring out the functional significance of the lesioned pathways. This approach has already proven of value in the recent dorsal column literature (e.g. Melzack and Southmayl, 1974; McCormack and Dubrovsky, 1979), and a case for its value in the study of locomotion is no less compelling (e.g. Udo *et al.*, 1980).

Definition of reflex reactions during stepping

As more groups begin to study the nature of reflex movements during stepping, it is inevitable that the movements will be defined differently in different laboratories. Dialogue and agreement on definitions is important if progress is to ensue in an orderly rather than a circular fashion. For example, the study of spasticity has been impeded by the lack of agreement on its clinical manifestations (Landau, 1974; Lance and McLeod, 1975), which, in turn, leads to a lack of agreement on an appropriate animal model for the syndrome (*see*, however, Burke *et al.*, 1972; Rymer *et al.*, 1979). As an example of this issue, Wand *et al.* (1980) have recently argued that the 'placing' reactions they studied are similar in most respects to the 'stumbling corrective reaction' of Forssberg (1979—*see also* Andersson *et al.*, 1978) and the 'non-specific, non-nociceptive flexion reflexes' of Duysens and Stein (1978), if the motor events are defined in terms of limb movement (Bard, 1938) rather than electrophysiology. It would have been fitting for there to have been dialogue on this viewpoint at the symposium, because, far from being simply an exercise in semantics, it would have brought out the need to distinguish between afferent and efferent activities induced by electrical (as opposed to the more naturally occurring mechanical) stimulation, and, further, the caution required when making the claim that a naturally occurring reflex is of purely spinal origin in the mammal.

Definition of 'fictive' locomotion

Much has been said at the symposium about 'fictive' locomotion. The term was apparently first used by Perret *et al.* (1972a, b; *see also* Perret, 1973, and, for a possible immediate precedent, Viala and Buser, 1971). Originally, it denoted a pattern of peripheral efferent neuronal activity in a paralysed preparation that bore close resemblance to the pattern in a locomoting animal, as if to suggest that the 'same' interneuronal machinery was in operation but deprived of supportive effects from locomotor-induced sensory feedback.

Subsequently, the term appears to have been used less precisely such as to include any evidence in a paralysed preparation of rhythmic efferent discharge in even a single ipsilateral hindlimb muscle nerve. However, it is well known that non-paralysed decorticate and decerebrate cats can exhibit several forms of rhythmical discharge that bear varying degrees of relationship to the locomotor rhythm. Similarly, it should not be construed that rhythmical discharge in a hindlimb nerve of a low spinal paralysed cat as induced by Nialamide and DOPA is necessarily closely related to the locomotor rhythm, because when the same drugs are given to the non-paralysed preparation, locomotor-type activity may or may not ensue.

It would appear that 'ground rules' are required for use of the term 'fictive' locomotion along the lines initially suggested by Perret (1973; *see also* Grillner and Zangger, 1979 and Cabelguen *et al.*, 1980) that will presumably include close attention to the patterns of efferent discharge in nerves to select muscles on both sides of the body.

NEUROPHYSIOLOGICAL ISSUES

Definition of the 'stepping generator'

We heard at the symposium of several interesting attempts to unravel the extent to which phase-dependent reflex effects are controlled by the interneuronal machinery that produces the stepping rhythm (the 'stepping generator') as opposed to reflex pathways with direct access to the motoneurons themselves. In my opinion, this effort, while technically commendable, is somewhat misplaced. In contrast to work on lower vertebrates and invertebrates (reviewed in Herman *et al.*, 1976), precise delineation of the nature and operation of the neuronal machinery that controls stepping in a mammal defies current technology and theory. Even its anatomical definition must await more detailed kinesiological studies because, at this stage, it is premature to emphasise spinal control of the rhythmicity of stepping to the exclusion of brain stem and cerebellar control, not only with regard to interlimb co-ordination which has facultative propensities (Stuart *et al.*, 1973; English, 1979), but also for the 'fine tuning' of a single limb's step (*see*, for example, figure 21 in Wetzel and Stuart, 1976).

In recent years, much has been said about the reciprocal half-centre hypothesis (Lundberg, 1969) and the ring hypothesis (Shik and Orlovsky, 1976) which were elaborated to explain the remarkable spinal contribution to at least the hindlimb component of stepping. While these hypotheses have stimulated much interest in stepping research and its rich history, they should serve today to warn us of the immensity of the task ahead. Even a cursory examination of the extensive literature on the genesis of respiratory rhythmicity (e.g. von Euler, 1977; Wyman, 1977) which, at least at first glance, appears far 'simpler' than the four-legged locomotor rhythm, is sufficient to illustrate that the CNS could control the locomotor rhythm in a multiplicity of ways. Given such complexity, what would be particularly valuable at this stage, and, in the light of what we have heard at this symposium from Prochazka and Wand (1981) and from Loeb and Hoffer (1981), what appears technically feasible in the immediate years ahead is the generation of detailed kinesiological data on how muscles actually contribute to the execution of locomotion. Despite valuable beginnings (Engberg and Lundberg, 1969; Gambarjan *et al.*, 1971; Walmsley *et al.*, 1978; English, 1979), much is to be learned in this area which has so often been neglected in the motor-control literature to this point. Only when the subtle details of the muscle timings and forces are fully understood will it be possible to elaborate a kinesiology-based hypothesis that gives insight into the step-generating mechanisms.

Selective activation of muscle afferents

My final concern is that in an area as promising as phase-dependent reflexes, any discussion of a reflex attributable to the input of a single afferent species (e.g. Schomburg and Behrends, 1978) should acknowledge the selective activation issue.

In my opinion, the only unequivocal method for studying the central actions of a single afferent species is to use either the conventional spike-triggered averaging (STA) approach of Mendell and Henneman (1968; for reviews, *see* Fetz *et al.*, 1979a and Kirkwood and Sears, 1980) for the analysis of monosynaptic connections, a modified STA approach that involves high-frequency intracellular stimulation of the reference afferent (Willis *et al.*, 1968; Fetz *et al.*, 1979a) for the analysis of monosynaptic and polysynaptic connections, or, in the case of Ia afferents, a 'quick' stretch or muscle vibration (Lundberg and Winsbury, 1960; Brown *et al.*, 1967; Stuart *et al.*, 1971; Fetz *et al.*, 1979b). This viewpoint is not meant to downgrade the importance of a large body of data collected over the last 30 years in which motoneuron responses have been documented to input evoked by either (1) graded electrical stimulation of muscle nerves, either with or without a concomitant nerve block, or (2) muscle contraction and/or vibration, or (3) a combination of (1) and (2) as in Coppin *et al.* (1970; *see also* Jack, 1978). In all such instances, with the possible exception of the aggregate monosynaptic Ia EPSP (Burke *et al.*, 1976) as evoked from *select* nerves, there is a large margin of doubt as to the selectivity and reproducibility of the afferent activation that was obtained (e.g. Eccles *et al.*, 1957; Eccles and Lundberg, 1959). The selective activation problem is well known (Hunt and Perl, 1960) and there are many allusions to it in recent reviews and papers (e.g. Stuart *et al.*, 1971, 1972; Matthews, 1972; McIntyre, 1974; Stauffer *et al.*, 1976; Jack, 1978; Fetz *et al.*, 1979a, b; Rymer *et al.*, 1979; Baldissera *et al.*, 1981). In addition, I have reservations about the Coppin *et al.* (1970) technique for selective activation of Ib afferents. It involves increasing the thresholds for excitation of Ia afferents to electrical stimulation above the threshold for Ib afferents by prolonged muscle vibration and then activating the Ib pathways by electrical shocks to the muscle nerve. While this technique might provide a selective activation of Ib afferents for a sufficient period of time to establish a central Ib connection (e.g. Baxendale and Rosenberg, 1977; Fetz *et al.*, 1979b), its use appears limited to a select number of muscles 'amenable' to longitudinal vibration (e.g. the triceps surae). Furthermore, it would be difficult and time consuming to test repeatedly for when the Ia afferents resume their activity. As such, the technique is somewhat impractical for detailed tracking experiments (e.g. Botterman *et al.* 1980) and for studies like those described at this symposium on polysynaptic muscle receptor input to a variety of muscles. In contrast, the modified STA technique (Willis *et al.*, 1968; Fetz *et al.*, 1979a) is indeed an unequivocal method for selectively activating and driving at high frequency single afferent axons supplying any muscle of interest.

In view of the technical perseverance and virtuosity displayed in the work of Schomburg *et al.* (1981) and Andersson *et al.* (1978), it does not seem unreasonable to propose to these investigators that the time is ripe for them to apply the original (Mendell and Henneman, 1968) and modified (Willis *et al.*, 1968; Fetz *et al.*, 1979a) STA techniques to the study of phase-dependent reflexes (*see also* Roscoe *et al.*, 1980).

In summary, my critique has focused on the current need for detailed kinesiological studies of locomotion in general and phase-dependent reflexes in particular. It has further emphasised that agreement is needed on the definition of reflex reactions during stepping, and that the selective activation issue should be taken into account when studying the central connections of muscle afferents. It is fitting to close with a note of commendation to the groups who have presented their technically demanding and imaginative work at this symposium, and, in the historical sense, to emphasise how much progress they have made in this area which, after the pioneering efforts of Marey (1874) and Graham Brown (1916), was revitalised by Bernstein (1947, 1967) and his students (reviewed in Gurfinkel and Shik, 1973; Shik and Orlovsky, 1976) and by Lundberg and his co-workers (reviewed in Lundberg, 1969).

ACKNOWLEDGEMENTS

This work was supported by U.S.P.H.S. Grants NS 07888 and HL 07249 and a travel award from the American Physiological Society.

Received on October 8th, 1980.

REFERENCES

Andersson, O., Forssberg, H., Grillner, S. and Lindquist, M. (1978). Phasic gain control of the transmission in cutaneous reflex pathways to motoneurones during 'fictive' locomotion, *Brain Res.*, **149,** 503–507

Baldissera, F., Hultborn, H. and Illert, M. (1981). Integration in spinal neuronal systems. In *APS Handbook of Physiology* series (ed. V. B. Brooks), Volume 4: *Motor Systems,* Washington, American Physiological Society

Bard, P. (1938). Studies on the cortical representation of somatic sensibility. In *The Harvey Lectures,* Series 33, Baltimore, Williams and Wilkins, 143–169

Baxendale, R. H. and Rosenberg, J. R. (1977). Crossed reflexes evoked by selective activation of tendon organ afferents axons in the decerebrate cat, *Brain Res.,* **127,** 323–326

Bernstein, N. A. (1947). *On the Construction of Movements* (monograph in Russian), Moscow

Bernstein, N. A. (1967). *The Co-ordination and Regulation of Movements,* New York, Pergamon

Botterman, B. R., Hamm, T. M., Reinking, R. M. and Stuart, D. G. (1980). Extent of monosynaptic Ia reflex localization in cat hamstring muscles, *Proc. XXVIII Int. Congr. Physiol. Sci.,* **XIV,** 336

Brown, M. C., Engberg, I. and Matthews, P. B. C. (1967). The relative sensitivity to vibration of muscle receptors of the cat, *J. Physiol., Lond.,* **192,** 773–800

Burke, D., Knowles, L., Andrews, C. and Ashby, P. (1972). Spasticity, decerebrate rigidity and the clasp-knife phenomenon: An experimental study in the cat, *Brain,* **95,** 31–48

Burke, R. E., Rymer, W. Z. and Walsh, J. V., Jr. (1976). Relative strength of synaptic input from short-latency pathways to motor units of defined type in cat medial gastrocnemius, *J. Neurophysiol.,* **39,** 447–458

Cabelguen, J. M., Orsal, D., Perret, C. and Zattara, M. (1980). Central pattern generation of forelimb and hindlimb locomotion activities in the cat, *Proc. XXVIII Int. Cong. Physiol. Sci.,* **XIV,** 74–75

Coppin, C. M. L., Jack, J. J. B. and McLennan, R. C. (1970). A method for the selective activation of tendon organ afferents from the cat soleus, *J. Physiol., Lond.,* **210,** 18–20P

Duysens, J. and Stein, R. B. (1978). Reflexes induced by nerve stimulation in walking cats with implanted cuff electrodes, *Expl Brain Res.,* **32,** 213–224

Eccles, J. C., Eccles, R. M. and Lundberg, A. (1957). Synaptic actions on motoneurones caused by impulses in Golgi tendon organ afferents, *J. Physiol., Lond.,* **138,** 227–252

Eccles, R. M. and Lundberg, A. (1959). Synaptic actions in motoneurones by afferents which may evoke the flexion reflex, *Arch. Ital. Biol.,* **97,** 199–221

Engberg, I. and Lundberg, A. (1969). An electromyographic analysis of muscular activity in the hindlimb of the cat during unrestrained locomotion, *Acta physiol. scand.,* **75,** 614–630

English, A. W. (1979). Interlimb coordination during stepping in the cat: an EMG analysis, *J. Neurophysiol.,* **42,** 229–243

von Euler, C. (1977). The functional organization of the respiratory phase-switching mechanisms, *Fed. Proc.,* **36,** 2375–2380

von Euler, C. (1980). The contributions of sensory inputs to the pattern generation of breathing, *Can. J. Physiol., Pharmacol.,* to be published

Fetz, E. E., Henneman, E., Mendell, L., Stein, R. B. and Stuart, D. G. (1979a). Properties of single cells in vertebrate motor systems revealed by spike-triggered averaging. In *Society for Neuroscience,* 8th Annual Meeting. Summaries of Symposia (BIS Conference Report No. 49), UCLA, Los Angeles, Brain Information Service/BRI Publications Office, 11–32

Fetz, E.E., Jankowska, E., Johannson, T. and Lipski, J. (1979b). Autogenetic inhibition of motoneurons by impulses in Group Ia muscle spindle afferents, *J. Physiol., Lond.,* **293,** 173–195

Forssberg, H. (1979). The 'stumbling corrective reaction'—a phase dependent compensatory reaction during locomotion, *J. Neurophysiol.*, **42,** 936–953

Forssberg, H., Grillner, S. and Rossignol, S. (1977). Phasic gain control of reflexes from the dorsum of the paw during spinal locomotion, *Brain Res.*, **132,** 121–139

Gambarjan, P. P., Orlovskii, G. N., Protopopova, T. Y., Severin, F. V.and Shik, M. L. (1971). The activity of muscles during different forms of locomotion of cats and the adaptive function of the (hindlimb) musculature in the family Felidae, *Proc. Inst. Zool. Acad. Sci. USSR*, **48,** 220–239

Graham Brown, T. (1916). Die reflex Functionen der Zentralnerven mit besonderer Berücksichtigung der rhythmischen Tätigkeiten beim Säugetier, *Ergebn. Physiol.*, **15,** 480–790

Grillner, S. (1975). Locomotion in vertebrates—central mechanisms and reflex interaction, *Physiol. Rev.*, **55,** 247–304

Grillner, S. (1979). Analogies between pattern generation in locomotion and breathing. In *Central Nervous Control Mechanisms in Breathing* (eds C. von Euler and H. Lagercrantz), New York, Pergamon

Grillner, S. and Zangger, P. (1979). On the central generation of locomotion in the low spinal cat, *Expl Brain Res.*, **34,** 241–261

Gurfinkel, V. S. and Shik, M. L. (1973). The control of posture and locomotion. In *Motor Control* (eds. A. A. Gydikov, N. T. Tankov and D. S. Kosarov), New York, Plenum, 217–234

Herman, R. M., Grillner, S., Stein, P. S. G. and Stuart, D. G. (1976). *Neural Control of Locomotion,* New York, Plenum

Hunt, C. C. and Perl, E. R. (1960). Spinal reflex mechanics concerned with skeletal muscle, *Physiol. Rev.*, **40,** 538–579

Jack, J. J. B. (1978). Some methods for selective activation of muscle afferent fibres. In *Studies in Neurophysiology.* Presented to A. K. McIntyre (ed. R. Porter), Cambridge University Press, 155–176

Kirkwood, P. A. and Sears, T. A. (1980). The measurement of synaptic connections in the mammalian nervous system by means of spike-triggered averaging. In *Spinal and Supraspinal Mechanisms of Voluntary Motor Control and Locomotion* (ed. J. E. Desmedt), *Progress in Clinical Neurophysiology*, Vol. 8, Basel, Karger, 44–71

Lance, J. W. and McLeod, J. G. (1975). *A Physiological Approach to Clinical Neurology,* London, Butterworths

Landau, W. M. (1974). Spasticity: The fable of a neurological demon and the emperor's new therapy, *Arch. Neurol.*, **31,** 217–219

Loeb, G. E., Bak, M. J. and Duysens, J. (1977). Long-term unit recording from somatosensory neurons in the spinal ganglia of the freely walking cat, *Science,* **197,** 1192–1194

Loeb, G. E. and Hoffer, J. A. (1981). Muscle spindle function during normal and peturbed locomotion in cats. This publication

Lundberg, A. (1969). *Reflex Control of Stepping,* Nansen Memorial Lecture to Norwegian Academy of Sciences and Letters, Oslo, Universitetsforlaget

Lundberg, A. and Winsbury, G. (1960). Selective activation of large afferents from muscle spindles and Golgi tendon organs, *Acta physiol. scand.*, **49,** 155–164

McCormack, M. and Dubrovsky, B. (1979). Impairments in limb actions after dorsal funiculi section in cats, *Expl Brain Res.*, **37,** 31–40

McIntyre, A. K. (1974). Central actions of impulses in muscle afferent fibres. In *Handbook of Sensory Physiology,* III/2 (ed. C. C. Hunt), New York, Springer-Verlag, 235–288

Marey, E. J. (1874). *Animal Mechanism. A Treatise on Terrestrial and Aerial Locomotion,* New York, Appleton–Century–Crofts

Matthews, P. B. C. (1972). *Mammalian Muscle Receptors and Their Central Actions*, London, Arnold

Melzack, R. and Southmayl, E. (1974). Dorsal column contributions to anticipatory motor behaviour, *Expl Neurol.*, **42**, 274–281

Mendell, L. M. and Henneman, E. (1968). Terminals of single Ia fibers: Distributions within a pool of 300 homonymous motoneurons, *Science*, **154**, 96–98

Perret, C. (1973). Analyses des mecanismes d'une activité de type locomoteur chez le chat, *These Doct. Sci.*, CNRS A08342, Paris, 1–259

Perret, C., Cabelguen, J. M. and Millanvoye, M. (1972a). Caracteristiques d'une rhythm de type locomoteur chez le chat spinal aigu, *J. Physiol., Paris*, **65**, 472A

Perret, C., Millanvoye, M. and Cabelguen, J. M. (1972b). Messages spinaux ascendants pendant une locomotion fictive chez le chat curarise, *J. Physiol., Paris*, **65**, 153A

Prochazka, A. and Wand, P. (1980). Independence of fusimotor and skeletomotor systems during voluntary movement. This publication

Prochazka, A., Westerman, R. A. and Ziccone, S. P. (1976). Discharge of single hindlimb afferents in the freely moving cat, *J. Neurophysiol.*, **39**, 1090–1104

Roscoe, D. D., Botterman, B. R., Cameron, W. E., Reinking, R. M. and Stuart, D. G. (1980). Use of a synchronization test in studies on segmental motor control. In Plenary and Symposia Lectures, *Proc. XXVIII Int. Congr. Physiol. Sci.* (ed. J. Szentagothai), Budapest, Hungarian Academy of Sciences

Rymer, W. Z., Houk, J. C. and Crago, P. E. (1979). Mechanisms of the clasp-knife reflex studied in an animal model, *Expl Brain Res.*, **37**, 93–113

Schomburg, E. D. and Behrends, H.-B. (1978). The possibility of phase dependent monosynaptic and polysynaptic Ia excitation to homonymous motoneurones during fictive locomotion, *Brain Res.*, **143**, 533–537

Schomburg, E. D., Behrends, H.-B. and Steffens, H. (1981). Changes in segmental and propriospinal reflex pathways during spinal locomotion. This publication

Sears, T. A. (1971). Breathing: A sensori-motor act, *Sci. Basis Med. Ann. Rev.*, 129–147

Shik, M. L. and Orlovsky, G. N. (1976). Neurophysiology of locomotor automatism, *Physiol. Rev.*, **56**, 465–501

Smith, O. A., Jr., Astley, C. A., Hohimer, A. R. and Stephenson, R. B. (1981). Behavioral and cerebral control of cardiovascular function. In *Neural Control of the Circulation* (eds C. D. Barnes and M. J. Hughes), New York, Academic Press

Stauffer, E. K., Watt, D. G. D., Taylor, A., Reinking, R. M. and Stuart, D. G. (1976). Analysis of muscle receptor connections by spike-triggered averaging. 2: Spindle secondary afferents, *J. Neurophysiol.*, **39**, 1393–1402

Stuart, D. G., Binder, M. D. and Botterman, B. R. (1979). Features of segmental motor control revealed in the single-unit recordings during natural movement. In *Posture and Movement* (eds R. E. Talbott and D. R. Humphrey), New York, Raven Press, 281–294

Stuart, D. G., Mosher, C. G. and Gerlach, R. L. (1972). Properties and central connections of Golgi tendon organs with special reference to locomotion. In *Research in Muscle Development and the Muscle Spindle* (eds B. Q. Banker, R. J. Pryzbylsky, J. P. Van Der Meulen and M. Victor), Amsterdam, Excerpta Medica, 437–464

Stuart, D. G., Willis, W. D. and Reinking, R. M. (1971). Stretch-evoked excitatory post-synaptic potentials in motoneurons, *Brain Res.*, **33**, 115–125

Stuart, D. G.,Withey, T. P., Wetzel, M. C. and Goslow, G. E., Jr. (1973). Time constraints for inter-limb co-ordination in the cat during unrestrained locomotion. In *Control of Posture and Locomotion* (eds R. B. Stein, K. G. Pearson, R. S. Smith and J. B. Redford), New York, Plenum, 537–560

Udo, M., Matsukawa, K., Kamei, H. and Oda, Y. (1980). Cerebellar control of locomotion: Effects of cooling cerebellar cortex in high decerebrate and awake walking cats, *J. Neurophysiol.*, **44,** 119–134

Viala, D. and Buser, P. (1971). Modalités d'obtention de rhythmes locomoteurs chez le lapin spinal par traitements pharmacologiques (DOPA, 5-HTP, d'amphetamine), *Brain Res.*, **35,** 151–165

Walmsley, B., Hodgson, J. A. and Burke, R. E. (1978). The forces produced by medial gastrocnemius and soleus muscles during locomotion in freely moving cats, *J. Neurophysiol.*, **41,** 496–508

Wand, P., Prochazka, A. and Sontag, K.-H. (1980). Neuromuscular responses to gait perturbations in freely moving cats, *Expl Brain Res.*, **38,** 109–114

Wetzel, M. C. and Stuart, D. G. (1976). Ensemble characteristics of cat locomotion and its neural control, *Prog. Neurobiol.*, **7,** 1–98

Willis, W. D., Letbetter, W. D. and Thompson, W. M. (1968). A small system of neurons in the mammalian spinal cord, *Brain Res.*, **9,** 152–155

Wyman, R. J. (1977). Neural generation of the breathing rhythm, *Ann. Rev. Physiol.*, **39,** 417–448